Oxide Thin Films and Nanostructures

Oxide Thin Films and Nanostructures

Falko P. Netzer

Surface and Interface Physics, Institute of Physics, Karl-Franzens University Graz, Austria

Claudine Noguera

Institut des NanoSciences de Paris, CNRS-Sorbonne Université, Paris, France

OXFORD

UNIVERSITY PRESS

Great Clarendon Street, Oxford, OX2 6DP,
United Kingdom

Oxford University Press is a department of the University of Oxford.
It furthers the University's objective of excellence in research, scholarship,
and education by publishing worldwide. Oxford is a registered trade mark of
Oxford University Press in the UK and in certain other countries

© Falko P. Netzer and Claudine Noguera 2021

The moral rights of the authors have been asserted

First Edition published in 2021

Impression: 1

Published in the United States of America by Oxford University Press
198 Madison Avenue, New York, NY 10016, United States of America

British Library Cataloguing in Publication Data
Data available

Library of Congress Control Number: 2020946227

ISBN 978–0–19–883461–8

DOI: 10.1093/oso/9780198834618.001.0001

Printed and bound by
CPI Group (UK) Ltd, Croydon, CR0 4YY

Preface

Why are scientists still writing books in the 21st century? This question formed the title of a commentary by Jeremy Matthews, acquisitions editor at MIT Press in Cambridge, Mass., in the January 2018 issue of *Physics Today*. And Matthews answered the question in that article in the sense that books are still part of the culture of our time, and that people still read books. He argues that textbooks and professional monographs are useful guides and references for students and professional scientists, whether they want to enter a new scientific area or broaden their knowledge in a practising research field. And, it appears that in spite of the prevalence of the electronic media, the majority of people still reach for a printed copy.

When we were asked to write this book on oxide nano-materials, we were aware of the fact that a book does not provide high-impact points or large citation records, but we considered it as an interesting project, which requires the reader to step back and reflect upon our field of scientific activity from a somewhat distant observation point. Writing a book is a different endeavour from writing an original research paper or even a review article. A book has a broader scope, it is less focussed, requires more research of the published literature, touches on areas that may be at the boundaries of one's immediate detailed expertise, and is expected to comply to pedagogic aims. A book is the authors' personal journey through the chosen topic or field, which the readers are invited to follow; the journey is affected by the personal experiences and preferences of the authors.

Charles Day, the editor in chief of *Physics Today*, reported in his monthly column in June 2019 about robot science writers: algorithms developed by computer scientists combining two subfields of artificial intelligence (AI), natural language processing and machine learning. *Springer Nature* had published a book 'Lithium Ion Batteries: a Machine-Generated Summary of Current Research (2019)'. The book summarizes the content of 1,086 scientific papers, analysing, simplifying, and synthesising text from the papers. Day noted that the book lacks a machine-written preface and concluded: providing a useful overview of an entire field is beyond the AI ability, because writing a good overview requires making choices of what to include and what to exclude. This choice requires a personal touch.

The personal touch of the present two authors is based on their different scientific backgrounds. We both have been working in the field of oxide surfaces, oxide nano-systems, oxide thin films and nano-structures, for more than two decades, but Claudine Noguera is a theoretical physicist with a touch of geochemistry culture and expertise in the many body problem, surface science, and computational science, whereas Falko Netzer is an experimental physicist with a strong background in surface physics and chemistry. We have both worked closely with the other 'domain', i.e. combining experimental work and theoretical analysis in our research, yet we have different approaches to the topics treated in this book. We hope that this creates a field of tension that is able to generate a particular charm in this book. The scientific topics contained in this book should introduce the reader to the field of oxide nano-structures; they have been selected to highlight areas of science and technology in which oxide materials at the nano-scale have an impact and represent crucial developments. The topics are selected views and reflect the personal interests and preferences of the authors. We hope that they lead to an interesting and entertaining journey through the world of oxide nano-science.

As for any undertaking, this book has greatly benefited from every-day contacts, work, and discussions with the colleagues of our groups in Graz and Paris. We also wish to particularly thank Svetlozar Surnev, Jacek Goniakowski, Bertrand Fritz, Jean Pierre Jolivet, Alain Baronnet, and Alain Meunier, who, in one way or another, have helped to improve the quality of this book. We are grateful to our home institutions, the Karl-Franzens University Graz and the CNRS and Sorbonne Université in Paris, for their support in terms of resources and space.

Falko Netzer and Claudine Noguera
May 2020

Contents

1
Introduction

Because our atmosphere contains oxygen and water vapour, because aqueous fluids are omnipresent at the Earth's surface as groundwaters (springs, wells) or as run-off waters (rivers, lakes or oceans), oxides, among which magnetite, quartz, or the various types of silicates (feldspars or pyroxene) form the major components of the Earth's crust. Oxides are also contained in the soil and strongly interact with the biosphere. Thus, it is fair to say: oxides are everywhere around us.

The diversity of oxide structures and compositions is astonishing, especially in comparison with metals or elemental semiconductors. Even simple binary oxides with a single type of cations display bulk structures which range from the simplest to more complex ones: rock-salt, wurtzite, spinel, corundum, fluorite, or antifluorite. Moreover, many oxides may exist under several polymorphic forms depending on thermodynamic conditions. As soon as several cations are present, as in perovskites or aluminosilicates (clays or zeolites), the structural complexity increases, concomitant with multiple possibilities of cation substitutions and multiple types of structural defects. Physicists have long considered oxides as uninteresting, classical objects with properties only driven by simple electrostatic forces. However, their electronic characteristics are as varied as their structures. Apart from the well-known large-gap insulators, some oxides are semi-conducting, others are metallic or even superconducting, and all flavours of magnetism may be encountered. This complexity represents a challenge for their controlled synthesis and to a precise engineering of their characteristics, but it also offers a tremendous richness of applications.

The interest of the field lies in its interdisciplinary nature and in the diversity of questions it raises, both on a fundamental and on an applied level. It gathers researchers from various horizons: geophysicists and geologists as far as the formation and weathering of the Earth's rocks are concerned; mineralogists interested in the structure and composition of complex minerals; toxicologists who decipher the interactions between small oxide particles and the biological medium; and chemists because oxides are good catalysts, largely used, for example, in petrochemistry. Oxides are also often present, although in an uncontrolled way, whenever an object is in contact with the ambient atmosphere. They play a fundamental role in corrosion processes and in the surface properties of real materials. Finally, they offer a large field of investigations to surface scientists.

The last three decades have witnessed important advances in the control of their synthesis, so that reproducible and well-characterized samples may now be obtained.

Oxide Thin Films and Nanostructures. Falko P. Netzer and Claudine Noguera, Oxford University Press (2021). © Falko P. Netzer and Claudine Noguera. DOI: 10.1093/oso/9780198834618.003.0001

The present situation bears no resemblance with that which prevailed at the start of the so-called 'high-T_c' era, which, quite unexpectedly, revealed that superconductivity exists in some copper oxide compounds with transition temperatures above that of liquid nitrogen. Because, at that time, the importance of the oxygen stoichiometry had not been perceived, for several years most experimental results turned out to be unreliable. Today, this is no longer the case. Playing with thermodynamic conditions and, in particular, the oxygen partial pressure (but also the water partial pressure) during synthesis is a lever to obtain oxide samples of precise predefined stoichiometry, a particular crucial requirement for studies at the nanoscale. In parallel, sophisticated means of characterization have been developed, in particular diffraction, microscopy, and spectroscopy methods available in synchrotron facilities, but also calorimetric methods for the study of nanoparticle phase diagrams, or various types of scanning microscopy/spectroscopy methods. Thanks to the ever-increasing computing power relevant simulations may now be performed, with a better account of strong electron correlation effects, of large numbers of atoms, or of processes on long time scales.

For all these reasons, oxides are no longer the 'dirty' materials of the past, but have integrated the technosphere as active and passive components in our technological surroundings. Understanding the structure and properties of oxides is thus crucial for many modern technologies. Oxide materials are present as dielectric elements in microelectronic devices (Lorenz *et al.*, 2016; Coll *et al.*, 2019), as transparent layers in optical coatings and solar energy harvesting devices, they form corrosion protective layers on metals, act as catalysts, and they confer biocompatibility to medical implants in the human body. Technological progress since the second half of the 20th century has been based to a large extent on the progress in thin film technology, amongst which oxide thin films have played and are playing a most prominent role. For example, the microelectronic revolution that led to the computer age and to our present-day digital society is essentially rooted in the metal-oxide-semiconductor field effect transistor (MOSFET) device element, in which the dielectric properties of thin films of SiO_2 as gate dielectrics and of other oxide layers in capacitive elements are essential for its functioning. With the ongoing trend towards miniaturization of all kinds of device elements and the progression from the microtechnology to the nanotechnology area in the last two decades, new challenges in the fabrication and characterization of materials have arisen. Ultra-thin films of nanometre thickness (nanolayers) with novel properties are more and more replacing the previous 'thick' micrometer films, and other forms of nanostructured materials with well-defined shape and behaviour (nanoparticles, nanosheets, nanorods) have become of increased interest in many scientific and technological fields. In this book, we want to address some of these developments and challenges as related to metal oxide materials at the nanometre scale.

It is necessary at this starting point, to define what we actually understand within the framework of a nanomaterial, a nanostructure. The definition is not unambiguous and has been used in the literature in various ways. Some authors define it as a material that has at least in one dimension a size of < 100 nm. Here in this book, we want to use a more restricted definition and reduce this length scale to the order of or < 10 nm. In this latter range, the scalable regime of properties with size is no longer valid and novel behaviour as compared to that of the macroscopic bulk phase has to be considered, due to quantum confinement effects and the significance of the interfaces to the environment. However, we intend to use this nanostructure definition in a somewhat pragmatic and flexible way, depending on the property to be investigated and discussed, and we will not exclude scales of the crossover region from the scalable to the non-scalable regimes.

In this book, we present concepts and phenomena of metal oxide materials in nanostructured forms, that is ultra-thin films of thickness $\sim < 10$ nm and nanoparticles of various shapes. Even in bulk form, oxides are characterized by a great variety of stoichiometries, structures, and thus, diverse physical and chemical properties. In oxide nanosystems, additional degrees of freedom are provided by the variable size dimension, the significance of interfaces to substrates and/or the environment, and by the morphology of surfaces and particles; these extra degrees of freedom may be used for modifying properties and for designing desired functionalities. In nanostructured systems, a large proportion of atoms is at the surface; thus, surface properties become particularly important. This is also reflected in the methodology that is employed for the experimental characterization: surface science techniques are widely used for obtaining atomic scale information of nanosystems and they are therefore prominently discussed here in this book. The progress in the fabrication of epitaxial oxide thin films on crystalline substrates by vacuum-based deposition methods during the last two decades has opened up the way to prospective applications in an all-oxide electronics (Ramirez, 2007), in all-oxide epitaxial thin film batteries (Liang *et al.*, 2019), and in the fabrication of epitaxial oxide heterostructures (Biswas and Jeong, 2017), the latter similar to the well-known heterostructures of semiconductor technology materials. Progress in the wet chemical preparation methods has enabled the reproducible fabrication of oxide nanoparticles of various forms and shapes. Their shape-dependent properties are finding use in chemical applications such as catalysis or biological recognition systems as well as in optoelectronic devices. The discussion of epitaxial oxide ultra-thin films and oxide nanoparticle ensembles thus forms a focus in this book.

The surface science of oxides has been treated in two previous seminal books of Henrich and Cox (1996) and Noguera (1996). Henrich and Cox's book is focussed on the properties of single crystal surfaces of bulk oxides with a view towards the adsorption of molecules as a first stage of catalysis. Noguera's book develops concepts of structure and electronic behaviour of oxide surfaces from a more theoretical angle. Both books are recommended to readers interested in the fundamental properties of oxide surfaces. More recently, Pacchioni and Valeri (2012) have edited an account of the science and technology of oxide ultra-thin films in a number of important technological applications. The collection of chapters in the book edited by Netzer and Fortunelli (2016) considers oxide materials at the two-dimensional limit, that is in (quasi-)two-dimensional form, not only concentrating on more fundamental aspects, but also pointing out relations to applications.

The approach we have adopted in this book attempts to serve a dual purpose: on the one hand, it introduces the reader to the fundamental concepts of oxide nanosystems incorporating both theoretical and experimental aspects; on the other hand, it highlights the characteristic properties of some prototypical oxide nanosystems from a phenomenological viewpoint, with sidelong glances on selected benchmarking applications. The book is organized along the following lines. We begin in Chapter 2 with methods for the fabrication of oxide thin films and nanoparticles, from the surface oxidation of bulk metals via thin film deposition methodologies to nanoparticle fabrication methods in the gas phase and liquid environments. An introduction into nucleation and growth concepts rounds off this chapter on preparatory aspects. In Chapter 3, experimental characterization techniques and quantum theoretical methods for geometric and electronic structure determination of oxide systems are presented. This chapter underlines the fact that the progress in the characterization of nanostructures and low-dimensional systems, as we have witnessed during the last two decades, has been made possible only by the close interaction of experimental and theoretical studies. Chapter 4 is devoted to fundamental properties of oxide thin films. Here, intrinsic properties of oxide films and the environmental effects introduced by the presence of substrates, that is interfacial phenomena, are discussed. Oxide thin films at the two-dimensional limit, i.e. oxides as two-dimensional materials and their characteristic novel physical and chemical properties, constitute the topics of Chapter 5. Some experiments, for which the 'proof of concept' exploratory stage for applications has just been overcome, are cited as examples for potential future applications. In Chapter 6, the basic physical concepts governing the properties of oxide nanoparticles—structure, electronic behaviour, and environmental effects—are investigated, both in the non-scalable and scalable size regimes. Clay minerals are agglomerations of natural oxide nanolayers and nanosheets; they have been used for thousands of years in human cultures as pottery, building materials and in pharmaceutical applications. Chapter 7 gives an introduction into these natural oxide nanomaterials, discussing structure and chemical reactivities of clay minerals and their interaction with the humid environment. Chapter 8 gives an impression of the various interdisciplinary fields in which oxide nanosystems are active and investigated in science and technology. The common theme of the diverse topics in this chapter is surface chemistry in its widest sense, involving electron transfer processes (redox reactions) as basic ingredients. Heterogeneous catalysis, one of the mature applications of oxide nanoparticles in the form of high surface area oxide powder catalysts is mentioned, together with the more recent photocatalysis, solid oxide fuel cells, solar energy materials, and corrosion protection. The current trend to the development of 'green' nanotechnologies is reflected in these sections. A prominent place in Chapter 8 is given to biological applications of oxide nanosystems, including the oxide-mediated biocompatibility of materials, diagnostic and therapeutic properties of oxide nanoparticles, biosensing, and the response of biosystems to electric fields in ferroelectric oxides. The book is concluded with a general summary, some conclusions and a cautious glimpse into possible future developments in Chapter 9. Three appendices are added. They give a simple introduction to the principles underlying the Wulff construction and to the Frenkel–Kontorova model of incommensurable interfaces, as well as a list of the acronyms used in the book.

The references cited in this book refer often to review articles for introductory purposes, but important key publications are mentioned as well. Due to the wide range of different topics and aspects treated in this book, an exhaustive bibliography is beyond its scope. Further specialist literature can be found in the cited review publications and books.

2

Growth of oxide thin films and nanoparticles: Methods of fabrication

The fabrication of oxide materials in the form of thin films and nanoparticles is the key step to enable the scientific study of oxides at the nanoscale and their applications in the fields of nanoscience and nanotechnology. As in other areas of condensed matter physics and chemistry, the availability of good quality samples is a prerequisite for the experimental characterization of the fundamental properties of nanomaterials, and progress in improving sample quality has often led to the discovery of novel and emergent phenomena. The discovery of the quantum Hall effect may be cited in this context, which was only possible after the quality of semiconductor heterostructure samples had reached a critical level of low defect densities (Von Klitzing, 1993; Von Klitzing, 2005). Good quality in samples of oxide thin films means well-ordered homogeneous crystalline layers with few impurities and few local and extended defects, whereas it is a controllable and uniform size and shape distribution in the case of oxide nanoparticles.

The first section of this chapter is devoted to the formation of oxide thin films by oxidation of the outer layers of bulk solids, the second section addresses methodologies of oxide thin film deposition, whereas the third section presents methods of fabrication of oxide nanoparticles. The forth section contains an introduction to the fundamental processes of thin film and nanoparticle nucleation and growth.

2.1   Oxidation of bulk solids

Oxidation in the present context means the chemical reaction of a solid (metal, alloy or semiconductor) with an oxidant. This involves the transfer of electrons and the formation of a new compound, the oxide, with a concomitant change in the bonding state from metallic/covalent to (partly) ionic. Formation of an oxide layer requires the creation of cations from the metallic (electropositive) element, the dissociation of the oxidant molecule (assuming molecular oxidants such as O_2 or H_2O) at the surface, and the formation of oxygen anions. After the direct oxidation of the outermost first atomic layers of the solid, diffusion processes in the solid are required for the mass transport necessary for further oxide growth. The diffusion may involve metal cations

Oxide Thin Films and Nanostructures. Falko P. Netzer and Claudine Noguera, Oxford University Press (2021). © Falko P. Netzer and Claudine Noguera. DOI: 10.1093/oso/9780198834618.003.0002

Table 2.1 Comparison of metal and respective oxide crystal structures. hcp and fcc structures stand for hexagonal close packed and face centred cubic.

Metal	Structure	Oxide	Structure	Symmetry
Mg	hcp	MgO	Rock-salt	Cubic
Al	fcc	$\alpha\text{-}Al_2O_3$	Corundum	Hexagonal
Ti	hcp	TiO_2	Rutile	Tetragonal
Co	hcp	CoO	Rock-salt	Cubic
Ni	fcc	NiO	Rock-salt	Cubic
Zr	hcp	ZrO_2	Fluorite	Cubic

or oxygen anions, depending on the material, and electrons and holes to maintain charge neutrality in the growing film. The following treatment will be restricted to the gas-solid oxidation reaction, i.e. the oxidants are applied from the gas phase. The electrochemical anodic oxidation in aqueous environments, which is of major interest in corrosion processes (Marcus and Maurice, 2011), will not be addressed here.

While it may appear that the oxidation of the outer region of an elemental bulk solid is a straightforward way to form an oxide film, this procedure has several limitations. First, on the materials side, it is not flexible and is restricted to an oxide film grown on its own metal (semiconductor). Thus, oxide films on different support materials, as desirable for many purposes, require deposition transfer techniques (see Section 2.2). Second, it has shortcomings in terms of the oxide film quality: the oxidation of bulk solids tends to yield imperfect films with poor long-range order, high defect densities, and rough film morphologies. However, for many applications of oxide thin films, flat layers with epitaxial long-range order and low defect densities are desired. Epitaxial film growth, i.e. the growth of a crystalline layer on top of a crystalline substrate, requires that the lattices of the growing film and the substrate match at their interface in terms of symmetry and lattice parameters. In general, the lattice match between metals and their own oxides is poor, preventing the epitaxial growth of oxide films on their parent metals. This point is illustrated in Table 2.1, where a few selected examples of metal and oxide lattice structures are compared. For ultrathin oxide films, the strict overlayer-substrate lattice matching condition in terms of coherent interfaces may be relaxed due to the flexibility of (quasi-)two-dimensional oxide lattices and the stability of novel interface oxide phases, and epitaxial growth may still be enabled in cases of significant mismatch of the bulk lattices of oxides and substrates (Obermüller *et al.*, 2017). This will be further discussed in Chapter 5.

Oxide thin films on metals may be the result of unintended corrosive processes, i.e. oxidation reactions at metal surfaces taking place in our environmental atmosphere containing oxygen, water, halogens, or other aggressive oxidants. The formation of rust on iron surfaces is a well-known phenomenon. The effect of these corrosion reactions limits the stability of metals in every-day practical applications and is an important technical problem, but corrosion can be inhibited by protective oxide layers (Macdonald, 1999). Passivating oxide layers of limited thickness can be grown conveniently as a result of self-limited growth; characterized by high thermal and mechanical stability with good adherence to the parent metal, passivating oxide layers may protect the bulk of the metallic substrate from the corrosive environment (Olsson and Landolt, 2003). Aluminium and chromium oxide for example can be used for passivating oxide layers.

2.1.1 Thermal oxidation of bulk metals and semiconductors

The gas-solid oxidation reaction at room temperature commonly stops, after a thin oxide layer has been formed, due to kinetic constraints (passivation). At elevated temperatures, thicker oxide films can be grown slowly. It is convenient to separate the thermal growth of oxides into two regimes (stages): high-temperature and low-temperature growth; these two regimes are synonymous with thick and thin film growth, respectively. In the high-temperature regime, the thermal energy is sufficient for ionization and diffusive transport, as described by Wagner (1933), resulting in a parabolic growth law for thick films. In the low-temperature stage, an electric field across the growing oxide film has been invoked by Cabrera and Mott (1949) to promote diffusion transport. This leads to a logarithmic growth law for thin films. The question of what is a thick or a thin oxide film, or of what is a high- or a low-oxidation temperature, depends on the materials and on their crystallographic state and will now be discussed.

The elementary processes occurring during the growth of an oxide film are depicted schematically in Fig. 2.1. Oxygen molecules as oxidants from the gas phase are adsorbed at the surface of the solid, dissociate and become ionized forming O^{2-} anions at the gas-solid interface. Concomitantly, metal cations are created at the metal-oxide interface. The ions and the liberated electrons (holes) diffuse in a gradient of oxygen chemical potential μ_O and electric field, eventually forming oxide nuclei and then further contributing to the oxide film growth. The following discussion of theoretical considerations is a shortened version of the treatment given by Atkinson (1985) in his excellent review on oxide film growth. These 'classical' theories of oxide film growth do not address the atomistic processes at the very beginning of the oxidation process. A short discussion of the latter as revealed by a modern density functional theory approach will be given in the last part of this section.

Theory of thick film growth. The theory of thick oxide film growth formulated by Wagner (1933) is a phenomenological approach, which relates the growth rate of the

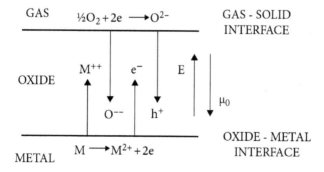

Fig. 2.1 Schematic of ion formation and transport in gradients of chemical potential of oxygen μ_O and electrical potential across a growing oxide film. Adapted from Atkinson (1985).

oxide film to transport properties such as diffusion coefficients and electrical conductivities. Thick films here means thicknesses of the order of or greater than $\approx 1~\mu\text{m}$, as will be discussed. Although such thick films are not the major topics of this book, which is intended to concentrate on nanoscale objects, Wagner's theory provides a useful starting point for a discussion of oxide film growth. Previous experimental observations have revealed that the thickness of oxide films X follows a parabolic kinetics:

$$X^2 = k_p t \tag{2.1}$$

with k_p the parabolic rate constant and t the time. Such parabolic kinetics is consistent with a transport mechanism in a gradient of a driving force, which becomes smaller as film thickness increases. This becomes apparent by differentiating eqn 2.1:

$$\frac{dX}{dt} = \frac{k_p}{2X} \tag{2.2}$$

with increasing X, the growth rate dX/dt slows down.

Wagner's theory is based on the assumption that diffusion in the film is the rate limiting step; both ionic and electronic transports across the film are necessary to provide the material transport through the film (see Fig. 2.1). Note that electronic transport (electrons and holes) is required to create oxygen anions and metal cations at the respective interfaces. Since the diffusing species are electrically charged, the diffusion fluxes are caused by gradients of the chemical potential and electric fields, the latter created by the separation of charge. The diffusion current density J_i of a particle i is given by

$$J_i = \frac{D_i C_i}{k_B T} \left(\frac{-d\mu_i}{dX} + q_i E \right) \tag{2.3}$$

where C_i is their concentration (in particles/unit volume), μ_i the chemical potential ($\mu_i = k_B T \ln C_i + \text{const}$), D_i the diffusion coefficient, q_i the charge, and E the electric field; k_B and T have their usual meaning, Boltzmann constant and temperature. The validity of eqn 2.3 depends on the validity of the Nernst–Einstein relation, which describes the diffusion of charged particles in terms of their electric mobility m_q and the diffusion coefficient D:

$$D = \frac{m_q k_B T}{q} \tag{2.4}$$

This in turn relies on the hypothesis of a small electric field E: $qEa \ll k_B T$, where a is the ionic jump distance.

The electric field during thick film growth may be caused by ambipolar diffusion (i.e. diffusion of particles with both polarities). If metal cations are more mobile in the oxide than oxygen anions, a new oxide is formed at the oxide/oxygen interface (gas-solid interface, see Fig. 2.1). Electrons must also diffuse outward with the cations (or holes diffuse inward) to ionize the oxygen atoms at the interface that become subsequently incorporated into the oxide. Electrons are more mobile than ions, and an electric field develops that speeds up the ions and slows down the electrons; the gas-solid interface develops a negative electric potential with respect to the oxide-metal interface. The same is also true, if the film grows mainly by diffusion of oxygen anions.

In Wagner's theory, it is assumed that the system is in a pseudo-steady-state, i.e. that there is no net electrical current. Further, it is assumed that a local chemical equilibrium exists throughout the film. Eliminating the electric field term from transport equations like eqn 2.3 and all chemical potentials but μ_O, the chemical potential of oxygen, the parabolic rate constant k_p can be expressed by:

$$k_p = \frac{k_B T}{4e^2 \alpha N_0} \int_{a(O_2)_I}^{a(O_2)_{II}} t_e t_{ion} \sigma_{tot} d[\ln a(O_2)] \qquad (2.5)$$

e is the modulus of the electric charge, α the stoichiometry coefficient and N_0 the number of MO_α oxide molecules per unit volume. σ_{tot} is the total electrical conductivity of the oxide with t_e and t_{ion} the fractions of the total conductivity provided by electrons and ions (transport numbers), respectively. The integration limits in eqn 2.5 are the activities of molecular oxygen at the metal-oxide interface $a(O_2)_I$ and the oxide-gas interface $a(O_2)_{II}$.

For most oxides $t_{ion} \ll t_e \approx 1$, i.e. the ionic conductivity is much less than the electronic conductivity, and eqn 2.5 can be recast into

$$k_p = \int_{a(O_2)_I}^{a(O_2)_{II}} \left(\alpha \frac{D(M)}{f_M} + \frac{D(O)}{f_O} \right) d[\ln a(O_2)] \qquad (2.6)$$

D are the self-diffusion coefficients of cations (M) and anions (O), and f_M and f_O are the correlation factors for metal and oxygen ion self-diffusion ($f_M \approx f_O \approx 1$). The significance of eqn 2.6 is that it expresses a quantitative relation between k_p and the tracer (self-)diffusion coefficients of ions in the oxide film, which are experimentally accessible quantities. This relation is phenomenological in that it does not disclose the atomistic diffusion mechanism (apart from the correlation factors $f_{M/O}$, which are weakly dependent on it). The dependence of the D data on the oxygen activity $a(O_2)$ contains information on the defect structure of the oxide film, and this is a useful way of obtaining such information.

The range of validity of Wagner's model requires, first, the validity of the Nernst-Einstein relation and thus small electric fields relative to the thermal energy $k_B T$. The E field from ambipolar diffusion is of the order $\approx 100 k_B T/eX$ (see Atkinson (1985) for details), $X \gg 100a$, and thus film thickness $X \gg 20$ nm. Second, the claim of local chemical equilibrium requires electric charge neutrality throughout the film; this, however, is not valid close to the interface boundaries. The surface charges at the interface cause a space-charge region of opposite polarity extending into the film. The extent of the space-charge region is of the order of the Debye–Hückel screening length L_D

$$L_D = \frac{\epsilon \epsilon_0 k_B T}{e^2 C_d} \qquad (2.7)$$

where ϵ and ϵ_0 are the dielectric constants of the oxide and vacuum, and C_d is the total number of elementary charges per volume due to charged defects at equilibrium in the bulk oxide. The assumption of charge neutrality is reasonable only if $X \gg L_D$. Unfortunately, C_d is not well known, but it can be estimated from conductivity data. An educated guess for a typical oxide with low defect concentrations suggests that at

$\approx 500°C$, L_D is probably less than 1 μm. The ambipolar diffusion is a further source of space charge: the additional charges at the interfaces create the electric field in the film, but also a space charge due to non-uniformity effects. Estimates of local variations of the defect concentration indicate that $X \gg L_D$ should be a valid assumption for local electrical neutrality. Taken together, for most oxides and growth temperatures larger than $500°C$, Wagner's theory should be valid for film thickness $X > 1$ μm.

Theory of thin film growth.   For oxide films of $X < 1$ μm thickness, the electrical neutrality concept within the film is unreliable, and for $X < 20$ nm the Nernst–Einstein relation is no longer valid. Thus, the presence of large E fields and large space charges have to be considered. Several theories have been formulated, with a corresponding number of kinetic expressions for film growth: logarithmic, inverse logarithmic, cubic, or parabolic (Atkinson, 1985). Cabrera and Mott's theory ((Cabrera and Mott, 1949)) describes the oxidation in atomic terms and is illustrative for the present discussion: electrons pass freely from the metal to the adsorbed oxygen in an equilibrium electrochemical potential (= Fermi level) of the metal and the adsorbed layer, a uniform electric field is created in the film by positive surface charges on the metal and negative charges from excess O^{2-} ions at the gas-oxide interface. This field drives the slow ionic transport across the growing film. The adsorbed layer of oxygen ions is in equilibrium with the gas phase according to the reaction:

$$\frac{1}{2}O_2(g) + 2e(metal) \longrightarrow O^{2-}(surface) \tag{2.8}$$

The equilibrium constant of this reaction is

$$K = \frac{a(O^{2-})}{a(O_2)^{1/2}a(e)^2} \tag{2.9}$$

$$K = \exp\left(\frac{-\Delta G}{k_BT}\right) \tag{2.10}$$

$$a(O^{2-}) = \frac{n_0}{N_s} \tag{2.11}$$

n_0 is the number of excess O^{2-} ions, i.e. those created by the gas-surface reaction, N_S the total number of O^{2-} ions per unit surface area. The activity of electrons with respect to the Fermi level is:

$$a(e) = \exp\left(-\frac{e\Delta\phi}{k_BT}\right) \tag{2.12}$$

with $\Delta\phi$ the voltage across the film. Combining these equations and solving for n_0 gives:

$$n_0 = N_s a(O_2)^{1/2}\exp\left(-\frac{\Delta G + 2e\Delta\phi}{k_BT}\right) \tag{2.13}$$

a relation between the excess O^{2-} ions n_0 and the voltage across the film. The film and the surface charges may be regarded as a simple capacitor:

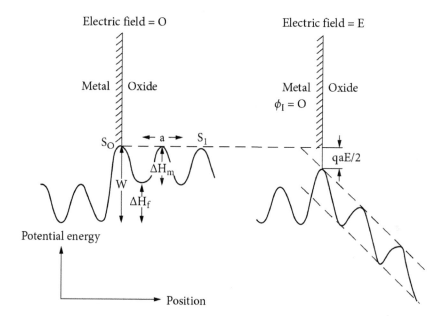

Fig. 2.2 Schematic diagram of the potential energy of an interstitial metal ion as a function of position at the metal-oxide interface. The electric field in the oxide film, generated by the transfer of electrons from metal to oxygen, lowers the energy barrier for ions moving into the oxide film. Reprinted from Atkinson (1985) with permission. Copyright 1985 by the American Physical Society.

$$n_0 = \frac{\epsilon\epsilon_0 \Delta\phi}{2eX} \tag{2.14}$$

Eliminating n_0 by setting eqn 2.13 = eqn 2.14 and reformulating yields:

$$\frac{2e\Delta\phi}{k_B T} + \ln\left(\frac{2e\Delta\phi}{k_B T}\right) = -\frac{\Delta G}{k_B T} + \ln\left(\frac{4e^2 N_s a(O_2)^{1/2} X}{k_B T \epsilon\epsilon_0}\right) \tag{2.15}$$

Since typically $e\Delta\phi/k_B T \gg 1$, the left side of eqn 2.15 is dominated by the first term and the second term can be neglected; solving for $\Delta\phi$ gives:

$$\Delta\phi = -\frac{\Delta G}{2e} + \frac{k_B T}{2e} \ln\left(\frac{4e^2 N_s a(O_2)^{1/2} X}{k_B T \epsilon\epsilon_0}\right) \tag{2.16}$$

The voltage across the oxide film $\Delta\phi$ is thus mainly dependent on the Gibbs free energy ΔG of reaction 2.8 and on the temperature, and less on the activity of oxygen in the gas phase $a(O_2)$ and the film thickness X. For typical values, $\Delta\phi = 1$ V, $\epsilon = 10$, $X = 10$ nm, the number of excess O^{2-} ions can be calculated (from eqn 2.14) to $n_0 = 2.8 * 10^{16}$ m^{-2}, which is only a small fraction of a monolayer.

Cabrera–Mott assumed that the rate controlling step of oxide film growth is the injection of defects into the oxide at one of the interfaces. Here we consider only the simplest case of oxide defects injected at the metal-oxide interface, in the form of an interstitial metal ion M^{2+}. As illustrated on the left side of Fig. 2.2, the activation energy for a jump of a metal atom into the oxide is W. Under the influence of an electric field during oxide film growth, the barriers at the interface are reduced by $qaE/2 = qa\Delta\phi/2X$ (see Fig. 2.2, right). In Wagner's model, the interface is assumed to be in equilibrium, i.e. the jump frequencies to the right and to the left of the barrier S_0 are equal. For Cabrera–Mott, the interface is far from equilibrium, i.e. if the E field is large enough, jumps occur predominantly to the right. The condition that reverse jumps (to the left) are negligible is $qaE/2 \gg k_BT$ (i.e. the thermal energy is insufficient to overcome the field effect). This is the same condition as the Nernst–Einstein relation becoming invalid, and it is likely to be true for films $X < 20$ nm.

In this high field limit, the jump rate corresponds to a growth rate

$$\frac{dX}{dt} = a\nu \exp\left(-\frac{W - qa\frac{\Delta\phi}{2X}}{k_BT}\right) = a\nu \exp\left(-\frac{W}{k_BT}\right)\exp\left(\frac{qa\Delta\phi}{2k_BTX}\right) \tag{2.17}$$

with ν the vibrational frequency of metal atoms at the interface. This may be rewritten into the form:

$$\frac{dX}{dt} = \frac{D_i}{a}\exp\left(\frac{X_1}{X}\right) \quad X_1 = \frac{qa\Delta\phi}{2k_BT} \quad D_i = a^2\nu\exp\left(-\frac{W}{k_BT}\right) \tag{2.18}$$

X_1 corresponds to the upper limit of thickness within the validity of the basic assumptions, D_i has the dimension of a diffusion coefficient. Equation 2.18 predicts an oxidation rate that decreases exponentially as the oxide thickness X increases. If $X \ll X_1$, eqn 2.18 may be approximately integrated:

$$\frac{X_1}{X} \approx -\ln\left(\frac{D_i X_1 t}{aX_L^2}\right) \tag{2.19}$$

This is the inverse logarithmic kinetic equation of the Cabrera–Mott theory. X_L has the meaning of a limiting thickness, above which the growth rate falls below an arbitrary low rate. With the Cabrera–Mott criterion of a negligible rate of 10^{-15} ms^{-1}:

$$X_L \approx \frac{X_1}{\frac{W}{k_BT} - 39} \tag{2.20}$$

for $X_L = X$, the upper limit of the validity of the model, $T = (W/40k_B)$ defines a critical temperature: if $T < W/40k_B$, the oxide film grows to a limiting thickness of $X_L < X_1$, but if $T > W/40k_B$, the film continues to grow beyond X_1 and reaches the parabolic regime.

The theories of Wagner, for thick films, and of Cabrera–Mott, for thin films, rely on assumptions that are probably valid only at the limits of very thick and very thin oxide films. To illustrate the range of the Cabrera–Mott and Wagner theories, Fig. 2.3 displays the growth rate of a hypothetical oxide as a function of thickness, using Cabrera–Mott for thin ($X < X_1$) and Wagner for thick ($X > L_D$ and X_1) films. The parameters used have been estimated for a NiO film growing by lattice diffusion at 500°C (Atkinson, 1985). As seen from the plots, at the limits of their validity the Cabrera–Mott and Wagner theories are mutually compatible.

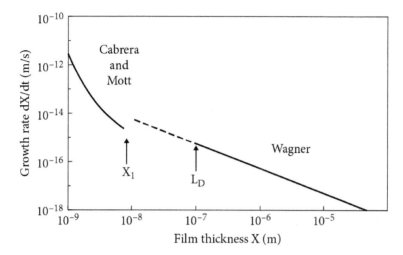

Fig. 2.3 Growth rate of a hypothetical oxide film as a function of thickness, calculated using the Cabrera–Mott theory when thin ($X < X_1$), and of Wagner when thick ($X > L_D$ and X_1). The space charge is significant only for $X > X_1$. Electrical neutrality is achieved in most of the film in the Wagner regime. The parameters used are appropriate for NiO growing at 500°C. After Atkinson (1985) with permission. Copyright 1985 by the American Physical Society.

Possible tests of various oxide growth theories in terms of their experimentally verifiable quantities are discussed in detail by Atkinson (1985). Wagner's theory can be readily compared with experiments and it is in agreement with fast growing oxides at high temperature (e.g. NiO, Fe_2O_3, Cr_2O_3, Al_2O_3, CoO). Oxide growth occurs by outward metal diffusion, with the exception of Al_2O_3, which appears to involve inward diffusion of oxygen (possibly along grain boundaries). At intermediate temperatures, the growth predicted from the lattice diffusion model is several orders of magnitude slower than experimentally observed. A possible reason may be diffusion along grain boundaries, which is much faster than diffusion within the lattice. The Cabrera–Mott theory qualitatively explains many observations, but there are also many anomalies. In general, it appears that it is difficult to fit experimental growth data to theoretical kinetic expressions to test the various theories: on the one hand, the predicted differences between the theoretical kinetic expressions for parabolic (Wagner), logarithmic (Cabrera–Mott), cubic, and other theories of growth are not as significant and the accuracy of experimental data is insufficient to allow a safe distinction between the various growth models.

Thermal oxidation of silicon. The growth of SiO_2 on silicon has been enormously important for the evolution of the present-day silicon device technology and it is worth separate treatment here. Thin layers of SiO_2 are used as a gate dielectric for metal-oxide-semiconductor (MOS) devices. SiO_2 as a dielectric material on silicon is characterized by a number of outstanding properties for electronic device fabrication: it is a good insulator with high stability, a good barrier layer for dopant atoms, and the

SiO$_2$-Si interface features perfect electrical behaviour with a low density of electronic trap states (the latter is due to the saturation of all Si bonds at the interface). The SiO$_2$-Si interface may be regarded as a gift of nature, on which the whole silicon-based digital revolution that we have been witnessing over the last 60–70 years has relied. An additional benefit of the SiO$_2$/Si system is that good quality oxide films can be readily grown by thermal oxidation.

The oxidation of silicon is significantly different from the oxidation of metals, in that the oxide grows in an amorphous state. This ensures the planarity of the oxide layer with no grain boundaries, which again is beneficial for device fabrication. Since the diffusion of Si atoms in SiO$_2$ is much slower than the diffusion of O$_2$ molecules, the oxide growth process occurs at the oxide-silicon interface. The accepted growth model of SiO$_2$ on Si has been proposed by Deal and Grove (1965) and it results in a combined linear-parabolic growth kinetics. The derivation of the Deal–Grove model is based on a steady-state condition of three different fluxes. The gas transport flux F_1 is given by:

$$F_1 = h_G(C_G - C_S) \tag{2.21}$$

where C_G and C_S refer to the gas and surface concentrations of oxygen, respectively, and h_G is a mass transfer coefficient derived from Henry's law. The diffusion flux through the oxide F_2 is derived from Fick's first law:

$$F_2 = \frac{D(C_S - C_i)}{X} \tag{2.22}$$

with D the diffusion coefficient, X the oxide film thickness, and C_i the concentration of oxygen at the oxide-silicon interface. The reaction flux at the oxide-Si interface F_3 is:

$$F_3 = k_i C_i \tag{2.23}$$

k_i is the interface reaction constant. The steady-state condition $F_1 = F_2 = F_3$ then leads to the Deal–Grove rate equation:

$$X^2 + AX = B(t + \tau) \tag{2.24}$$

A and B are rate constants and τ takes into account the existence of an oxide layer at $t = 0$; τ is the time corresponding to the growth of a pre-existing oxide, e.g. the so-called 'native' oxide layer (1–2 nm in thickness) that is used to stabilize (passivate) Si wafers.

For a thick oxide film, $X^2 \gg AX$ and $X^2 = B(t + \tau)$:

$$X \approx \sqrt{B(t + \tau)} \tag{2.25}$$

This is the parabolic growth regime, with B the parabolic rate constant. For a thin oxide film:

$$X^2 \ll AX \quad X \approx \frac{B}{A}(t + \tau) \tag{2.26}$$

this is the linear regime with B/A the linear rate constant. A and B are determined experimentally; A is proportional to the diffusion coefficient D, and B/A is proportional to the interface reaction constant k_i. The Deal–Grove equation thus describes an oxide growth, which starts fast (linear) and slows down (parabolic) with increasing oxide thickness.

What is the transition region between thick ($X > A$) and thin ($X^2 < AX$) in the Deal–Grove model? Experimental data for the dry oxidation of Si(100), i.e. using gas phase O_2, at 1000°C, yield $A = 165$ nm (Deal and Grove, 1965). Interestingly, the so-called wet oxidation of silicon using water vapour (Si + 2 $H_2O \longrightarrow SiO_2$ + 2 H_2) is much faster: the corresponding $A = 226$ nm (Deal and Grove, 1965). This has been associated with the faster diffusion of H_2O as compared to O_2 in SiO_2. However, the oxide resulting from dry oxidation is denser and of higher quality and the dry oxidation procedure is thus used for device fabrication. In the ultrathin linear regime of dry oxidation (< 50 nm), it has been observed that the SiO_2 growth is faster than predicted by the linear-parabolic description of the growth of thicker layers (Massoud *et al.*, 1985*a*; Massoud *et al.*, 1985*b*). The Deal-Grove model has thus been revisited and a more complex behaviour has been recognized (Massoud *et al.*, 1985*c*).

2.1.2   Oxidation of alloy single crystal surfaces

The selective oxidation of alloy single crystal surfaces is a practicable way to produce well-ordered ultrathin oxide films. This method of preparation has been used mainly for fundamental studies on oxide model surfaces under ultrahigh vacuum surface-science type conditions. The interaction of oxygen with binary alloy surfaces leads to the preferential segregation of one of the metallic components and the formation of an oxide surface layer; the metallic component that is more easily oxidized forms the oxide overlayer. The oxide film is initially disordered (amorphous), but high-temperature annealing promotes crystallization and long-range ordering in the oxide layer.

Employing this method, the transition metal aluminides (NiAl, Ni_3Al, FeAl) in the form of single-crystal surfaces have been primarily used to fabricate ultrathin ordered aluminium oxide layers (see the review of Franchy (2000) and references therein). The aluminides themselves have interesting engineering applications, e.g. as metallization layers in semiconductor heterostructure devices or high-temperature resistant environmental coatings. The latter application is based on the very effect discussed here, namely the formation of a compact passivating oxide layer by selective oxidation of the Al component. All low-index surfaces of NiAl and Ni_3Al (100), (110), (111) have been shown to form ultrathin, well ordered crystalline Al_2O_3-type films after exposure to oxygen and high-temperature annealing (1000–1200 K) (Franchy, 2000), though with some differences in the detailed atomic structures. The formation of Al oxide is facilitated by thermodynamics, the heat of formation of Al_2O_3 being much larger than the one for ($\Delta H_f(Al_2O_3) = -1690.7$ kJmol^{-1} versus $\Delta H_f(NiO) = -240.8$ kJmol^{-1} (Lide *et al.*, 2012)). The growth of Al-oxide under ultrahigh vacuum conditions (that is at typical oxygen pressures $\leq 10^{-5}$ mbar) is self-limited to 0.5–1 nm thickness, which corresponds roughly to two Al-O bilayers. It was originally thought that the structures of the Al-oxide films are derived from one of the polymorphs of Al_2O_3 (e.g. α-, γ-, or δ-Al_2O_3) consisting of essentially hexagonal O and Al planes, but the complex atomic structures have remained a riddle for more than ten years. The Al-oxide structure on NiAl(110) was finally solved by a combination of atomically resolved scanning tunnelling microscopy data and extensive density functional theory calculations (Kresse

et al., 2005). The idea of hexagonal oxygen planes, assumed previously from the structure of practically all known Al_2O_3 bulk phases, had to be given up and has been
replaced by a square and triangular arrangements of oxygen atoms, with the Al atoms
in between, slightly below the oxygen layer, giving an $Al_4O_6Al_6O_7$ stoichiometry of
the two Al-O bilayers, in contrast to the usual Al_2O_3 stoichiometry. The registry of
the substrate and the oxide is provided by the interfacial Al atoms, which are bonded
to the Ni rows of the NiAl part of the interface. Subsequently, using their experience with the alumina model on NiAl(110), Schmid *et al.* (2007) have also solved the
structure of the Al-oxide on $Ni_3Al(111)$ with a similar Al-O-Al-O stacking sequence,
involving square and triangular arrangement of atoms at the surface, but with a hole
in the unit cell reaching down to the Ni_3Al substrate. These holes are arranged in
an ordered sublattice and provide anchoring centres for the growth of monodisperse
metal clusters, making this alumina overlayer an excellent nanotemplate for the growth
of regular arrays of nanoparticles (Gragnaniello *et al.*, 2012). The ultrathin alumina
nanolayer films on NiAl alloy surfaces formed by high-temperature oxidation are thus
structurally and chemically very different from the bulk-terminated surfaces of the
various Al_2O_3 phases, and they are prototypical examples of so-called surface oxides,
which are discussed in the next subsection.

As mentioned previously, the growth of alumina films on all NiAl alloy surfaces
is self-limited in thickness to about two Al-O bilayers (0.5–1 nm), if the oxidation is
performed at low oxygen pressures. Under atmospheric oxygen pressure conditions,
the thickness of the oxide layer can be increased by roughly a factor of two (Franchy,
2000). Interestingly, an increase of the alumina thickness on $Ni_3Al(111)$ has also been
observed at low oxygen pressures under the influence of metal nanoparticles at the
oxide surface, which catalyse the additional growth of Al-oxide at the oxide-alloy
interface (Gragnaniello *et al.*, 2012). This indicates that the dissociation of the oxygen
molecules at the outer Al-oxide surface is the rate-limiting step for the oxidation: metal
nanoparticles can easily dissociate O_2 at their surface forming adsorbed O species,
which can readily diffuse through the oxide, in the case of the Al-oxide on $Ni_3Al(111)$
through the holes in the oxide lattice (as previously mentioned), and oxidise another
Al layer at the oxide-alloy interface.

The formation of Ga-oxide by oxidation of a CoGa(100) alloy surface (Franchy,
2000) is another example of the success of the 'alloy oxidation route' for oxide ultrathin
film growth. Amorphous Ga_2O_3 formed after oxygen exposure of CoGa(100) at 300 K
crystallizes to ordered β-Ga_2O_3 after heat treatment at 700 K. Again, the thermodynamic heat of formation condition, $\Delta H_f(Ga_2O_3) = -1,815$ kJmol$^{-1} > \Delta H_f(Co_3O_4)$
$= -905$ kJmol^{-1} (Lide *et al.*, 2012), enables the selective surface segregation of Ga
and the formation of Ga oxide.

The advantage and success of the 'alloy oxidation route' for the formation of highly
ordered quasi-2-D oxides as compared to the oxidation of the pure metals is based on
the high-temperature annealing step, which promotes crystallization and long-range
ordering. Metal alloys are more stable at high temperature than the oxide forming
metal constituents, and higher annealing temperatures can be employed to induce
ordering without melting the substrate. This is exemplified by the comparison of the
melting points: $T_m(\text{NiAl}) = 1955$ K; $T_m(\text{Al}) = 933$ K; $T_m(\text{Ni}) = 1728$ K.

2.1.3 Formation of surface oxides

The term 'surface oxide' has been coined to designate an ultrathin oxide phase that is formed under thermodynamic conditions, which are insufficient for the formation of a bulk oxide. A surface oxide may be regarded as a metastable intermediate between a chemisorbed oxygen layer on a substrate surface and the corresponding bulk oxide, thus representing a precursor stage to bulk oxidation. The atomic arrangement of a surface oxide is distinctly different from the respective bulk oxide and may be highly complex (see e.g. the discussion of the Al-oxides in the preceding subsection). Surface oxides as distinct phases have been described mostly on surfaces of the late transition and noble metals, such as Ru, Rh, Pd or Pt (Lundgren *et al.*, 2006). Different surface oxide stoichiometries and structures have been reported as a function of thermodynamic growth parameters (oxygen pressure, substrate temperature). Since these late transition metals are excellent oxidation catalysts, surface oxides may form under catalytic reaction conditions and may actually be the catalytically active phases. This has caused major interest in the catalytic community to elucidate the nature of these surface oxide phases.

A commonly observed structure motif of surface oxides is a hexagonal trilayer stack, which forms under oxygen pressure and temperature conditions of 10^{-5}–10^{-6} mbar O_2 and 500–700 K on the (100) and (111) surfaces of Rh, Pd, and Pt (O-Rh-O; O-Pd-O; O-Pt-O) with various registries with respect to the substrates. These surface oxide trilayers differ in structure and stoichiometry from their respective Rh_2O_3, PdO and PtO_2 bulk oxides.

As mentioned, the formation of a surface oxide may be considered as the initial step in the oxidation of a bulk metal. This step has not been addressed in the older theories of the thermal oxidation of metals as discussed in the previous subsections, but the more recent work of Reuter *et al.* (2002), using an extensive density functional theory (DFT) approach, has given an atomistic description of these initial stages of oxygen incorporation into the first metal layers of an Ru surface and the subsequent nucleation of surface and bulk oxides. The pathway from oxygen adsorption on the Ru(0001) surface to rutile bulk RuO_2 is characterized by three metastable precursor stages: i) adsorption of a dense oxygen monolayer, ii) formation of the O-Ru-O trilayer surface oxide, and (iii) transformation of the original trilayer via a lateral shift into a stacking fault geometry. With increasing concentration of oxygen at the surface, oxygen is incorporated into subsurface sites between the first and the second Ru layer, forming patches of an O-Ru-O trilayer. With the growth of trilayer islands, a registry shift of the trilayer into a stacking fault geometry with respect to the underlying Ru substrate becomes energetically favourable and thus a lateral displacement of the trilayer is predicted. With continuing oxidation, successive trilayers are formed, which are only loosely coupled to each other, leading to an expansion of the outer Ru layer distances. Beyond a critical thickness of the trilayer phase (> 2 trilayers), the transition to the RuO_2 (110) rutile bulk structure can easily occur.

It appears that this tendency to form subsurface oxygen islands, that destabilize the metal surface and form metastable surface oxides as precursors for a structural phase transition to the bulk oxide, is a more general phenomenon on metal surfaces. This mechanism provides a rational explanation of the initial stages of the formation of bulk oxide layers on top of a metal, whose further growth in thickness has been investigated by the 'classical' theories (Wagner, Cabrera–Mott, and so on).

2.2   Thin film deposition methodology

The fabrication of thin films by deposition requires the transfer of atomic (or molecular) species from a target material to a substrate surface. This transfer can be stimulated thermally, by photons or ions, or effectuated by transport via a carrier medium; in the latter case, precursor compounds of the film material are typically transported to the substrate, where a chemical reaction forms the desired final film compound. In this reactive deposition method, the molecular transport can occur through the vapour or the liquid phase. A formal and somewhat arbitrary classification of thin film deposition methodologies may distinguish between physical and chemical methods (Valeri and Benedetti, 2012). In physical methods, the deposition is achieved via a ballistic transport in the vapour phase and typically in a vacuum environment. Chemical methods employ carrier gases for the molecular transport or a liquid environment.

2.2.1   Physical methods

Physical vapour deposition (PVD), molecular beam epitaxy (MBE).   In PVD or MBE, an atomic or molecular beam is generated by thermal evaporation of an elemental or compound solid and made to impinge on a (heated) substrate. The technique of MBE was introduced by John Arthur and Alfred Cho from the Bell Telephone Laboratories in the late 1960s (Cho and Arthur, 1975; Arthur, 2002) to fabricate epitaxial thin films of compound semiconductors such as GaAs by evaporating Ga and As_4 species (thus the term 'molecular beam'): the As_4 units dissociate on the heated substrate surface and react with Ga atoms to form the GaAs film. Ultrahigh vacuum (UHV) environment and clean crystalline substrate surfaces are required to enable thin-film growth with quasi-ideal parameters, i.e. epitaxial single-crystalline order, growth in a layer-by-layer fashion forming an atomically sharp interface to the substrate while maintaining the bulk structure and stoichiometry of the film material. In practice, however, many mitigating factors may prevent this ideal outcome (Chambers, 2000). Today, the MBE term is used synonymously with PVD to denote vapour phase deposition techniques involving thermal evaporation of substances under UHV and well controlled growth parameters. The latter include an atomically defined state of the substrate surface, control of the substrate temperature and the evaporation rate, and a programmable deposition protocol. 'Reactive' PVD is performed by evaporation of an element in the presence of a gaseous reactant (oxidant), e.g. a metal in a controlled oxygen atmosphere. An interesting variant of the MBE technique has been reported recently using cluster molecules as precursors for oxide ultrathin film growth. A beam of $(WO_3)_3$ molecules, created by thermal sublimation of WO_3 powder, has been directed onto a metal single-crystal surface and the adsorbed $(WO_3)_3$ clusters have been made to condense by thermal treatment into a well-ordered epitaxial WO_3 film (Li *et al.*, 2011). This is a very gentle method of preparing ultrathin WO_3 films, in view of the high melting temperature of W metal, which makes the standard PVD method less convenient. It appears that this method is more generally applicable by using other volatile molecular oxide clusters (e.g. $(MoO_3)_3$ for MoO_3 films (Du *et al.*,

2016), and possibly other late transition metal maximum valency compounds). PVD is the method of choice for the preparation of ultrathin oxide films for fundamental surface science type studies, due to the UHV environment and the ability to achieve atomic precision in the film deposition procedure. An important issue in oxide thin film growth is the choice of the substrate. The crystallographic properties of the substrate such as lattice symmetry and lattice constant are essential ingredients for ordered epitaxial film growth. The lattice mismatch between the substrate and the film as defined by $m = (a_f - a_s)/a_s$, with a_f and a_s the film and substrate in-plane lattice constants, respectively, should be as small as possible to reduce interfacial strain. For $a_f < (>) a_s$ the film will be under tensile (compressive) stress before relaxation. A small amount of strain can be maintained in epitaxial films before it is released by the creation of defects (misfit dislocations) or morphological transitions (generation of mosaics, tilting of the surface, growth mode change, e.g. from layer-by-layer to island growth). Since the lattice strain can modify the physical and chemical properties of materials (e.g. magnetic anisotropy, surface chemistry), the engineering of strain in thin films by epitaxial growth on appropriate substrates is an interesting approach to design functional behaviour. In ultrathin films or two-dimensional (2-D) materials, the proper choice of the film lattice constant a_f has to be considered: since the reduction of dimensionality from 3-D to 2-D typically causes a reduction of the in-plane lattice constant, the use of the bulk lattice constant of the film material is inappropriate to determine the lattice mismatch m (Thomas and Fortunelli, 2010; Netzer and Surnev, 2016). An estimate of a_f of a free-standing layer of the film material is required, which is generally not available experimentally but is accessible by theoretical calculations (Thomas and Fortunelli, 2010; Barcaro *et al.*, 2010).

The substrate can play a passive role as an inert support or an active role in oxide thin film growth. In this context, the choice of metal versus oxide substrate surfaces has to be considered for oxide PVD film growth. On the one hand, the crystal symmetry match is easier to find between two oxides and thin oxide films grown on oxide substrates are more likely to retain a bulk-like structure (Chambers, 2000). On the other hand, the preparation of ordered metal surfaces is well established and the characterization of the grown oxide films is easier, since charging effects resulting from the application of charged-particle probes are less prominent. However, the metal surface may play an active role in the oxide growth, in particular in ultrathin films, supporting kinetically stabilized meta-stable phases as a result of interactions at the oxide-metal interface (Netzer and Fortunelli, 2016). For very reactive metals (e.g. Fe), sharp oxide-metal interfaces may not be achievable, and intermixing may occur at the metal-oxide interface (thermodynamic bulk phase diagrams of solubility may give an indication of its likeliness). The growth of intermediate inert buffer layers may be necessary to achieve phase separation between the metallic substrate and the desired oxide overlayer.

The choice of the gaseous oxidant in reactive PVD is determined by the condition that the oxidation of metallic deposits on the support surface should be much faster than the film growth (Chambers, 2000). For many metals, molecular O_2 can be easily dissociated at the support surface and the growth front and fulfils therefore the previously mentioned condition, whereby the chemical potential of oxygen μ_O determines the oxidation state of the oxide (in case of the existence of several stable oxidation states). Species with higher oxidation potential such as atomic oxygen, ozone O_3, or NO_2 (forming $NO + O$ at the surface) are alternative choices of more aggressive oxidants that have been successfully used in reactive PVD growth. The postoxidation procedure of metallic PVD films with one of the mentioned oxidants gives reasonable film qualities for ultrathin oxide films, but oxidation may remain incomplete for thicker films. Cycling of metallic layer deposition followed by postoxidation has been used to obtain thicker films. Finally, high-temperature annealing in a vacuum or in an oxygen atmosphere has to be emphasized as an important step to improve long-range ordering and single-crystalline quality in PVD grown oxide thin films.

Ion assisted deposition–sputter deposition.   Sputtering is a process, in which atoms are ejected from a target or source material that are deposited on a substrate as a result of bombardment of the target with high energy particles. Sputter deposition in its simplest configuration is carried out in a vacuum chamber, which is backfilled with a low pressure of rare gas such as argon, a dc voltage is applied between the target and an electrically conducting substrate on which the film is deposited. The voltage ionizes the gas to form a glow discharge plasma, consisting of Ar^+ and electrons. The Ar^+ ions are accelerated to bombard the target, and via momentum transfer, target atoms are sputter-ejected, some of which are deposited on the substrate. In reactive sputtering, mostly employed for oxide film growth, an oxidant gas is added to the vacuum chamber that reacts with the target material cloud creating a molecular compound, which then becomes the thin film coating. The primary reaction controlling the film composition occurs thus at the film-growth surface. In magnetron sputtering, a glow-discharge sputter deposition occurs with a closed, crossed electric and magnetic field. Various designs with different field configurations have been described (Kelly and Arnell, 2000). The magnetic field influences the electron trajectories enhancing ion formation in the plasma region and preventing them from bombarding the substrate. Faster deposition rates can thus be achieved. For a comprehensive treatise of thin film sputter deposition, we refer to the seminal review of Greene, which gives both a historical account and an overview of recent developments of the technique (Greene, 2017).

Pulsed laser deposition (PLD).   A high-power pulsed laser beam is focused inside a vacuum chamber onto a target material that is vaporized in a plasma plume, containing neutrals, ions, and electrons, and is deposited as a thin film on a substrate. For depositing oxide films, this process is commonly performed in an oxygen background atmosphere to ensure stoichiometric transfer of the material from the target to the substrate. The physical phenomena of laser-target interaction are complex, consisting of electronic and thermal excitations and evaporation, ablation, and plasma formation processes. Key deposition parameters are substrate temperature, laser energy, laser pulse repetition rate, and the ambient gas pressure, which have to be finely tuned to achieve good quality films. The outstanding feature of PLD is the ability to ablate practically any target material and to control the materials' stoichiometry. Since no surface segregation in the target material takes place during the short evaporation phase, a congruent transfer of the target compound material to the substrate surface is possible. PLD is thus particularly suited for the fabrication of thin films of complex multicomponent oxides. PLD has also been used for depositing complex oxide films in a new high-throughput method for screening local epitaxy and film properties: combinatorial substrate epitaxy (Zhang *et al.*, 2012; Havelia *et al.*, 2013). The strategy uses a well-characterized, polished polycrystalline ceramic surface (e.g. complex oxide) as a substrate, which consists of a large number of different grains. The orientation and structural quality of the deposited films is investigated locally on each grain, e.g. by using scanning probe microscopies or electron backscattering diffraction (Pravarthana *et al.*, 2014). High throughput correlations between film-substrate pairs, epitaxial crystal orientation, and functional properties can thus be generated. This approach has the potential to become a rapid and cost-effective screening method for the search of novel oxide phases (Lorenz *et al.*, 2016).

2.2.2 Chemical methods

Chemical vapour deposition (CVD), atomic layer deposition (ALD). In CVD, a substrate surface is exposed to volatile precursor compounds, which react and decompose on the heated substrate to produce the desired film. The transport of the precursors to the substrate is typically effectuated via a carrier gas. Volatile byproducts of the surface reaction can be removed by the gas flow through the reaction chamber (Choy, 2003). CVD is a very flexible technique as a wide range of chemical precursors, typically metal-organic compounds, are available. A distinct advantage of CVD is that it is a non-ballistic deposition technique, which enables the coating of complex-shape objects. Drawbacks of the CVD method are chemical and safety hazards due to the use of toxic and inflammable precursor gases and the difficulty to deposit multicomponent materials with well-controlled stoichiometry.

ALD is a CVD method based on sequential self-limiting reactions via alternating pulses of gaseous chemical precursors that react with the substrate surface, thereby producing one monolayer at the surface (this first step is called 'half-reaction'). Subsequently, the reaction chamber is purged with an inert gas to remove unreacted precursor, or reaction by-products. Then, a counter-reaction precursor pulse is applied, reacting with the monolayer of the first 'half-reaction' and producing one layer of the desired film material. This process of two consecutive 'half-reactions' is cycled until the appropriate film thickness is achieved (Johnson *et al.*, 2014). ALD processes are typically conducted at moderate temperatures ($<350°$C).

Spray pyrolysis, sol-gel processes. The techniques described here employ liquid precursor compounds and processing under ambient conditions. The methods are relatively cheap and suitable for large-scale production, but at the expense of the film quality (higher roughness, less crystallinity) and with reduced control over the deposition process (as compared to the PVD derived vacuum methods). Spray pyrolysis is a process in which a thin film is deposited by spraying a solution of precursor compounds onto a heated surface, where the constituents react to form the desired compound of the film. The chemical reactants are selected such that the reaction products other than the desired compound are volatile at the temperature of the deposition (Perednis and Gauckler, 2005). Spray pyrolysis is a common method for the fabrication of metal oxide thin films, in particular transparent conducting oxide films, on glass substrates. The quality and morphological properties of the film depend largely on the process parameters, the most important one being the substrate temperature. The other important parameter affecting the morphology of the deposited film is the precursor solution, which forms the aerosol during the spraying process.

Sol-gel processes involve the conversion of monomers into a colloid solution (sol) that acts as a precursor for an integrated network (gel). For oxide films, typical precursors are metal alkoxides. The precursor sol can be deposited on the substrate by dip-coating or spin-coating. Removal of the solvent is achieved by drying and thermal treatment (Brinker and Scherer, 2013). An advantageous feature of sol-gel processes is that small quantities of dopants can be introduced in the sol, which end up uniformly dispersed in the final film.

To finish this section, a brief comment on the choice of the method for thin film fabrication is added: it depends largely on the desired properties of the film, such as the film material and the required quality of the film and the film thickness and scale; ultimately the most appropriate method is determined by the targeted application. For studies of fundamental scientific character, the UHV PVD techniques are the methods of choice, providing atomic control over cleanliness, epitaxial order, and the process parameters. For large-scale applications where commercial considerations are important, the cheaper chemical methods are more suitable. For semiconductor device technology, techniques based on sputter deposition have provided both the required high film qualities and large throughput capacities.

2.3   Oxide nanoparticle fabrication methods

Nanoparticles (NPs) represent an important state of matter. They are the elementary building blocks of nanostructured materials and devices in bottom-up approaches, and they display spectacular structural, electronic, optical, magnetic and catalytic properties compared to bulk materials. Because their importance has long been recognized in many fields of science and technology—metallurgy, electronics, catalysis, biology, medicine, the food industry, energy, materials chemistry, chemical and drug manufacturing—intense efforts have been put to develop and refine synthesis protocols which allow a control of their shape, crystallinity, and size distribution. Aside from industrial interests, understanding the conditions under which NPs are formed in the environment is also of key relevance in the Earth sciences, due to their role in many geochemical and mineralogical processes.

Many methods of nanoparticle (NP) synthesis closely resemble those of thin film deposition, although no substrate is used on which condensation takes place. For an overview, a number of excellent reviews on the subject are available, including (Swihart, 2003; D'Souza and Richards, 2006; Stankic *et al.*, 2016). Depending on the authors, the chosen classification makes the distinction between physical and chemical methods, or between synthesis in the gas and the liquid phase. Considering the importance of oxide NP synthesis in liquids in laboratories and industries, as well as the prevalent NP formation in natural environment fluids, it is the latter presentation that we adopt in the following. However, to be exhaustive, a third class of methods should be mentioned, which are based on biogenic or biomimetic concepts. We will not cover them here, but refer the reader to recent reviews (Stankic *et al.*, 2016; Deravi *et al.*, 2007). Finally, one should also keep in mind that, in order to limit the environmental

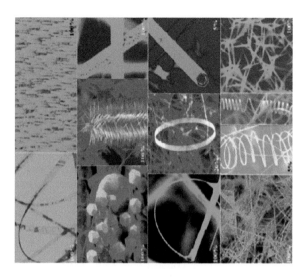

Fig. 2.4 ZnO nano-objects: rods, rings, tetrapods, belts, spirals, helices, combs obtained by synthesis in the gas phase. Reprinted from Wang (2004a).

impact of NP production methods, green chemistry routes are being researched, using lower-temperature conditions and safer reagents (Diodati *et al.*, 2015; Duan *et al.*, 2015).

2.3.1 Synthesis in the gas phase

NPs are produced in the gas phase when a vapour becomes supersaturated with respect to the solid phase (metal vaporization or combustion methods), or when gaseous molecules chemically react to form a condensed phase (chemical vapour synthesis). The general requirement to prepare small particles is to create a high degree of supersaturation, induce a high nucleation density, and rapidly quench the system to prevent the particles from growing.

Metal vaporization and combustion methods. The starting point for these methods is heating a metal to obtain a metallic vapour, which is then mixed with a cold inert gas (He for example) to induce condensation. As early as the 80s, this was the standard method for producing free metal clusters in laboratories. To obtain stoichiometric or non-stoichiometric oxide NPs, an oxidizing gas (O_2 or N_2O) was added to the inert gas (Ziemann and Castleman Jr, 1991). More recently, a similar method was used to produce ZnO NPs (Wang, 2004a) with a wide variety of shapes, such as nanorods, nanorings, nanobelts, nanospirals, nanohelices, nanocombs, etc. (Fig. 2.4). The processing parameters which fix the NP characteristics are the temperature, the gas pressure, its flow rate related to the pressure in the heating chamber, and the evaporation time.

Variants of this method use other means to produce the metallic vapour and its oxidation. In the *spark discharge* method, electrodes made of the desired metal are charged up to the breakdown point. An electric arc forms between the electrodes, which carries a small amount of metallic NPs. The *pulsed laser deposition technique* (PLD), described previously for the growth of thin films (Section 2.2), can also be used to produce NPs. The material of interest is vaporized by a laser beam and strongly confined in the plasma plume. Supersaturation results which induces NP nucleation. The advantage of this technique is the formation of NPs made of materials which cannot be easily evaporated.

Metal or metal oxide NPs can also be obtained by the *combustion method* (Patil, 2008). A metal wire is resistively heated in a mixture of inert and oxidizing gases until oxidation starts. The heat delivered by this reaction fosters the combustion and the production of NPs. As in the previous methods, the control of the particle size distribution may be obtained by playing with the total gas pressure and the relative pressure of the oxidizing gas (Stankic *et al.*, 2011).

Chemical vapour synthesis (CVS). In analogy with the chemical vapour deposition (CVD) method (Section 2.2), chemical vapour synthesis (Swihart, 2003; Djenadic and Winterer, 2012) uses chemical precursors under a vapour state, which are introduced into a hot-wall reactor to react or decompose under conditions that allow particle nucleation. The precursors can be solid, liquid, or gaseous at ambient conditions but are vaporized before being introduced in the reactor. This method allows a large range of nanomaterials to be produced thanks to the flexibility of using the many chemical precursors conceived for the CVD. Mixing several precursors may produce complex or doped compounds and adopting multi-step protocols allows synthesizing core-shell particles.

There are several variants of CVS. In the *flame synthesis* methods, the energy to decompose the precursors is in-situ provided inside a flame (Strobel and Pratsinis, 2007). Depending upon whether the precursors are in a vapour or liquid state, one distinguishes the *vapour-fed aerosol flame synthesis* from the *liquid-fed aerosol flame synthesis*. They allow large amounts of particles to be obtained in short times, for example, in the industrial production of silica, alumina, or titania NPs. While the basis of these methods have been known and used since antiquity, advances in a more rigorous control of the NP characteristics have been made in the last two decades.

2.3.2 Precipitation in a liquid medium

NPs are produced in a liquid medium when the solution becomes supersaturated with respect to the solid phase. Compared to the standard precipitation methods, many variants have been developed in order either to provide additional energy to the system (sonochemistry, microwave assisted methods), or to use confinement effects to better control the NP size distribution (sol-gel method, microemulsion method, template-based method).

Co-precipitation in water. The co-precipitation method relies on the dissolution of a salt precursor, generally a metal nitrate or chloride, in water leading to the formation of oxides, hydroxides, or oxyhydroxide NPs. In water, metallic cations M^{z+} may form aquo-hydroxo complexes $[M(OH)_h(OH_2)_N]^{(z-h)+}$ or oxo-hydroxo complexes $[MO_{N-h}(OH)_h]^{(2N-z-h)-}$, which, by association (olation and oxolation, respectively), give polyanions, polycations, or neutral entities. Since the condensation of charged species remains limited due to the charge increase as growth proceeds, polyanions or polycations are unable to lead to the formation of a macroscopic solid. At variance, the condensation of neutral complexes may proceed without limitation up to the solid phase. Generally, water is expelled during the precipitation of neutral aquo-hydroxo complexes, leading to hydroxide NPs, while oxide NPs are produced by condensation of oxo-hydroxo complexes involving the elimination of hydroxo ligands (Henry *et al.*, 1992; Jolivet, 2019). A wide range of oxide NPs has been synthesized in that way, with crystalline structures, sizes and shapes which can be fine-tuned by playing with the experimental parameters, such as the degree of supersaturation, the acidity of the solution, the presence of specific anions, the temperature, and so on. The method is simple, low cost, and corresponds to mild reaction conditions.

While specific protocols are needed in the lab to obtain a supersaturated solution, mineral nanoparticles are commonly encountered in nature, as a result of various geochemical processes (Banfield and Zhang, 2001). Among those which involve inorganic reactions is *chemical weathering*. Rain or underground waters progressively dissolve the rocks through which they travel, which results in supersaturation and NP nucleation in the pores of the rocks or in the bulk fluid. Granite dissolution, for example, yields quartz particles (sand) and clay minerals, the latter being composed of nanometre-thick aluminosilicate sheets (see Chapter 7). High degrees of supersaturation are also locally encountered close to deep ocean vents, in zones of mixing between waters, or in evaporites.

Under high-temperature conditions which foster the solubilization of reagents and the crystallization of the solid phase, the synthesis is said to be *hydrothermal* (*solvothermal* if the liquid medium is not water). In the lab, it is performed in an autoclave, while in nature, it may take place in deep regions of the Earth's crust or in the vicinity of magmatic chambers. Hydrothermal conditions allow the synthesis of compounds in a large range of crystalline phases, compositions, or dopings. A recent review (Hiley and Walton, 2016) presents selected examples of oxides with structures or compositions never reported before (ruthenates, iridates), unusual dopings (e.g. CeO_2 with Fe, Bi or Pd, or TiO_2 with Sn, Ru, La), and NPs with unusual/controlled morphologies. The stabilization of specific habits may be obtained by playing with the solution concentration, temperature, heating rate, pH, and additives (Wu *et al.*, 2016*b*).

Hydrothermal synthesis is commercially used to grow synthetic quartz, or gems like emeralds, rubies, alexandrite, and so on. Natural growth in aqueous solutions under high-T/high-p conditions can produce giant crystals, like beryls, quartz, or gypsum. The Naica cave in Mexico, for example, contains giant gypsum crystals which have formed at an extremely slow rate during more than half a million years. The largest ones reach 12 m in length, 4 m in diameter, and 55 tons in weight.

Precipitation in ionic liquids.   Most synthesized NPs exhibit low energy surfaces characterized by high atomic densities and slow growth rates (Section 2.4). However, crystal surfaces with a high surface energy are expected to be more catalytically active due to their larger proportion of undercoordinated sites and/or due to charge uncompensation in the case of polar surfaces. Strategies to stabilize them during crystal growth consist of providing charges by interaction with an ionic liquid, which may be a strong acid, like hydrofluoric acid in water, or a molten salt like $LiNO_3$ (Jiang *et al.*, 2010).

ZnO, TiO$_2$, or MgO NPs synthesized in that way display a large proportion of polar facets.

Sonochemistry.   The principle of *sonochemistry* relies on an ultrasonic irradiation of an aqueous or organic solution (Bang and Suslick, 2010) which, by cavitation, induces the formation, growth, and implosion of bubbles. Hence, a high energy is released in ultra-short times, leading to a local heating at a rate which may exceed 10^{10} K.s^{-1}, and to transient temperatures and pressures of the order of 5000 K and 1000 bars. In water, free OH$^{\cdot}$ and H$^{\cdot}$ radicals are formed, which can combine to produce H$_2$, H$_2$O$_2$ or HO$_2^{\cdot}$ by association with O$_2$. These oxidants or reductants are used for NP synthesis. This method has proved to be successful for oxide NP synthesis, leading to uniform size distributions, high surface areas, fast reaction times, and improved phase purity and crystallinity.

Ultrasonic spay pyrolysis.   *Spray pyrolysis* also makes use of ultrasound but at lower intensity and higher frequency. The solution is nebulized in microdroplets in which the chemical reaction takes place, assisted by gas heating. This method has been largely used in industry, especially for the production of multicomponent or composite materials.

Microwave assisted methods.   The principle of these methods consists in replacing the slow and rather inefficient conventional heating by furnaces or oil bath by microwave heating which is particularly appropriate when the materials or solutions contain charges or dipoles able to move and rotate under the effect of the electric field (Bilecka and Niederberger, 2010). Dedicated ovens are necessary for this purpose, with a tight control of the irradiation power and reaction temperature. While mainly applied to organic synthesis, microwave assisted methods are used more and more for the synthesis of a whole range of functional inorganic materials.

Sol-gel methods.   In the sol-gel methods (D'Souza and Richards, 2006), a colloidal suspension of hydroxide or oxo-hydroxide particles (the 'sol') is first produced by hydrolysis of precursors in an alcoholic solution. Then centrifugation or sedimentation may be used to remove a part of the liquid and initiate the 'gel' formation which is due to the interactions between the colloidal particles. The gel consists of a dense network of polymerized material. After the removal of the remaining solvent, a drying treatment yields ultra-fine powers. For this purpose, the use of supercritical gases may allow a better penetration in the porous medium and thus facilitate the NP formation (Sui and Charpentier, 2012). The sol-gel method had been initially conceived for the production of high-quality SiO$_2$ but is now currently used for the synthesis of many metal oxide NPs.

Microemulsion method.   This is an alternative method for the fabrication of NPs, in which the chemical reaction takes place in nanosized water droplets dispersed in oil, which represent confined reaction media (D'Souza and Richards, 2006). Surfactant molecules are added, with polar heads which point toward water and non-polar tails which face the oil surface. When the density of surfactant exceeds a critical value, the molecules associate to form micelles into which the reactants are introduced. In order to synthesize metal-oxide nanoparticles, usually two microemulsions are prepared and mixed, one with the metal salt of interest and the other with the reducing or oxidizing agent. During the collision of water droplets, a fast exchange of reactants and the chemical reaction take place. This method allows obtaining NPs with narrow size distribution, fixed by the droplet size in the microemulsion. However, large quantities of solvents and surfactants are needed, which is not environment friendly. For this reason, techniques based on the reverse configuration, i.e. oil droplets dispersed in water have been developed and shown to allow NP formation under milder conditions (Sanchez-Dominguez *et al.*, 2009).

Template-based methods. The idea of having confined reaction media to better control the NP size distribution also underlies template-based methods (Huczko, 2000). The most commonly used templates include porous membranes obtained by irradiation with nuclear fission fragments, porous alumina obtained by anodic oxidation of aluminium metal in an acidic solution, mesoporous silica prepared by sol-gel methods, zeolites, lamellar materials like clay minerals, or carbon nanotubes. The latter, for example, were used in the synthesis of various oxides (PbO, Bi_2O_3, V_2O_5, SiO_2, Al_2O_3, MoO_3, MnO_2, Co_3O_4, ZnO, and WO_3). Depending upon the shape of the confined space—pores, channels, interlayer spaces—spherical particles, individual or arrays of nanotubes, nanowires, etc. may be formed. The use of templates may be associated with various methods, electrodeposition, chemical, sol-gel, or CVD synthesis.

2.4 Nucleation and growth concepts

The kinetics of NP or thin film formation results from the combined effect of various elementary processes which take place in the liquid or gas phases, at the surface of the particles, or on a substrate. In a first step, called nucleation, the new phase appears with nano- or sub-nanometric dimensions. Growth then proceeds which displays specific features according to whether particles are formed in an homogeneous system (gas phase or aqueous solution), or whether a thin film grows on a substrate. Growth can coexist with other processes such as dissolution, ageing, Ostwald ripening, aggregation, fragmentation, and/or recrystallization, depending upon external conditions. Playing with thermodynamic parameters, such as temperature, pressure, acidity, and addition of counter-ions in the case of aqueous solutions, nature and structure of the substrate, is a way to control their size, shape, and stoichiometry.

2.4.1 Elementary processes

During nucleation and growth, atoms, or more generally growth units, leave the ambient phase, meet each other, and incorporate at the surface of the newly formed phase. Figure 2.5 sketches the major elementary atomistic processes which take place when thin films are grown on a substrate by vapour deposition (Venables *et al.*, 1984). They are also relevant for NP formation if what is represented as a substrate in Fig. 2.5 is understood as a facet of a growing particle.

Individual atoms first land on the substrate, with a rate proportional to the gas phase pressure, adsorb on it (process 1) and diffuse on it (process 2). Their net sticking coefficient is the result of a steady state which settles between the two reverse processes

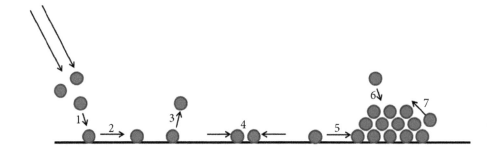

Fig. 2.5 Elementary atomistic processes taking place during the formation of a thin film on a substrate: 1): adsorption on the substrate; 2) surface diffusion; 3) desorption; 4) dimer formation; 5) incorporation into a pre-existing particle; 6) adsorption on-top a pre-existing particle; 7) diffusion on a forming particle.

of adsorption and desorption (process 3). Adsorption and diffusion on the substrate are controlled by specific energies E_{ads} and E_{Dsubs}. The adsorption energy E_{ads} is always larger on defect sites (step edges or kinks) than on flat surfaces. E_{Dsubs} represents the barrier that has to be overcome for diffusion to take place, which is a function of the local structure of the substrate (orientation, roughness). When a diffusing atom meets another one and binds to it, a dimer is formed (process 4) which represents a seed for further growth. Alternatively, individual atoms may bind to a pre-existing particle or an island in formation (process 5), directly land on it (process 6), and/or diffuse on it (process 7). The transport of atoms between different layers is controlled by step edge barriers of the Ehrlich–Schwoebel type (Schwoebel and Shipsey, 1966).

This simple atomistic description directly applies to the formation of a film made of a single type of atoms. The growth of oxide films, by post-oxidation of a metallic deposit or by dissociation of a molecular precursor on the heated substrate (see examples in Section 2.2), involves additional steps, among which are the precursor dissociation and/or the diffusion of the two (or more) species. For oxide NP synthesis in an aqueous medium, dehydration or dehydroxylation of the growth units also have to take place prior to incorporation into the growing particle.

All these processes are associated to characteristic energies and times which are functions of the atom concentration on the surface, and whose combined effect controls the kinetics of formation of the new phase.

2.4.2 Nucleation

Nucleation in systems which deviate from thermodynamic equilibrium occurs when the concentration of a solute in a solution exceeds the solubility of the solid phase (nanoparticle formation), when the partial pressure of a gas in a vapour exceeds the saturating vapour pressure of its liquid phase (droplet formation), when a melt is undercooled, when an electrolyte is submitted to an overvoltage, etc. (Adamson, 1960; Markov, 1995). In all cases, the emergence of the new phase necessarily requires a transient formation of small clusters (particles, droplets, bubbles) in the volume of the ambient phase, as first recognized by Gibbs (Gibbs, 1876a; Gibbs, 1876b). It represents a first-order phase transition close to the line of coexistence of the two phases. The nucleation stage is of key importance since it determines the number of nuclei which are created and will thus subsequently grow.

More precisely, the deviation from thermodynamic equilibrium is measured by the difference in chemical potential $\Delta\mu$ of one growth unit between the two phases present (the dilute and the solid phases, for example the vapour and the solid). The saturation state of the system I is defined as $I = \exp(-\Delta\mu/k_B T)$ (k_B the Boltzmann constant and T the temperature). It is equal to 1 at thermodynamic equilibrium where $\Delta\mu = 0$. $I - 1$ is thus the quantity which drives the instability of the system and its kinetic behaviour. If $I - 1 > 0$ condensation/precipitation of the solid phase takes place, while $I - 1 < 0$ indicates that the dilute phase is more stable so that dissolution may take place. When gas phase condensation occurs at constant volume and temperature, I is equal to the ratio between the gas pressure and the equilibrium pressure on the transition line. When nucleation takes place in an aqueous solution, I is equal to the ratio between the ion activity product IAP associated to the solid phase, divided by its solubility product K. For example, in the case of a simple mineral, like $Mg(OH)_2$, $IAP = [Mg^{2+}][OH^-]^2$, K is the value of IAP at equilibrium with the solution and the symbol [A] refers to the activity of ion A in the solution.

The capillary approach (Nielsen, 1964), which makes use of the bulk properties of the forming phase as an approximation, gives a good qualitative description of the phenomenon and has been widely used in various contexts, in particular when oxide nanoparticles are formed in an aqueous medium or in the gas phase. The nucleation step of thin film rather relies on an atomistic description involving elementary processes of addition or subtraction of monomers from the particles. Finally, in the last decade, multiple-step nucleation processes have been considered involving precursor phases or pre-nucleation clusters.

Capillary approach.   The simplest description of nucleation relies on a capillary approach which recognizes that the solid phase does not immediately gather all the excess of solute. Small nuclei first appear, which contain a few growth units (monomers of the same composition as the solid phase) in coexistence with the gas or aqueous phase. In the capillary approach, the number of growth units n in a nucleus is treated as a continuum variable and the nuclei are assumed to display the bulk atomic structure. These nuclei, being small, have a very large surface-to-volume ratio (typically equal to $4\pi\rho^2/(4\pi\rho^3/3) = 3/\rho$ for spherical particles of radius ρ), which means that there is a very large contribution of the interface energy (assumed equal to that of the bulk phase) to their free energy of formation. It is the competition between surface and volume effects which determines the kinetics of nucleation.

The change in Gibbs free energy in the nucleation process contains two contributions:

$$\Delta G = -nk_B T \ln I + X n^{2/3} v^{2/3} \overline{\gamma} \tag{2.27}$$

The first term $(-nk_B T \ln I)$ is proportional to the difference $\mu_B - \mu_A$ between the chemical potentials of the n growth units in the solid (B phase) and in the mother A phase (gas or aqueous solution). It represents the driving force to nucleation when negative $(I > 1)$, which means that the solid phase is thermodynamically stable under

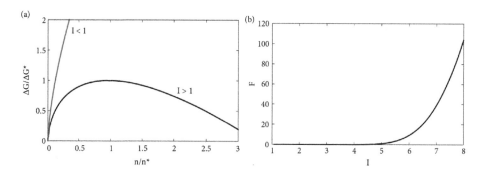

Fig. 2.6 Left: typical size dependence of the change in Gibbs free energy during nucleation, in the case of under-saturation ($I < 1$) and supersaturation ($I > 1$). Right: typical variation of the nucleation rate with the saturation state I.

the experimental conditions and the system is said to be supersaturated. However, for finite size particles, one has to take into account surface/interface effects, included in the second term which is positive and proportional to the particle surface area ($X n^{2/3} v^{2/3} = 4\pi \rho^2$ for spherical particles, v being the volume occupied by a growth unit in the solid and X a geometrical factor which depends on the particle shape) and the mean interface energy $\bar{\gamma}$, between the particle and the environment.

In the gas or aqueous phase, growth units constantly associate and dissociate. These fluctuations are driven by the dependence of ΔG on size. In the under-saturated regime ($I < 1$), ΔG is positive for all sizes and increases with n, so that the probability of having large clusters $p(n) \propto \exp(-\Delta G/k_B T)$ is small. At variance, when the system is supersaturated ($I > 1$), ΔG presents a maximum ΔG^* as a function of n, which, by analogy with the concept of activated complex in chemical reactions (Eyring, 1935), can be considered as the barrier that nuclei have to overcome during their size fluctuations (Fig. 2.6). This point defines the characteristics of the *critical nuclei*, in particular their number n^* of growth units, equal to:

$$ n^* = \frac{2u}{\ln^3 I} \qquad \frac{\Delta G^*}{k_B T} = \frac{u}{\ln^2 I} \qquad u = \frac{4 X^3 \bar{\gamma}^3 v^2}{27 (k_B T)^3} \qquad (2.28) $$

The maximum point of the ΔG curve represents an *unstable* equilibrium with the surrounding phase. The addition or subtraction of one growth unit to the critical nucleus irreversibly drives it towards increasing size or dissolution, respectively. The nucleation barrier ΔG^* is equal to one third of the surface term whatever complex the particle shape is. Nucleation occurs more easily (smaller ΔG^*) if v is small (smaller growth units nucleate more easily), $\bar{\gamma}$ is small (softer materials), or I is high (larger departure from equilibrium), and this is associated to smaller nuclei sizes n^*.

The previous derivation applies to both homogeneous and heterogeneous nucleation. In the latter case, a third medium intervenes, which can be a substrate in the case of thin film formation, the walls of a container in laboratory or industrial processes, or the pore walls inside rocks in the natural environment. To a first approximation, the two cases only differ by the value of the mean interface energy $\bar{\gamma}$.

For example, for tetragonal particles with a-, b- and c-oriented facets, $\overline{\gamma} = (\gamma_a \gamma_b \gamma_c)^{1/3}$ for homogeneous nucleation and $\overline{\gamma} = (\gamma_a \gamma_b (\gamma_c - W_{adh}/2))^{1/3}$ when the particle nucleates with its c-oriented facet on a substrate with which it interacts (adhesion energy W_{adh}). The role of a substrate is thus to decrease the average surface energy and, as a consequence, the nucleation barrier. Heterogeneous nucleation is indeed known to be always more efficient than homogeneous nucleation.

From eqn 2.28, one can deduce the Gibbs–Thomson equation, which, in the case of spherical particles reads $k_B T \ln I = 2\overline{\gamma} v / \rho$ (Markov, 1995). Its inversion yields the value of the saturation index $I_{\text{eff}}(n)$ of the environment with which a particle containing n growth units is in equilibrium:

$$I_{\text{eff}}(n) = \exp\left(\frac{2u}{n}\right)^{1/3} \tag{2.29}$$

In an aqueous solution, $I_{\text{eff}}(n)$ may be written $K(n)/K$, which defines the solubility product $K(n)$ of a particle of finite size containing n growth units, and shows that it is always larger than the solubility K of the infinite crystal. The relative increase of solubility becomes significant only below the micrometre size (for example, 5% for 0.1 μm and more than 50% for 10 nm NaCl particles). Alternatively, it defines the Laplace excess pressure $P(n) = P_0 \exp\left(\frac{2u}{n}\right)^{1/3}$ inside a particle which condenses in the vapour phase.

Specifically in the case of thin film formation, it is often assumed that the critical nuclei are dimers, which means that once two atoms meet and bind on the substrate, no further dissociation will take place. In some cases (layer growth of flat faces), 2-D nucleation processes take place. Its description in the capillary approximation is very close to that of 3-D nucleation just developed. Only the 'surface term' is modified, to take into account the border length (rather than the particle area) which scales as a 1/2 power of the particle area and the mean step energy $\overline{\gamma}$ (rather than the mean surface energy $\overline{\gamma}$). The change in Gibbs free energy then reads:

$$\Delta G = -n k_B T \ln I + X' n^{1/2} a^{1/2} \overline{\gamma} \tag{2.30}$$

with X' a constant which depends on the island shape ($X' \neq X$) and a the elementary area occupied by a (planar) formula unit. The nucleation barrier height and the number n^* of formula units in the critical nucleus are changed accordingly:

$$n^* = \frac{u}{\ln^2 I} \qquad \frac{\Delta G^*}{k_B T} = \frac{u}{\ln I} \qquad u = \frac{X'^2 a \overline{\gamma}^2}{(k_B T)^2} \tag{2.31}$$

Whatever the dimensionality of the nuclei (2-D or 3-D), they condense and dissociate randomly with time, with large size fluctuations not accounted for in the present approach. After some transient time, a steady state regime of nucleation is established. As in the transition state theory for chemical reactions, nucleation is an activated process. Its stationary rate F, i.e. the number of nuclei per unit time and volume, varies exponentially with the nucleation barrier:

$$F = F_0 \exp\left(-\frac{\Delta G^*}{k_B T}\right) \tag{2.32}$$

The pre-factor F_0 is not precisely specified in this approach. For nucleation in an aqueous medium, besides slowly varying factors, it contains the effect of an additional kinetic barrier due to the dehydration of the growth units before incorporation into the nuclei. However, in many instances, it is considered as an empirical constant. For moderate supersaturations, F monotonically increases with I (Fig. 2.6). However, there exists a metastable zone (sometimes called latency zone) delimited by a critical saturation state I_c, in which F is vanishingly small. I_c is arbitrarily defined by the condition that $F(I_c) = 1$ particle par second in the reference volume:

$$\ln I_c = \sqrt{\frac{u}{\ln F_0}} \quad (3 - \mathrm{D}) \qquad \ln I_c = \frac{u}{\ln F_0} \quad (2 - \mathrm{D}) \tag{2.33}$$

In the range $1 < I < I_c$, nucleation is thermodynamically allowed, but it cannot be observed in reasonable time scales. When I exceeds I_c, the nucleation rate strongly increases, and many particles are produced.

Atomistic approach.   The capillary nucleation theory just presented uses a continuous description of the solid phase, relying on the properties of the infinite phase. It is thus suited for the formation of large nuclei at small or moderate degrees of supersaturation. Alternatively, nucleation is better described at an atomistic level, when the saturation state is large, because the nuclei under formation then contain a small number of growth units (Adamson, 1960; Markov, 1995).

The atomistic approach accounts for the size fluctuations of particles A_n containing n formula units, through equations of the type:

$$A_n + A_1 \longleftrightarrow A_{n+1} \tag{2.34}$$

which assumes that growth or desorption occurs via addition or subtraction of a monomer A_1. The rate F_n of reaction (2.34) is the result of processes of association proportional to the A_n concentration C_n, and dissociation proportional to the A_{n+1} concentration C_{n+1}:

$$F_n = a_n C_n - b_{n+1} C_{n+1} \tag{2.35}$$

The coefficients a_n and b_n represent the probabilities of attachment and detachment of a monomer on a cluster of size n. The balance equation for A_n particles is:

$$\frac{\partial C_n}{\partial t} = F_{n-1} - F_n \tag{2.36}$$

The resolution of these coupled equations, made in the literature under various approximations for the a_n and b_n coefficients, leads to a complete time description of the state of the system. In particular, Zeldovich has proposed a continuum resolution allowing a link between the capillary and atomistic approach, which yields the following expression for the stationary nucleation rate (Markov, 1995):

$$F = aC_1 Z \exp\left(-\frac{\Delta G^*}{k_B T}\right) \tag{2.37}$$

Z is called the Zeldovich factor. It is inversely proportional to the curvature of ΔG at its maximum:

$$Z \propto \left(\frac{\Delta G^*}{3k_B T n^{*2}}\right)^{\frac{1}{2}} \tag{2.38}$$

Attempts to numerically estimate the prefactor of the exponential function in eqn 2.37 have usually resulted in huge (several orders of magnitude) discrepancies with measured values. However, the Zeldovich approach is interesting because it stresses that the stationary nucleation rate F is proportional to the association rate of the critical nucleus F_{n*}, which itself is the smallest of all the F_n. The formation of a critical nucleus thus represents the bottleneck of the nucleation process.

The capillary and atomistic theories of nucleation both pertain to the classical nucleation theory (CNT). Beyond it, statistical methods, most of which relying on Monte Carlo simulations, have been developed which account for entropic effects.

Beyond the classical nucleation theory: non-scalable regime; precursor phases; prenucleation clusters. While in the classical nucleation theory, the nuclei are assumed to have the same atomic structure and properties as those of the final solid phase, this assumption may break down at different levels.

First, many past experimental and theoretical studies of free clusters have shown that, from a structural point of view, two size regimes can be identified: the so-called scalable regime at large sizes in which the clusters may be considered as bulk pieces modified by small relaxation effects close to under-coordinated atoms, and the non-scalable regime below a critical size which depends on the compound. In the non-scalable regime, clusters adopt specific shapes and structures which are highly size- and temperature-dependent (see Chapter 6). Molecular dynamics (MD) or Monte Carlo (MC) simulations can be used to simulate their structure and calculate their free energy. The latter has no direct relationship with the bulk one, and the energy lowering which results from these structural changes increases the C_n probabilities of small clusters compared to those predicted by bulk properties. Applied to the critical nucleus, it yields an effective lowering of the nucleation barrier. One should note that the growth unit volume v and the interface energy $\bar{\gamma}$ also lose their significance in this regime.

Second, polymorphism, which is the ability of a compound to exist under several crystalline forms, is very frequent in oxides. For example, TiO_2 crystallizes in rutile, anatase, or brookite structures depending upon thermodynamic conditions. The same is true for Al_2O_3 which, depending on temperature and pressure, may adopt various crystallographic structures (α, γ, θ, κ, etc.) in addition to an amorphous one. The bulk polymorph which is thermodynamically preferred under the experimental conditions will determine the structure of the final product. However, at small sizes but still in the scalable regime, it may not be the most stable. This is, for example, the case for TiO_2 NPs which display an anatase structure rather than the rutile most stable bulk phase (Ranade *et al.*, 2002). This suggests that in the condensation process, anatase NPs will first form, but will have to eventually transform into rutile, either via a solid-to-solid

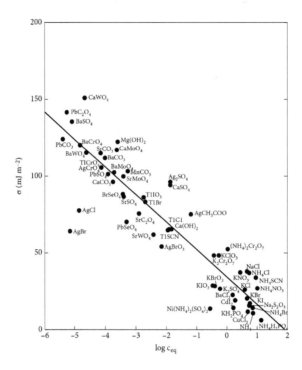

Fig. 2.7 Correlation between mean surface energies σ ($\bar{\gamma}$ in eqns 2.27 and 2.28) and the logarithm of equilibrium concentrations (equivalent of solubility products K). Reprinted from Söhnel (1982). Copyright 1982, with permission from Elsevier.

phase transformation or via a dissolution-recrystallization process. In this example, anatase thus represents the precursor of the rutile phase.

Third, while the previous considerations rely on a *thermodynamic* basis, some polymorphs of low stability may be *kinetically* favoured during the first steps of nucleation, due to their larger solubility. When nucleation takes place in an aqueous solution, it has been experimentally established (Söhnel, 1982) for many minerals that larger solubilities K are correlated to lower mean surface energies $\bar{\gamma}$ (Fig. 2.7). A lowering of the nucleation barrier ensues (eqn 2.28), which promotes the nucleation of the less stable polymorphs. This is why amorphous or hydrated more soluble phases often form as precursors of the most thermodynamically stable phases. However, in later stages of the precipitation process, a solid-to-solid phase transformation or a dissolution-recrystallization process has to take place to produce particles with the most stable polymorph structure.

Finally, another non-classical nucleation pathway in aqueous solutions has been revealed in the last decade, based on the existence of stable pre-nucleation clusters (PNCs) (Gebauer *et al.*, 2008; Gebauer *et al.*, 2014; Van Driessche *et al.*, 2016). Indeed, solute species which are substantially larger than ion pairs have been evidenced even in under-saturated solutions for many biominerals, calcium carbonates and phosphates, iron oxy-hydroxides, and silica. At variance with classical nuclei, PNCs are not solid particles, but rather poly-nuclear solute complexes exhibiting a highly dynamic character and a strongly hydrated composition. For example, in silica rich aqueous solutions, poly-nuclear entities with various linear, branched, rings, or cage structures are present, whose abundance and degree of protonation depends on the pH and the concentration of the solution. However, a number of these PNCs are polyanions or polycations which are known to be unable of leading to the formation of macroscopic solids. At variance, the condensation of neutral complexes may proceed without limitation (Jolivet, 2019) (Section 2.3).

2.4.3   Nanoparticle growth

As nucleation, crystal growth is a non-equilibrium kinetic process, which describes how the particle volume increases beyond that of the critical nucleus ($n > n^*$) (Markov, 1995). The incorporation rate of growth units scales with the adsorption energies on the various particle sites. It is the largest on kink sites, intermediate at step edges, and the lowest on flat surfaces. Consequently, the growth rates R of the three types of surfaces (flat 'F', stepped 'S', and kinked 'K') are usually in the order $R_F \ll R_S < R_K$, leading to a final particle habit in which, among all possible facet orientations, the most represented are those with the slowest growth rates.

More precisely, the elementary growth mechanisms are different on the different surface types. On rough surfaces (many kinks and steps), growth is limited by the incorporation at the particle surface, or by the diffusion of the elements in the mother phase. These are the most efficient processes at high saturation states. On flat surfaces, growth proceeds by 2-D nucleation and island spreading (step propagation) at moderate supersaturation. Finally, at low supersaturation, spiral growth may take place if screw dislocations are present (Burton *et al.*, 1951). These processes are schematized in Fig. 2.8.

When growth is limited by surface reactions (Fig. 2.8(a)), one may consider that the particles retain their equilibrium shape if surface diffusion is faster than the incorporation of growth units. The Wulff theorem then allows determining the particle shape as that which minimizes the total surface energy at constant volume (Zangwill, 1990; Muller and Kern, 2000) (see Appendix A). Critical nuclei displaying equilibrium shapes are those which are produced with the highest frequency since, in the framework of the nucleation capillary approach, the surface energy minimization implies a minimization of the nucleation barrier (Section 2.4.2). Simplified growth models assume that the overall growth rate is proportional to the deviation of the system from thermodynamic equilibrium I-1. However, in a closed system, this assumption implies that all nucleated particles grow and that their number never decreases, which is incompatible with the requirement of minimization of the total surface energy of the solid phase. To reach the true thermodynamic equilibrium, a size dependent growth rate is required, for example that given by the following expression (Wagner, 1961; Parbhakar *et al.*, 1995)(written in the case of spherical particles $n = 4\pi\rho^3/(3v)$):

$$\frac{d\rho}{dt} = \kappa \left(I - \exp\left(\frac{2u}{n}\right)^{1/3} \right) \tag{2.39}$$

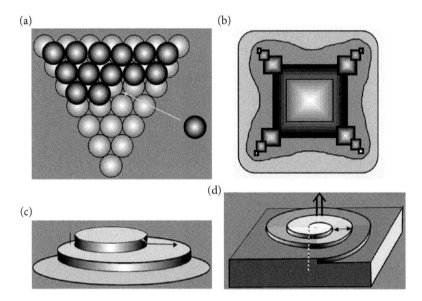

Fig. 2.8 Sketch of the four elementary processes of nanoparticle growth. a) growth of rough particles limited by surface processes; b) growth limited by diffusion in the mother medium; c) island growth by 2-D nucleation and spreading on flat facets; and d) spiral growth on flat facets driven by the emergence of a screw dislocation. Courtesy A. Baronnet.

On the right-hand side of this equation, one recognizes the quantity $I_{\text{eff}}(n)$, equal to the saturation state of an (hypothetical) environment with which a particle containing n growth units is in equilibrium (eqn 2.29). Moreover, the actual saturation state I is related to the number n^* of growth units in the critical nuclei by a similar relationship $I = \exp(2u/n^*)^{1/3}$. As a consequence, the time derivative of the particle size $d\rho/dt$ is positive or negative depending upon whether $n > n^*$ or $n < n^*$. This is the Ostwald ripening effect (Myhr and Grong, 2000; Ratke and Voorhees, 2002), also named coarsening, which describes the dissolution of sub-critical particles $(n < n^*)$ and the growth of super-critical ones $(n > n^*)$ simultaneously in the supersaturated medium. This exchange of matter occurs via the external medium and is controlled by the gradients of concentrations in the space between the particles. It is recognized as one of the most efficient mechanisms for decreasing the number of particles in the ageing phase close to thermodynamic equilibrium, but it may also occur in earlier stages of the precipitation process (Fritz *et al.*, 2009). In a closed system, where the saturation state I decreases with time due to the precipitation of particles and tends to 1 when the system approaches thermodynamic equilibrium, n^* constantly grows and tends to infinity at infinite time (eqn 2.28). Only the largest particle survives, which contains all the growth units initially available. This is the true thermodynamic equilibrium state of the *whole* system, including not only the aqueous solution $(I = 1)$ but also the solid phase (with a minimized total surface energy).

When growth is limited by diffusion in the ambient medium (Fig. 2.8(b)), the growth velocity is determined by the gradient of the growth unit concentration close to the surface. It is maximum close to the corners, and minimum on flat facets. The particles usually become hollow with dendrites growing at their corners. They display large surface-to-volume ratios which do not correspond to a minimum of the total energy. These growth shapes differ from the equilibrium shapes predicted by the Wulff theorem. The diffusion of atoms along the facets may counterbalance the dendrite formation if it is faster than the incorporation of atoms at the corners.

On flat surfaces, the adsorption of growth units is slower. Similar to what happens on a flat substrate, diffusion of adsorbed atoms is necessary to initiate an island formation (2-D nucleation), after which growth proceeds. Diffusing atoms which incorporate at the step edges of the island participate to the island spreading (lateral growth), while adsorption and diffusion on top of the island allow the formation of a second layer and thus induce vertical growth (Fig. 2.8(c)).

Finally, spiral growth can be observed around the emergence of screw dislocations on a flat surface (Fig. 2.8(d)). No nucleation barrier is involved to initiate growth which corresponds to a mere propagation of the ledge.

Aside from these individual growth mechanisms, aggregation processes occur when two or more particles coalesce to form a larger one. It involves the attraction, collision, and attachment between two NPs, facilitated by NP motion that brings the particles close to each other. As a whole, this process decreases the NP number. It usually takes place in solutions with a high concentration of small NPs, at high temperature and slow agitation speed. In the case of crystalline particles, an alignment of the crystallographic lattices has to take place via translation and/or rotation of the particles, and atomic diffusion allows a smoothing of the structural discontinuities at the contact. In the lab, aggregation is usually uncontrollable and thus considered undesirable. The use of capping agents or ways to modify the surface charges can prevent it.

2.4.4 Thin film growth

Thin film growth basically involves the same atomistic processes as those described previously, to which one has to add the competition between adsorption on the substrate and on growing particles/islands and strain effects associated with the difference in lattice parameters between the substrate and the particles/film when they are crystalline. The three primary growth modes of thin films on a substrate result from the combined effect of these processes (Bauer, 1958). They are schematized in Fig. 2.9.

In the Volmer–Weber mode (Fig. 2.9(a)), clusters of all sizes are formed, and a percolation occurs when the equivalent thickness of the deposit exceeds some critical value. This mode is found when the interaction between growth units is larger than their interaction with the substrate, for example when metal atoms are deposited on an oxide substrate. As in the case of NP growth described previously, the cluster shape depends on the rate-limiting process. If surface diffusion is fast enough, the clusters display equilibrium shapes which are given by an extension of the Wulff theorem (Muller and Kern, 2000) in which the surface energy of the face in contact to the substrate is replaced by the interface energy in the evaluation of the total surface energy (Wulff–Kaishev theorem) (see Appendix A).

The layer or Frank–van der Merwe mode (Fig. 2.9(b)) is a 2-D growth mode in which layers are formed one after the other. It allows obtaining thin films with a

(a)   (b)   (c)

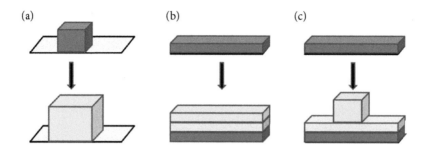

Fig. 2.9 Sketch of the three main growth modes of thin films on a substrate. a) the Volmer–Weber mode; b) the Frank–van der Merwe mode and c) the Stranski–Krastanov mode.

controlled thickness. Under some circumstances, a new layer may start forming before the previous one is completed, which leads to a monolayer-simultaneous-multilayer mode. These modes are encountered whenever the interaction of the growth units with the substrate is much larger than that between growth units. Ultra-thin oxide films on metallic substrates are obtained in that way, but also semiconductor films on semiconductors, rare gas films on graphite, and so on. Usually, the binding between subsequent layers decreases towards the value of the deposit bulk.

The Stranski–Krastanov mode, also named layer-plus-island growth mode (Fig. 2.9(c)) is an intermediate case between the two previous ones. It proceeds in two steps. After forming one or several layers in a layer-by-layer fashion (Frank–van der Merwe mode), subsequent layer growth becomes unfavourable and growth continues through the nucleation growth and coalescence of 3-D particles. This occurs when the interface energy between the last full layer and the deposit becomes high, due to some incompatibility between the lattice of the first layers and that of the bulk deposit (lattice mismatch, symmetry, orientation) leading to an accumulation of strain. At some critical thickness, a switch of the growth mode towards a Volmer–Weber mode occurs.

2.5   Summary

Contrary to the past opinion that oxides are 'dirty materials', there are now well-mastered protocols to fabricate oxide thin films with pre-determined thickness, long-range order, and precise stoichiometry. The same is true for nanoparticles, whether in the gas phase or in an aqueous medium, for which the control of temperature, oxygen partial pressure, solution pH, or ionic strength allows obtaining particle ensembles with a well-defined stoichiometry and a narrow size distribution function.

3
Methods of study

This chapter gives a description of the various experimental and theoretical methods which are mostly used for the characterization and modelling of oxide thin films and nanostructures.

3.1 Experimental characterization

In this section, the experimental methods are presented that allow us to study the physical and chemical properties of thin films, specifically oxide thin films and nanostructures. Given the wide variety of methods necessary to characterize thin oxide films and nanostructures, only a brief mention of the most important techniques, of their principles of operation, and of their information-giving capabilities will be discussed. This variety of characterization methods is also a reflection of the wide range of applications of oxide thin films, which encompass such diverse fields as heterogeneous catalysis, electronic device technology, optical and protective coatings, and magnetic nanostructures, to name just a few. For the purpose of presentation, we will divide the physico-chemical properties of oxides and classify their respective characterization tools into categories of general properties and specific properties. In the category 'general properties', we include geometry (atomic structure, morphology), chemical composition (including oxidation state and stoichiometry), and electronic structure (occupied-unoccupied electronic states). The more 'specific properties' refer to the magnetic and vibrational properties and to the chemical (catalytic) behaviour.

3.1.1 Oxide Geometry: Structure and Morphology

The tools for structure determination can be separated into real space and reciprocal space techniques. The former comprise microscopy imaging techniques such as scanning probe microscopies, electron microscopies, and photoemission electron microscopy. The latter include the diverse scattering methods, involving scattering particles such as photons, electrons or ions.

Real space imaging techniques. In *Scanning Probe Microscopy* (SPM) techniques, one introduces a parameter P which describes the tip-surface interaction. Accordingly, if P is the tunnelling current between the tip and the sample, the resulting instrument is the *Scanning Tunnelling Microscope* (STM); if P is the atomic force between the tip and surface atoms, the it Atomic Force Microscope (AFM) is generated. Likewise, the

Oxide Thin Films and Nanostructures. Falko P. Netzer and Claudine Noguera, Oxford University Press (2021). © Falko P. Netzer and Claudine Noguera. DOI: 10.1093/oso/9780198834618.003.0003

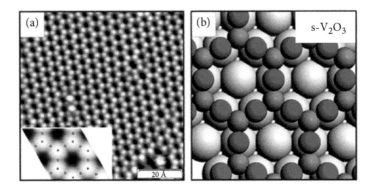

Fig. 3.1 STM image of the surface-V_2O_3 (2×2) phase on Pd(111) (a), and the corresponding DFT model (b) (green: V; red: O; grey: Pd). The insert in (a) displays the DFT simulated STM contrast. Reprinted from Surnev *et al.* (2003). Copyright 2003, with permission from Elsevier.

magnetic tip-surface force leads to the *Magnetic Force Microscope* (MFM), the surface potential is the probing parameter for the *Kelvin Probe Atomic Force Microscope* (KPAFM). The operating principle of all SPMs is similar: a sharp tip is scanned over a surface, the parameter P of the tip-surface interaction is measured and used to programme the tip-surface distance, generating an image of the surface in terms of the interaction parameter P, with atomic resolution in the best cases. For example, in STM, a bias is applied between the tip and a conducting surface, the tunnelling current between tip and surface is detected and used to control the tip-surface distance. Since the tunnelling current is exponentially dependent on the tip-surface distance, it is a very sensitive measure of the surface corrugation. By displaying the tip-sample distance at a constant tunnelling current along the scanning path, a topographic image of the surface in real space is generated. However, there is a caveat: the STM image is not purely of geometric origin. The tunnelling current depends not only critically on the tip-surface distance, but also on the Local Density of States (LDOS) of tip and surface, since it is proportional to the overlap of the wave functions of the tip and the surface atoms. STM images thus present a contour map at the constant surface LDOS close to the Fermi level, and electronic and geometric effects are intimately mingled up and difficult to disentangle, making the interpretation of STM images sometimes a difficult task. This is particularly true for oxide surfaces, where it is not always clear what the origin of the contrast (the maxima and minima) in the STM images (cations or anions) is. Theoretical simulation of the surface LDOS for the particular oxide structure is necessary to interpret the contrast in the STM images in terms of the atomic structure. This is illustrated in Fig. 3.1 (Surnev *et al.*, 2003). Panel (a) shows an atomic resolution topographic STM image, recorded in constant current mode, of a 2-D V_2O_3 layer on a Pd(111) surface displaying a so-called honeycomb structure. The DFT derived structure model is given in panel (b). The insert in (a) displays the DFT-simulated STM contrast, which reveals that the bright protrusions forming the honeycomb lattice are due to the V atoms, whereas the oxygen atoms are invisible.

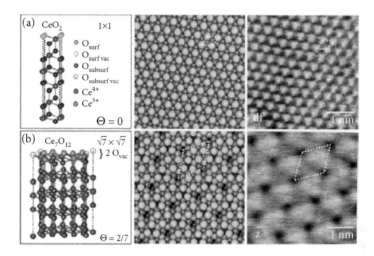

Fig. 3.2 Ceria phases observed with non-contact AFM (right column) and corresponding DFT structure models (left and middle columns). (a) (1×1) CeO_2; (b) $(\sqrt{7} \times \sqrt{7})R19.1°$ Ce_7O_{12}. Reprinted from Olbrich *et al.* (2017) with permission. Copyright 2017 American Chemical Society.

STM requires a minimum of conductivity of the sample or a very thin film of an insulating overlayer on a conducting substrate. In the latter case, electron tunnelling into the substrate modulated by the overlayer structure is possible and may provide meaningful STM images. For small band gap oxides, tunnelling into the conduction band at appropriate bias choices may also allow the recording of STM images. For further reading we refer to (Chen, 1993) for a general treatise of the STM technique, and to (Nilius, 2009) for a review of STM on oxide thin films.

For non-conducting samples, the AFM technique may be applied to image surfaces with atomic resolution. The tip on a cantilever brought into close proximity of a surface experiences interactions with surface atoms leading to a displacement that is recorded while the tip is scanned over the surface. Recording the height of the probe with respect to the sample that corresponds to a constant probe-sample interaction leads to an image displaying the atomic force contrast of the surface. A variety of forces (van der Waals, electrostatic, chemical bonding) may give rise to tip-sample interactions and the interpretation of AFM images is also not always straightforward, requiring theoretical simulations for the determination of the atomic structure of the surface. Because of the long-range nature of the relevant forces, the microscopic/mesoscopic shape of the tip is important; but as the shape of the tip is not always known, the potential of simulations is sometimes limited. Various modes of operation of the AFM have been established: contact, tapping, and non-contact AFM modes. In the latter non-contact mode, the cantilever is set into resonant frequency excitation and the modulation of the frequency due to the interaction with surface atoms is recorded and provides the image contrast. This non-contact AFM mode is used for high-resolution imaging, giving sub-atomic resolution in selected cases with tips that have been specially terminated with molecules (Giessibl, 2003). Figure 3.2 shows examples of non-contact AFM images from a thick ceria film: the right column gives the AFM images, the left and middle columns the DFT derived surface structure models. Figure 3.2 (a) shows the (1×1) structure of the $CeO_2(111)$ surface, whereas (b) is from the $(\sqrt{7} \times \sqrt{7})$ surface reconstruction of a slightly reduced Ce_7O_{12} surface.

STM and AFM are uniquely able to image atomic scale defects on solid surfaces, and this is most relevant for the characterization of oxide thin films, whose physical and particularly chemical properties are decisively influenced by the type and density of their defect population.

Electron microscopies allow a microscopic examination of surfaces and thin films either by electron-optical imaging with fixed and scanned electron beams, leading to the *(Scanning) Transmission Electron Microscopy* ((S)TEM) or *Secondary (Scanning) Electron Microscopy* (SEM) techniques. The basis of transmission electron microscopy is the de Broglie wavelength of electrons and the scattering of electron waves in matter. According to the diffraction limit, a point-to-point resolution comparable to the electron wavelength is possible in principle; for an electron wave with an energy of 300 keV, the de Broglie wavelength is ~ 2 pm. However, electromagnetic lens aberrations degrade the diffraction limited resolution by a factor of at least 10^2, and a resolution of order of 0.1 nm can only typically be reached in modern aberration-corrected TEM instruments. TEM is now a ripe technique with many sophisticated imaging modes, and we refer to the specialist literature for a more detailed discussion of the principles of image generation and of the various imaging modes (Williams *et al.*, 1998; Titchtmarsh, 2009; Spence, 2013; Goldstein *et al.*, 2017).

In STEM, a fine-focus electron beam is rastered across a thin sample and the transmitted electron intensity is detected as a function of position in the x-y plane. The resolution in modern STEM and (high-resolution) TEM instruments is comparable and better than 0.1 nm, i.e. subatomic point resolution can be achieved. The majority of modern (S)TEM instruments is designed to operate with a primary energy $E_p = 200$–300 keV. This range is the optimum compromise between spatial resolution, useful sample penetration thickness, and instrument size and cost. The electron-induced sample damage due to electron stimulated decomposition, atom displacement, and electron stimulated desorption processes is however an issue at these high electron illumination/penetration energies and has to be considered. For example, at $E_p \simeq 200$ keV the energy transfer between the electrons and sample atoms is of the order of 40 eV, which is typically above the threshold for atom displacement damage in solids. (S)TEM techniques can be applied to specimen of thickness up to 150 nm (due to the strong scattering/absorption interaction of electrons with matter). TEM images of a SnO_2 film deposited on an α-$Al_2O_3(012)$ substrate by the ALD method (a) and by conventional CVD (b) are shown in Fig. 3.3. High-resolution images displaying details of the SnO_2 surface (top) and of the SnO_2-Al_2O_3 interface (bottom) are also included in (a). The two chemical vapour deposition methods both produce epitaxial oxide films, but the TEM images clearly reveal a smoother morphology for the ALD-produced film (a) (see Chapter 2, Section 2.2). Figure 3.4 shows an example of a TEM image of oxide nanoparticles, MgO smoke particles; the figure also contains a plot of the particle size distribution.

(a)

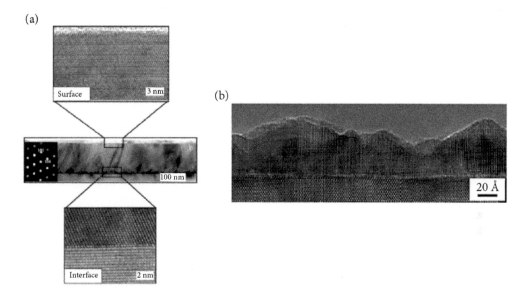

Fig. 3.3 Transmission electron microscopy images of SnO_2 films on α-Al_2O_3(012), grown by ALD (a) and CVD (b). High-resolution images of the film surface (top) and film/substrate interface (bottom) are included in (a). The insert on (a, middle) represents a selected area electron diffraction pattern. Reprinted from Sundqvist *et al.* (2006). Copyright 2006, with permission from Elsevier.

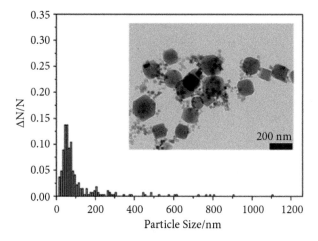

Fig. 3.4 TEM image and particle size distribution of MgO smoke nanoparticles obtained by burning Mg in a mixture of O_2 and Ar. Reprinted from Stankic *et al.* (2011). Copyright 2011, with permission from Elsevier.

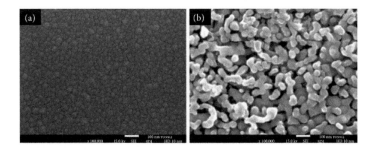

Fig. 3.5 SEM images of sputter-deposited ZnO films as deposited (a) and after annealing at 1000°C in ambient air (b). Reprinted from Ghosh *et al.* (2016).

SEM is a tool for the topographic visualization of the near-surface region of a sample, with a point-to-point resolution of 1–10 nm at best in secondary electron imaging mode. The focused incident electron beam ($E_p \simeq 0.5$–30 keV) is rastered over the surface of the sample, back-scattered primary electrons and secondary electrons, the latter generated by inelastic interactions of primary electrons with the solid are detected as a function of the (x, y) position revealing the surface topography. The contrast mechanisms rely on the dependence of the back-scattered electron intensity on the average atomic number Z of atoms in the interaction volume (primary electrons) and on the variations of the work function (secondary electrons) across the sample. SEM images of ZnO films deposited using a reactive sputtering technique are displayed in Fig. 3.5, showing the morphological properties of the films in the as-deposited state (a) and after high-temperature annealing (b). Nice examples of SEM images showing tetrapod-shaped ZnO nanorods, useful for high-resolution scanning probe tips or field-emission tip applications, are presented in Fig. 3.6.

Analytical energy-dispersive spectroscopy providing local chemical composition information can be combined with both SEM and (S)TEM-type techniques. While in the back-scattering mode of SEM instruments energy dispersive X-ray analysis (EDX) is mostly applied, electron energy loss spectroscopy (EELS) is used in the transmission modes. In EDX, the inelastically scattered incident electrons create core hole excitations, which decay by the chemically specific characteristic X-ray emission, which is detected. In EELS, the characteristic electron energy losses of primary electrons, due

Fig. 3.6 SEM images of tetrapod ZnO nanorods deposited by thermal evaporation of Zn chunks in air at around 850–1050°C. Reprinted from Ronning *et al.* (2005) with the permission of AIP Publishing.

to inelastic scattering via core hole excitation, are measured in an electron energy analyser after transmission through the sample. The spatial resolution in the spectroscopy modes is determined by the diameter of the incident electron beam (which can be < 1 nm), the energy resolution is higher in EELS, and of the order of a few tenths of an eV. Thus, modern aberration corrected STEM instruments with attached electron energy loss spectrometers are capable of chemical analysis at the atomic resolution level. The use of core-loss signals after energy selection to form energy-filtered atomic images with chemical specificity is an additional attractive feature of modern electron microscopy techniques.

(X-ray) Photoemission Electron Microscopy ((X)PEEM) is a microscopic and spectroscopic technique that uses emitted photoelectrons for image creation of surfaces with a lateral resolution of one to several tens of nanometres. The sample is irradiated with ultra-violet (UV) or X-ray photons stimulating the emission of photoelectrons, which are accelerated through electrostatic lenses and grids onto a 2-D detector, thus creating an image contrast via work function variations (UV photons) or chemical inhomogeneities (X-ray photons) across the sample surface. The spatial resolution of the technique is determined by the electron optics in the imaging column and limited mainly by the chromatic and spherical aberrations of the electron optics (as in TEM). Laboratory UV light sources based PEEM typically enables a somewhat higher spatial resolution ($\sim$ 10 nm) than X-ray PEEM utilizing synchrotron radiation sources (several 10 nm), but the ability of chemical sensitivity in XPEEM images, employing the photoelectron emission from specific core levels, makes XPEEM an attractive and powerful method for the analysis of heterogeneous surfaces (Grinter and Thornton, 2011). The variable polarization available at synchrotron radiation sources allows one to use magnetic linear and circular dichroism-based photoemission for contrast and image formation. The imaging of magnetic domain structures has thus become possible. As a full field microscopy technique, PEEM images may be recorded in $\sim$ 100 ms, allowing real-time observation of surface topography changes and thin film growth phenomena.

Reciprocal space techniques–Diffraction Methods. The reciprocal space methods do not directly reveal the position of atoms or of their electron distribution as the real space techniques, but instead measure the Fourier transform of the atom-atom correlation function, from which the direct space location of the atoms has to be retrieved by a suitable back-transformation. All diffraction techniques are reciprocal space imaging techniques. Here the treatment will concentrate on the most popular ones involving electrons and photons as scattering particles.

The analysis of the structural properties of (ultra-)thin films requires methods with a certain sensitivity to a thin layer of the material, preferably to a layer in the nanometre thickness range; that is, the surface sensitivity of the methods is an important property. Surface sensitivity can be achieved in two ways: i) by strongly interacting probe particles to reduce the transmission to deeper layers into the solid, or ii) by an appropriate scattering geometry for weaker interacting particles to minimize the scattering contributions of the bulk. In the first category, slow electrons are the most common probes, whereas photons belong to the second category. The strong interaction of electrons with matter can be expressed by the electron *inelastic mean free*

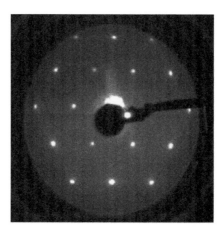

Fig. 3.7 LEED pattern of a 2-D V_2O_3 layer on Pd(111) showing a (2×2) surface structure. Electron energy 60 eV.

path λ, which follows a so-called 'universal curve' for most materials as a function of the electron energy, displaying a broad minimum in the energy range from 20 to 1,000 eV with $\lambda < 2$ nm, and $\lambda \sim 0.5$ nm around 100 eV (Seah and Dench, 1979). Photons interact weakly with matter and X-ray diffraction is the most common technique for the determination of the bulk structure of solids. To achieve the surface sensitivity required for thin film applications, surface X-ray diffraction employs a grazing incidence scattering geometry below the critical angle for total external reflection. Critical angles for typical X-ray wavelengths of a couple of Å (appropriate for revealing the atomic structure) are a few tenths of a degree. This in combination with the low scattering cross section of X-rays necessitates a brilliant (i.e. intense and highly collimated) X-ray source, as nowadays available in synchrotron radiation laboratories.

Low-Energy Electron Diffraction (LEED) is the oldest and the most common surface structure determination technique and has been developed in the early 1960s, when the vacuum technology had reached the state to enable the generation of ultrahigh vacuum. The strong interaction of slow electrons with matter requires a high vacuum in the experimental scattering chamber to avoid the scattering of electrons with gas phase particles in the residual atmosphere. In a typical LEED experiment, an electron beam in the energy range 10–200 eV is back-scattered from the surface, the diffracted electrons are displayed on a fluorescent screen or another type of detector that monitors their spatial distribution and records the diffraction pattern. The diffraction pattern is a projection of the surface reciprocal net at a magnification determined by the incident electron energy and gives an immediate impression of the symmetry of the surface. Figure 3.7 shows a simple LEED pattern of a 2-D V_2O_3 layer on Pd(111), displaying a (2×2) surface structure. The LEED reflections at the corners of the outer hexagon of the picture are from the Pd(111) substrate, whereas the additional spots are due to the (2×2) V-oxide surface structure (Netzer and Surnev, 2016). Note that this LEED pattern is from the same 2-D V_2O_3 /Pd(111) surface as shown in Fig. 3.1.

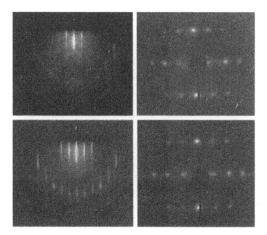

Fig. 3.8 RHEED (left panels) and LEED (right panels) patterns of a $TiO_2(110)$ surface (top) and of the (110) surface of a thick RuO_2 film grown on the TiO_2 substrate (bottom). Reprinted from Kim *et al.* (1997*a*). Copyright 1997, with permission from Elsevier.

The display nature and simplicity of a typical LEED optics instrument together with the easy assessment of a variety of surface parameters such as long-range order, translational symmetry, and surface unit cell dimensions have made LEED the most popular surface structure determination method in surface science (Ertl and Küppers, 1985; Woodruff, 2016; Conrad, 1996). Measurement of the intensities of the diffracted LEED beams as a function of the kinetic energy (= wavelength) of the electrons gives the so-called LEED (I–V) curve. From the (I–V) data, the atomic structure of the surface can be deduced with a theory that treats the multiple scattering of electrons in an appropriate way. Due to the strong electron-matter interaction, strong multiple scattering effects are prevailing, which have to be treated in a complex dynamic scattering theory (Vanhove *et al.*, 2012). The experimental simplicity of LEED is thus offset by the more complex data analysis (as compared to surface X-ray diffraction, see following section).

In *Reflection High-Energy Electron Diffraction* (RHEED), high-energy electrons of 10–100 keV primary energy are reflected from a surface in grazing incidence scattering geometry, the latter to ensure surface sensitivity. At these high electron energies, forward scattering is dominant and the diffraction pattern yields reflections on an arc that corresponds to the intersections of a large-radius Ewald sphere (determined by the wave vector k of the electron wave) with the reciprocal lattice rods of the surface (Woodruff, 2016). From the RHEED pattern and the intensities of the diffraction spots, the surface translational symmetry and the unit cell dimensions and, with an appropriate theoretical analysis, the position of the atoms within the unit cell can be deduced in much the same way as in LEED. Figure 3.8 shows a comparison of RHEED and LEED patterns from a TiO_2 (110) and RuO_2 (110) surface. The grazing angle of intersection of the Ewald sphere and the reciprocal net rods, the latter broadened by surface imperfections, leads to a *streaking* of the diffraction spots in RHEED, as seen in the respective patterns in Fig. 3.8. From the streaking and the shape of the RHEED spots, the nature and degree of surface imperfections can be deduced.

Although the multiple scattering of electrons at RHEED electron energies is reduced, RHEED has been rarely used for quantitative surface structure analysis (possibly due to the lack of commercial RHEED theory packages (Woodruff, 2016)). However, the great advantage of the RHEED technique lies in the experimental set-up, i.e. the scattering geometry of the experiment, which allows one to access the sample surface during observation of the diffraction pattern: for example, while growing a thin film by physical vapor deposition methods. RHEED is thus able to monitor in real-time the atomic layer-by-atomic layer growth of epitaxial films, by observing the oscillations in the intensity of diffracted beams in the RHEED pattern; this is of great value during MBE growth of thin films (Chambers, 2000).

In the *Surface X-ray Diffraction* (SXRD) technique, the diffraction of X-rays from a surface at grazing incidence angle conditions leads to lines of scattered intensity along the reciprocal lattice rods normal to the surface, corresponding to the Bragg peaks of the 3-D bulk structure. The intensity variation along the surface reciprocal lattice rods, called crystal truncation rods (CTR) in X-ray terminology, contains information on the atom positions in the surface layer, which can be retrieved by comparison with single scattering structure simulations. The great advantage of SXRD over electron based techniques is that the kinematic (single scattering) approximation of the scattering problem is adequate. Apart from the CTRs, the surface structure may give rise to superstructure rods, which are caused by fractional order diffracted beams due to a surface superstructure. The intensities along superstructure rods contain information only on the relative location of surface atoms within the surface layer, but no information on the registry of surface atoms to the bulk. Since surface scattering is weak, the scattered intensity along CTRs at Bragg peak positions is dominated by the scattering from the substrate bulk. However, well away from the 3-D diffraction conditions on the CRTs, the intensities of substrate and overlayer scattering are approximately equal and the overlayer scattering may therefore be assessed more accurately. As an example, Fig. 3.9 shows the scattered intensity along a CTR from the famous (7×7) reconstruction of the clean Si(111) surface. The diverging high intensity at the perpendicular momentum transfer positions -4, -1, $+5$ are the *Bragg peaks* due to the Si bulk structure. It is seen that the (7×7) surface reconstruction causes a significant modulation of the scattered intensity *in between* the Bragg peaks, as compared to a bulk terminated (1×1) simulation (dashed curve).

With the rececnt availability of synchrotron radiation sources all over the world, the inherent disadvantage of SXRD of low X-ray scattering yields has been overcome and SXRD has been established as an important tool for surface structure determination. The experimental complexity, as e.g. relative to LEED, is offset by the relative simplicity of the kinematic structure analysis. The possibility to measure structure phenomena under ambient pressure conditions and to study buried interfaces is another benefit of the SXRD technique.

In contrast to diffraction techniques such as LEED and SXRD, which rely on long-range sample order, *Photoelectron Diffraction* (PED) is a reciprocal space structure determination technique that is sensitive to short range order. In addition, it provides

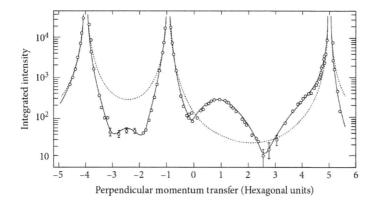

Fig. 3.9 The (1,0,l) CTR from the (7 × 7) reconstruction of the Si(111) surface. The dashed curve is a fit to the data using only the bulk CTR intensity. The solid curve is a fit to a 4-layer model of the interface. Reprinted from Robinson *et al.* (1986) with permission. Copyright 1986 by the American Physical Society.

atomic, chemical and site specificity. Photoelectrons emitted from a core level undergo elastic scattering from the atoms surrounding the emitter atom, the outgoing electron wave displays coherent interference effects and angular-dependent intensity variations as a function of the environment of the emitting atom, thus on the structure. The intensity variations are manifested in the scattering direction, i.e. the angular distribution, and also are a function of the kinetic energy of the photoelectrons, i.e. the photon energy of the exciting X-rays. This accounts for two detection modes: i) scanned angle PED, in which the photoemission intensities are measured as a function of the detection angle at constant photon energy (often called X-ray Photoelectron Diffraction (XPD)); and ii) a mode in which photoemission intensities are measured as a function of the photon energy. XPD has the advantage that laboratory X-ray sources can be used, whereas scanned photon energy PED requires the use of synchrotron radiation. The detection of photoemission from the core levels of particular atoms determines the chemical specificity of the technique. By comparing the measured intensity distribution with scattering simulations, the structural surroundings of and bond distances to the emitter atom can be determined.

Contrary to the scattering techniques mentioned previously, which are sensitive to the geometry of the atomic structure, the *Small Angle X-ray Scattering (SAXS)* method is sensitive to electron density variations at the nanoscale (1–100 nm) (Gerson *et al.*, 2009). It is used for the determination of nanoparticle shapes and size distributions, the sizes of pores in porous solids, and characteristic distances in partially ordered materials. If the experiment is performed at *grazing incidence (GISAXS)*, thus acquiring surface sensitivity, the method is suitable for the investigation of the morphology and preferential alignment of nanoscale objects on surfaces or inside a thin film (Renaud *et al.*, 2003).

3.1.2 Oxide composition–chemical analysis

The spectroscopy of electrons emitted from solids is the basis of the major tools for the compositional analysis of surfaces and thin films. The detection and energy analysis of the electrons emitted with kinetic energies in the range 5–2,000 eV gives information on the electronic energy levels in the solid, more specifically from the surface and from surface-near regions. The latter is the result of the high probability of inelastic scattering of such electrons, which causes energy losses during the transmission through the solid and reduces the useful sampling depth for the detection of electrons that are characteristic of a particular quantum excitation. The sampling depth is usually described by the inelastic mean free path of electrons or the electron attenuation-length λ, which takes into account both inelastic and elastic scattering processes. The attenuation of an electron beam after transmission through a distance d in a solid is proportional to $exp(-d/\lambda)$, with $\lambda \leq 2$–3 nm in the previously mentioned energy range and only weakly dependent on the material (Seah and Dench, 1979). The emission of electrons from matter may be stimulated by photons and electrons, and this forms the basis of the two outstanding electron spectroscopy techniques, photoelectron spectroscopy and Auger electron spectroscopy. These are the major spectroscopies for the chemical analysis of atoms at surfaces, and this discussion will be centred mainly on these two techniques. Two related spectroscopies, X-ray absorption and electron energy loss spectroscopy will also be mentioned shortly. Finally, an ion-based element specific spectroscopy, ion scattering spectroscopy (ISS), which is less common but has some merits for the investigation of oxide thin films, will be described briefly at the end of this subsection.

Both *X-ray Photoelectron Spectroscopy* (XPS) and *Auger Electron Spectroscopy* (AES) for the chemical analysis of surfaces are core level spectroscopies, which involve the ionization of core levels by an X-ray or a high-energy electron beam. The emitted electrons are detected outside the solid and their kinetic energy is analysed (Hüfner, 2013; Woodruff, 2016). Since the core level binding energies are characteristic of the atomic species, the respective photoelectrons and Auger electrons bear information on the chemical identity of the emitting atoms.

The photoemission process, i.e. the absorption of a photon and the emission of a photoelectron is described by the modified Einstein relation:

$$E_{kin} = h\nu - E_B - \phi \tag{3.1}$$

where E_{kin} is the kinetic energy of the photoelectron with reference to the vacuum level (i.e. outside the solid), $h\nu$ is the energy of the absorbed photon, E_B is the binding energy of the core level electron with respect to the Fermi level, and ϕ is the work function of the solid. Equation 3.1 is essentially a manifestation of the law of energy conservation during the photoemission process. Figure 3.10 shows an XPS spectrum of a thin WO_3 layer on an Ag surface, stimulated by the 1,253 eV photons of the Mg Kα X-ray emission line.

The spectrum displays a number of peaks due to the photoemitted electrons from the W and O atoms of the oxide overlayer as well as from the Ag atoms from the underlying substrate. The XPS peaks are superimposed on a background of secondary

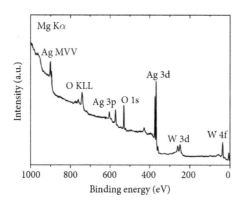

Fig. 3.10 XPS survey spectrum of a thin WO₃ layer (approximately 3 monolayer-thick) on an Ag(100) surface; excitation by Mg Kα X-rays ($h\nu = 1{,}253$ eV).

electrons, which are created by inelastic scattering processes of the emitted photoelectrons. In addition to the photoemission peaks, features due to Auger electron emission (see the following text) are also visible in the XPS spectrum (e.g. O KLL). The core-level binding energies, as measured in the XPS spectrum, are not only element specific, but are also sensitive to the chemical environment as a result of the influence of the valence electron distribution on the respective inner core level energies. These chemical shifts of core electron energies can amount to several eV, e.g. between an atom in a metal and in the corresponding oxide, and they are a sensitive measure of the oxidation state of an atom. For a further discussion of XPS and of the origin of chemical shifts of core-electron binding-energies we refer to (Hüfner, 2013; Woodruff, 2016).

The emission of an Auger electron is the result of the decay process of an excited atom after the ionization of a core level. The excited atom with a hole in the core state can decay by filling the core hole with an electron from an outer shell, the respective energy difference can be emitted in the form of an X-ray photon (X-ray emission process) or of another electron from an outer shell, the so-called Auger process. For low binding energy core-level ionization ($\simeq 2{,}000$ eV), Auger emission is the dominant decay process. The kinetic energy of an Auger electron is given by

$$E_{kin} = E_A - E_B - E_C - U \tag{3.2}$$

$E_{A,B,C}$ are the one-electron binding energies, as measured in XPS, and U is the hole-hole interaction energy in the two-hole Auger final state. Since the Auger process involves a combination of core-level binding energies, the energies of Auger electrons are also element specific. It is customary to designate the core level states involved in an Auger transition in X-ray spectroscopy notation: K; L_1, $L_{2,3}$; M_1, and so on. For example, for an Auger transition with an initial hole in the $1s$ level, a $2p$ electron filling the hole and emission of another $2p$ electron, the notation is $KL_{2,3}L_{2,3}$. Shallow electron levels in the valence band are designated with the symbol V. The initial core hole may be created either by a high-energy electron or an incident X-ray photon. Thus Auger electrons are also detected in the XPS spectra (see the O KLL or Ag MVV features

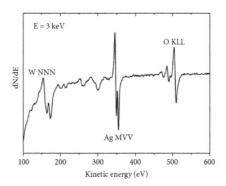

Fig. 3.11 AES spectrum of a thin WO$_3$ layer on a Ag(100) surface; excitation by 3 keV electrons.

in Fig. 3.10). Since the production of high-intensity electron beams with energies 1.5–5 keV is relatively easy and cheap (Rivière, 1990), AES is predominantly performed with incident electron beams. AES spectra are usually recorded by differentiation of the distribution curve of emitted secondary electrons (of which the Auger electrons are a part) with respect to energy, in order to emphasize the weak Auger peaks on the large secondary electron background. Figure 3.11 shows an AES spectrum of the thin WO$_3$ layer on Ag(100) (the same sample as shown in Fig. 3.10). The W NNN and O KLL Auger transitions of the oxide layer as well as the Ag MVV transitions of the substrate are clearly apparent. Note that the Auger features in the spectrum all consist of several individual Auger transitions (e.g. O KLL: KL$_1$L$_1$; KL$_1$L$_{2,3}$; KL$_{2,3}$L$_{2,3}$).

The primary excitation processes in XPS and AES, namely the absorption of an X-ray photon and the creation of a core hole by electron impact, respectively, form the basis of two other core level spectroscopies: *X-ray Absorption Spectroscopy* (XAS) and *Electron Energy Loss Spectroscopy* (EELS).

In XAS, the incident X-ray beam energy is scanned across an energy range, which includes the energy of excitation of a core electron to a continuum (i.e. unbound) state, giving rise to an abrupt increase of the absorbance, the absorption edge. The recording of an XAS spectrum requires an energy tuneable X-ray source, which is nowadays readily available with the use of synchrotron radiation. The intensity which is absorbed after transmission of the X-ray beam through the sample, as compared to the incident X-ray intensity, when plotted versus energy gives the XAS spectrum, which is traditionally a bulk measurement. However, the method of obtaining the XAS (X-ray absorbance) data determines the analytical information depth of the technique. If the XAS spectrum is measured via the number of electrons emitted from the sample, which is proportional to the absorption of X-ray photons, the technique can be made surface sensitive. Various methods of measuring the emitted electron yield have been developed: total yield (all the emitted electrons are considered, as most simply measured via the drain current, i.e. the current from ground to the sample, required to cancel the current of electrons emitted from the sample in response to the X-ray absorption); partial yield (integration of the electron yield over a chosen range of kinetic energies, measured with a simple band-pass filter); Auger yield (the Auger electrons emitted in a preset energy window, following the creation of a particular core level hole, being proportional to the number of absorbed photons, are detected; the measurement requires an Auger spectrometer, i.e. an electron energy analyser). The partial electron yield and Auger electron yield measurements of the X-ray absorbance are more surface sensitive than the total yield measurement, however at the cost of lower signal intensities in the former cases. This is because the slow secondary electrons ($\simeq 20$ eV), which are created by multiple inelastic cascade processes and which form the major part of the emitted electrons, have a larger escape depth than the higher energy electrons. The XAS spectrum at a core level threshold consists of the main absorption edge (called the white line, for historical reasons) and of distinct pre-edge and post-edge features, in a narrow energy range before and beyond the absorption edge. These features are due to electron excitation from the core level to *empty bound* states below and above the vacuum level. This energy region of the XAS spectrum close to the main absorption edge is called the *X-ray Absorption Near Edge Structure* (XANES) or *Near Edge X-ray Absorption Fine Structure* (NEXAFS) region and contains information on the unoccupied electronic energy levels (Stöhr, 2013). The XAS region beyond discrete one-electron excitations features intensity oscillations of the absorption coefficient due to the scattering of the emitted photoelectron waves; this is the Extended X-ray Absorption Fine Structure (EXAFS) region, from which structural information of the surroundings of the emitter atom may be derived. For further information on the XAS technique we refer to the comprehensive text of Koningsberger and Prins (1988).

The inelastic scattering of electrons in matter results in a characteristic energy-loss spectrum of an incident electron beam, as a result of well-defined quantum excitations. The energy losses of core electron excitations corresponding to the ionization of core levels are distinctly element specific. In EELS, the electronic excitation spectrum of the sample is measured via the characteristic energy losses of the primary electrons, whereby the surface sensitivity can be tuned by the appropriate choice of the electron primary energy. The energy-loss features in the secondary electron distribution curve can be distinguished from other features, e.g. Auger electrons, because they scale with the primary energy of the incident electrons (the Auger electrons occur at a fixed kinetic energy). Apart from core-energy loss peaks, electron energy losses due to valence electron excitations (one-electron inter-band transitions, collective plasmon excitations) are observed in the lower energy-loss region (1–30 eV) of the EELS spectra, which contains information on the electronic structure in the valence band region (see Section 3.1.3, ELS). The advantage of the EELS technique is that it can be performed with the same instrumentation as AES (electron gun, electron energy analyser) and thus can be applied as a complementary technique to AES, quasi for free (Rivière, 1990). The incorporation of EELS spectrometers into modern high-resolution STEM instruments has led to a recent revival of the EELS technique and has opened up possibilities for chemical analysis via core level spectroscopy with sub-nm spatial and good energy (< 1 eV) resolution. This is a very attractive feature for nanostructure research. Apart from electronic excitations, vibrational quantum excitations (in the 100 meV range) at surfaces can be probed by the EELS method: this high-resolution version of the EELS technique is presented in the section dealing with the study of vibrational properties of oxide thin films.

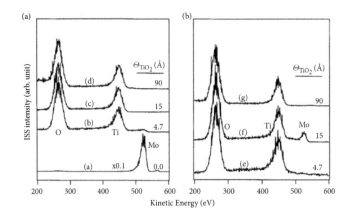

Fig. 3.12 ISS spectra of (a) clean Mo substrate; (b) and (e) 0.47 nm TiO$_2$ film; (c) and (f) 1.5 nm TiO$_2$ film; (d) and (g) 9 nm TiO$_2$ film. ISS spectra (b), (c), and (d) were taken after the growth of the TiO$_2$ film at 300 K. ISS spectra (e), (f), and (g) were obtained after annealing the film to 1,200 K. Reprinted from Oh *et al.* (1997) with permission. Copyright 1997, American Vacuum Society.

Ion Scattering Spectroscopy (ISS) (also referred to as *Low-Energy Ion Scattering (LEIS)*) is a technique in which a beam of ions (typically < 10 keV) is scattered from a surface. The kinetic energy of the scattered ions is measured: the peaks observed in the ion energy spectrum correspond to the elastic scattering of ions from specific surface atoms. Each element produces a peak at a different measured kinetic energy, which is caused by the momentum transfer between the incident ion and the surface atom. Scattered ions have an energy for a specific scattered angle that is defined by the masses of the scatterer and the scattered particle. Thus, an energy spectrum measured in an ISS experiment can be converted immediately into a mass spectrum, which provides an elemental analysis of the surface (Rivière, 1990; Woodruff, 2016). Since strong scattering signals are limited to the topmost atom layer, ISS is extremely surface sensitive. The high surface sensitivity is the result of several effects: *shadow cones* of the ion trajectories of the topmost scattering atoms and the high probability of the (relatively) slow ions of being neutralized after more than one collision. This surface sensitivity is of interest in the analysis of oxide materials to determine the chemical species that form the outermost surface layer. Figure 3.12 presents ISS spectra recorded during the growth of a TiO$_2$ film on a Mo(100) surface. The peaks of the Mo substrate and of Ti and O atoms of the growing film have been used to follow the growth process and the evolution of the oxide film morphology after annealing.

3.1.3 Electronic structure

The most direct method to probe the occupied electronic energy levels in the valence band region, which are responsible for most of the physical and chemical properties of matter, is the *Ultra-violet Photoelectron Spectroscopy* (UPS). The physical principle of UPS, absorption of a UV photon and emission of a photoelectron, is the same as in XPS and is described by the Einstein eqn (3.1), which determines the kinetic energy of

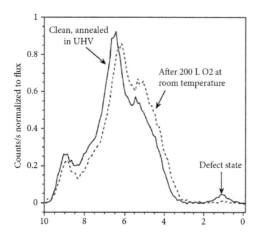

Fig. 3.13 Photoemission spectra ($h\nu = 35$ eV, normal emission) from the valence band region of a sputtered and UHV-annealed, clean $TiO_2(110)$ surface. After adsorption of oxygen at room temperature, the defect state in the bands gap region disappears. Reprinted from Diebold (2003). Copyright 2003, with permission from Elsevier.

the photoelectron as measured outside the solid. Experimentally, lower photon energies (than in XPS) in the UV region are typically used for valence band photoelectron excitation, which has the advantage of a high photoionization cross section and that good experimental resolution can easily be achieved. In home laboratories, rare gas discharge UV photon sources are used, yielding narrow resonance lines of UV radiation (e.g. $He_I = 21.2$ eV; $He_{II} = 40.8$ eV; line widths ~ 20 meV). Alternatively, continuously tuneable UV radiation is available in synchrotron radiation laboratories, where the photon energy can be selected by monochromators, with the additional benefit that the radiation is polarized (linear or circular). In addition to energy conservation (expressed by the Einstein relation (3.3)), momentum conservation has to be fulfilled upon photoexcitation. This means that a photoelectron detected at a particular angle outside the surface arises from a discrete initial energy and reduced momentum (i.e. $\vec{k}$ vector) state within the solid. This opens up the possibility to probe the electronic band structure of a solid $E(\vec{k})$ by UPS, specifically by *Angle-Resolved UPS (ARUPS)*. For a more detailed description of momentum conservation in ARUPS and its application to the mapping of band structures we refer to the more specialized literature (Ertl and Küppers, 1985; Woodruff, 2016; Kevan, 1992; Plummer and Eberhardt, 1982).

Here we note that an angular averaged UV photoelectron spectrum is essentially a replica of the joint density of states, weighted by the probability of transition between the initial (occupied) and final (empty) eigenstates involved in the photoexcitation process. The transition probability T is given by the Fermi Golden Rule (Ertl and Küppers, 1985):

$$T = \frac{2\pi}{\hbar} | < \psi_f | H | \psi_i > |^2 \delta(E_f - E_i - h\nu) \tag{3.3}$$

where ψ_f, ψ_i are the final and initial state wave functions, respectively, and H is the

Hamiltonian supporting the photoexcitation process. The last δ function term reflects the energy conservation of the process, namely that the photon energy must match the energy difference between initial and final state (Ertl and Küppers, 1985).

Figure 3.13 shows UPS spectra of the valence band region of $TiO_2(110)$ surfaces prepared by two different procedures: a vacuum (UHV) annealed surface and a surface after exposure to oxygen.

The TiO_2 valence band region is dominated by the O $2p$ emission from ~ 4 eV to $\sim 9 - 10$ eV binding energy. At the annealed surface, a peak at ~ 1 eV binding energy is apparent due to a Ti 3d derived point defect state, which is created by oxygen vacancies and a concomitant reduction of some surface Ti atoms from 4+ to 3+. After adsorption of oxygen, the defect state in the band gap region disappears as the oxygen vacancies become filled and the whole spectrum shifts by 0.2–0.3 eV to higher binding energy due to band bending.

The UPS technique also allows the measurement of the work function ϕ of the sample surface. The width of the UPS spectrum ΔE, i.e. from the Fermi energy (binding energy $= 0$) to the secondary electron cut-off (kinetic energy $= 0$) is given by $\Delta E = h\nu - \phi$. With rare gas resonance UV lines, where the photon energy is exactly known, measurement of the secondary electron cut-off is easily done and it gives an important physical characteristic of the surface, ϕ, *quasi for free.*

Resonant Photoemission Spectroscopy (RPES) is a variant of UPS, in which the photoemission from localized states in the valence region (d or f states) is excited with a photon energy of a core level resonance. This leads to a significant enhancement of the photoionization cross section of the respective valence level and thus to an increase in the experimental sensitivity. The process is illustrated with the photoemission spectra from the $4f$ level in a cerium oxide ultra-thin film (Fig. 3.14).

Cerium oxide can exist in two stable oxidation states (3+ and 4+, corresponding to Ce_2O_3 and CeO_2), with the two electronic ground state configurations $4f^1$ and $4f^0$, respectively. Photo-absorption $4d \longrightarrow 4f$, at a resonance energy of around 122–125 eV, leads to a $4d^9 4f^2$ intermediate state, which can decay via a Super–Coster–Kronig process:

$$4d^{10}4f^1 + h\nu \longrightarrow 4d^9 4f^2 \longrightarrow 4d^{10}4f^0 + e \tag{3.4}$$

The direct photoemission channel:

$$4d^{10}4f^1 + h\nu \longrightarrow 4d^{10}4f^0 + e \tag{3.5}$$

coherently interferes with the indirect absorption-decay channel, leading to an enhancement of the $4f$ photoionization cross section. Figure 3.14 displays valence band spectra of a ceria monolayer film on a Rh(111) surface after growth (as-laid: (a,b)) and after annealing at 600°C (c,d). Spectra (b,d) have been excited by a photon energy $h\nu = 125$ eV, corresponding to the maximum of the $4d \longrightarrow 4f$ resonance (on-resonance), whereas spectra (a,c) have been excited with a photon energy of 110 eV, i.e. in the pre-resonance region (off-resonance). The as-laid film shows no signs of $4f$ emission between E_F and 2.5 eV binding energy, indicating a stoichiometric CeO_2 film. However, after annealing the appearance of the $4f$ emission at ~ 1 eV binding energy in the on-resonance spectrum (d) signals the presence of Ce^{3+} species, generated by a temperature-induced reduction and the formation of oxygen vacancies. Note that this

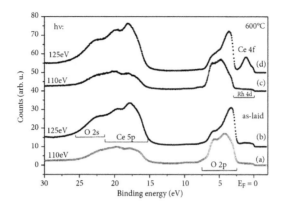

Fig. 3.14 Valence band photoemission spectra of a 1.3 monolayer Ce-oxide film on a Rh(111) surface, as-laid at 250°C (a,b) and after annealing at 600°C (c,d). Spectra (b) and (d) have been excited by a photon energy $h\nu = 125$ eV corresponding to the maximum of the Ce $4d \longrightarrow 4f$ resonance (on-resonance), while spectra (a) and (c) have been excited with a photon energy in the pre-resonance region $h\nu = 110$ eV (off-resonance). The Rh valence band emission is weak in all spectra, because these photon energies are at the Cooper minimum of the Rh $4d$ photoionization cross section. Reprinted from Eck *et al.* (2002). Copyright 2002, with permission from Elsevier.

Ce $4f$ signal is too weak and not apparent in the off-resonance spectrum (c). The peak at 3–3.5 eV, also seen in the on-resonance spectra (b) and (d), has been ascribed to O $2p$ states hybridized with Ce valence states. In conclusion, RPES is a powerful technique for the detection of a small concentration of minority species with localized states in the valence band region.

Among the *spectroscopies of unoccupied electronic states*, the *Inverse Photoemission Spectroscopy* (IPES) may be regarded as the time-reversed photoemission process: an incident electron loses a discrete amount of energy in falling from a continuum state into an unoccupied bound state of a target via emission of a photon, thereby providing information on the state into which it falls. The energy spectrum of emitted photons carries information of the density of unoccupied states. In practice, IPES is typically performed by detecting photons of a fixed energy (isochromat spectroscopy), electron primary energy is scanned such that the final state energy of the inverse photoemission process is swept through the empty state region of interest (Ertl and Küppers, 1985; Woodruff, 2016). The cross section of the radiative capture of an electron by a target surface is much lower than that of the direct photoemission process (by $\sim 10^{-5}$), so that a high current electron source and an efficient photon detection system is required. The latter has been realized in form of a Geiger–Müller-type detector with a simple band-pass filter (Dose, 1983).

Since both energy and momentum conservations rule the inverse photoemission process, $\vec{k}$-*Resolved Inverse Photoemission Spectroscopy* (KRIPES) can be performed by varying the angle of incidence of the incoming electron beam. Energy band mapping of unoccupied electronic states is thus possible (Woodruff, 2016).

The X-ray absorption spectrum may be separated into the near-edge region around the absorption threshold, which contains empty state information, and the extended far-edge fine structure region ($\sim$ 40–50 eV above threshold). The absorption intensity modulations in the EXAFS region, created by the interference of the outgoing electron waves by scattering from neighbouring atoms of the emitter atom, carry information of the local geometric structure and are used for fundamental structure analysis (Ertl and Küppers, 1985). EXAFS is an important method for the characterization of oxide nanoparticles, since it allows one to probe the local environment, e.g. the metal-oxygen bond distances, as a function of particle size. The absorption fine structure in the near-edge region (XANES or NEXAFS spectroscopy) contains mostly electronic information (Gerson *et al.*, 2009; Stöhr, 2013). Features in the X-ray absorption spectrum around the absorption edge of a core level (pre-edge and in the region immediately above the edge) are the result of the excitation of a core electron to unoccupied bound (conduction band) states, whereby the dipole selection rules have to be obeyed ($\delta l = \pm 1$; $\delta s = 0$; $\delta j = 0; \pm 1$; e.g. transitions $s \longrightarrow p$, $p \longrightarrow d$ states). A qualitative analysis of XANES structures may be based on a comparison of measured spectra with standard spectra, e.g. of bulk oxides. The following qualitative features of absorption in the XANES region have been recognized (Gerson *et al.*, 2009): i) the position of the absorption edge is sensitive to the elemental oxidation state, to first approximation; ii) the pre-edge features reflect the local coordination of the absorber atom, e.g. tetrahedral versus octahedral coordination; iii) the *white line* intensity, i.e. the peak intensity immediately above the absorption edge, may be indicative of the relative occupancy of the band (orbital), into which the electron excitation is occurring (the term *white line* is historic, the strong X-ray absorption appeared as a white line on the photographic film used as a detector).

Figure 3.15 reports Mn $L_{2,3}$ XAS spectra in the XANES region for various Mn oxides. The top spectrum is from a Mn-oxide monolayer on a stepped Pd(100)-type surface, with a formal Mn_3O_4 stoichiometry (Franchini *et al.*, 2009*b*). For display purposes, the spectra are presented with approximately equal size. The oxidation state of the Mn cations is indicated on the right hand side of the figure. The comparison of spectra shows a general similarity of the Mn-oxide monolayer on Pd (top curve) with Mn_3O_4 (second curve from bottom), indicating the presence of both Mn^{2+} and Mn^{3+} species, but there are differences in the details, as expected from the particular structure of the oxide nanowires (Franchini *et al.*, 2009*b*; Franchini *et al.*, 2012).

A quantitative interpretation of XANES spectra requires calculations of the multiple scattering of the excited photoelectron waves, which is a significant process at the low kinetic energies in the XANES region.

Scanning Tunnelling Spectroscopy (STS) can be performed in a scanning tunnelling microscope to probe the Local Density of States (LDOS) of both occupied and unoccupied states on a surface by measuring the tunnelling current response to a linear bias ramp at a constant tunnelling gap (Chen, 1993; Nilius, 2009). By keeping the tip-sample distance constant, the z-dependence of the tunnelling probability T is disabled, and the tunnelling current I at bias V is dominated by the LDOS of tip (ρ_T) and sample (ρ_S) :

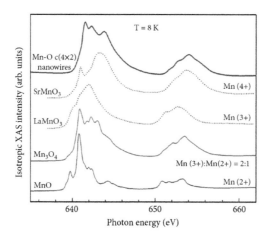

Fig. 3.15 Mn $L_{2,3}$ XANES spectra of various Mn-oxides (Courtesy, S. Altieri). The top spectrum is from a Mn-oxide monolayer on a stepped Pd(100) surface, forming nanostripes with a c(4 × 2) superstructure and a formal Mn_3O_4 stoichiometry (Franchini *et al.*, 2009*b*).

$$I \propto \int_{E_F}^{E_F+eV} \rho_S(z,E)\rho_T(E-eV)T(z,E,V)dE \tag{3.6}$$

Applying a voltage ramp across the tunnelling junction and recording the tunnelling current as a function of V, jumps in the tunnelling current I indicate that new electronic states enter the voltage window between E_F of tip and sample (Fig. 3.16 A, B).

While the I–V curve gives only an integral measure of the electronic state density in the probed energy window, energy resolved information is obtained by detecting the harmonic response of the tunnelling current to a bias modulation using a lock-in technique. The differential conductance dI/dV at bias V_0 is given by:

$$\frac{dI}{dV}(V_0) \propto \rho_S(z,V_0)\rho_T(E_F)T(z,V_0) \tag{3.7}$$

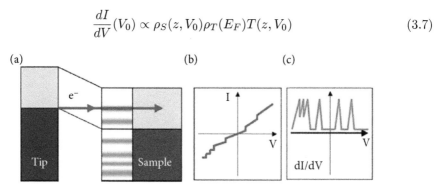

Fig. 3.16 (a) Potential diagram of an STM junction containing a system with discrete electronic states. (b) Current-voltage and (c) differential conductance curves modelled for the junction sketched in (a). Each energy level gives rise to a peak in the dI/dV spectrum. Reprinted from Nilius (2009). Copyright 2009, with permission from Elsevier.

Assuming ρ_T and T (at set-point z) are approximately bias independent, the dI/dV signal provides an expression of the sample LDOS (see Fig. 3.16, C). To account for bias dependent changes of the transmission probability, it is customary to divide dI/dV by the tunnel resistance V/I, which gives an even better representation of the sample LDOS:

$$\frac{dI}{dV}\frac{V_0}{I} \propto \rho_S(z, V_0) \tag{3.8}$$

However, the dI/dV signal is not a true reproduction of the LDOS, because the underlying model (leading to eqn (3.8)) is simplified and the spatial localization of the electronic states has to be considered. A large Density Of States (DOS) in the vacuum region above the surface and a concomitant strong overlap with tip wave functions gives extra weight in the tunnelling current. Therefore, s- and p-like wave functions tend to dominate over d- or f-like wave functions, which are more localized. Also, occupied electronic sample states are more difficult to probe than empty states, which are spatially more extended. Due to the spatial confinement and broken translational symmetry across the STM junction, momentum conservation is not obeyed during the tunnelling process, dI/dV spectroscopy therefore contains no $\vec{k}$-space information. Besides spectroscopy, conductance imaging can be performed by recording the dI/dV signal at fixed bias at each point of a topographic scan; the resulting dI/dV images contain information on the spatial distribution of the electronic states.

Conductance spectroscopy is used in oxide materials to investigate the LDOS in the vicinity of the Fermi energy and the gap size in semiconducting samples, the spectral characteristics of local defects and the electronic properties of single adsorbates on oxide surfaces (Nilius, 2009).

In *Electron Energy Loss Spectroscopy* (EELS or ELS, the latter acronym is mostly used for lower energy excitations), a monochromatic electron beam of energy E_p is directed onto a solid surface and the back-scattered secondary electron emission spectrum contains features due to characteristic energy losses at energies E_p-δE, which are the result of quantum excitations in the solid. The distinct electron energy losses δE can be caused by: i) excitation of core electrons (ionization losses), which are element characteristic and thus used for chemical analysis (e.g. in STEM instruments, see Section 3.1.1); ii) one-electron excitations of valence electrons, intra- and inter-band transitions (δE typically 3–20 eV); iii) collective bulk and surface plasmon excitations ($\delta E \sim 3$–30 eV); iv) vibrational excitations (δE in the meV range), see HREELS (Section 3.1.4) for vibrational analysis.

Since ELS spectra from surfaces and thin films are usually measured in reflection geometry, one elastic back–reflection event is necessary (prior or after the inelastic scattering event), since the momentum transfer perpendicular to the direction of the primary beam $q_\perp$ in one-electron and collective plasmon excitations is too small to cause back-reflection (Netzer, 1988). The intensities of energy losses may be described by the dielectric loss function (Netzer, 1988; Rivière, 1990; Ertl and Küppers, 1985):

$$\Im\frac{1}{\epsilon(\omega, \vec{q})} = \frac{\epsilon_2(\omega, \vec{q})}{\epsilon_1(\omega, \vec{q})^2 + \epsilon_2(\omega, \vec{q})^2} \tag{3.9}$$

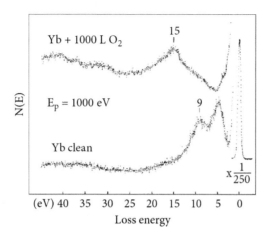

Fig. 3.17 Electron energy loss spectra of clean Yb and Yb exposed to 1000 L O_2 (1L = 10^{-6} torr.sec). Reprinted from Bertel *et al.* (1982). Copyright 1982, with permission from Elsevier.

where $\epsilon(\omega, \vec{q}) = \epsilon_1(\omega, \vec{q}) + i\epsilon_2(\omega, \vec{q})$ is the complex dielectric constant, $\vec{q}$ the momentum transfer and ω the frequency. Equation 3.9 is valid for energy loss events in the bulk. At the surface, the corresponding surface loss function is $\Im(1/(\epsilon(\omega, \vec{q}) + 1))$.

In specular reflection geometry, there is no momentum transfer perpendicular to the direction of the elastic primary beam (this is the optical limit $q_\perp = 0$) and the loss function made up from $\epsilon_1(\omega, 0)$ and $\epsilon_2(\omega, 0)$ can be obtained from optical data. Collective plasmon excitations produce features in both ϵ_1 and ϵ_2, whereas one-electron interband transitions correspond to features in ϵ_2. The latter are due to vertical transitions in the reduced band structure (for $q_\perp = 0$), as observed in optical (UV/Vis) spectroscopies. $\epsilon_2(\omega, 0)$ maps the Joint Density Of States (JDOS) for the transition according to:

$$\epsilon_2(\omega, 0) \propto |M_{fi}|^2 J_{fi}(\omega) \qquad (3.10)$$

where M_{fi} is the dipole matrix element for the transition between initial i and final f states at the excitation energy ($\hbar\omega$). The loss peaks due to collective electron excitations occur at the bulk plasmon energy, according to free-electron theory:

$$E_{bp} = 2eh \left(\frac{\pi n}{m}\right)^{1/2} \qquad (3.11)$$

with n the free-electron density in the conduction band, and e and m the electron charge and mass. The surface plasmon energy of a free electron metal bounded by vacuum ($\epsilon = 1$) is:

$$E_{sp} = \frac{E_{bp}}{\sqrt{\epsilon + 1}} = \frac{E_{bp}}{\sqrt{2}} \qquad (3.12)$$

The plasmon energies may deviate from free-electron-like behaviour in transition metals and in compounds, e.g. oxides, and are also influenced by the energetic vicinity of interband transitions, due to the interplay of ϵ_1 and ϵ_2 in the dielectric loss function.

The use of ELS to follow the oxidation of a metal surface is illustrated in Fig. 3.17. It presents ELS spectra, recorded in reflection geometry with an electron primary energy of 1,000 eV, of a clean Yb surface (bottom), and after oxidation (top curve). The clean Yb surface shows a plasmon peak at 9 eV, consistent with the free-electron value of the two $6s$ valence electrons of the divalent Yb metal. This peak disappears upon oxidation and is replaced by the oxide plasmon of trivalent Yb_2O_3 at 15 eV. The oxide plasmon at 15 eV conveys non-free-electron behaviour, it would be expected at significantly higher energy assuming eighteen free electrons per Yb_2O_3 unit.

In non-specular reflection geometry ($q_\perp \neq 0$), non-vertical (optically forbidden) interband transitions are possible. If angular resolution is employed in an ELS experiment, loss peak positions with various momentum transfer as selected by the scattering geometry may be used to obtain information about band shapes (interband transitions) and to derive the plasmon dispersion relation $E_{bp,(sp)}(\vec{k})$.

The *optical spectroscopies*, UV/Vis absorption and photoluminescence (PL), i.e. light emission after photoexcitation, are related to the ELS technique in that they probe the same electronic excitations. UV/Vis and PL are mostly applied in oxide nanoparticle research as they can be performed under ambient pressure conditions and require no vacuum environment. The determination of band gaps and localized electron-hole quasiparticle (exciton) energies in semiconductor nanoparticle ensembles are targeted experimental issues.

3.1.4 Specific properties

Magnetic properties.   The most common techniques to probe the magnetization structure of surfaces and thin films rely on the use of the polarization of incident photons in reflection or absorption processes. Depending on the wave length of the interacting radiation, visible/near IR or X-rays, two methods with different application potential and experimental sophistication have been developed: the magneto-optical Kerr effect and X-ray magnetic circular (linear) dichroism.

If polarized light interacts with a magnetic medium, its polarization state is influenced. In the *Magneto Optical Kerr Effect* (MOKE), the changes of light reflected from a magnetic surface is detected and analysed. If linear polarized light is incident on a ferromagnetic surface, the reflected light has s-and p-polarized components leading to elliptical polarization (the ratio of these is the *Kerr rotation*), which is measured. Depending on the relative orientation of the magnetic field and the plane of incidence of light, three MOKE configurations can be set up: polar (magnetization $\vec{M}$ perpendicular to the reflecting surface and parallel to the plane of light incidence); longitudinal ($\vec{M}$ parallel to surface and plane of incidence); transverse ($\vec{M}$ parallel to surface and perpendicular to plane of incidence). Depending on the configuration, the ellipticity (polar), the *Kerr rotation* (longitudinal), or the reflectivity (transverse) is measured. The MOKE set-up is experimentally relatively simple and is suitable for UHV studies. It involves a laser light source, an electromagnet at the sample position (providing a longitudinal, in-plane, out-of-plane field), polarisers and a photon detector.

The sensitivity of MOKE is sufficient to detect the signal from a single magnetic monolayer. But its drawback is its limited spatial resolution (due to the diffraction limit of the long wavelength visible light), which makes it unsuited to image magnetic structures at the nanoscale (< 200 nm). The latter is in contrast to the X-ray derived XMCD technique (see below), which can be used to 'see' the nanoworld. Nevertheless, Kerr microscopy is at the basis of the magneto-optical recording technology by employing small semiconductor lasers.

In *X-ray Magnetic Circular Dichroism* (XMCD), one measures the magnetic dichroism, i.e. the difference of the absorption spectra between right and left circularly polarized X-rays near core level absorption edges of magnetic atoms, allowing the determination of the magnetic anisotropy of a sample. Absorption at core level thresholds confers the method chemical specificity. The power of soft X-rays for magnetic studies arises from the fact that the most important absorption edges of interest in magnetic studies, the L-edges ($2p$) of Fe, Co and Ni and the M-edges ($3d$) of the rare earths fall into this soft X-ray range (700–$1{,}500$ eV). The respective absorption edges display large magnetic effects and through dipole allowed $2p \rightarrow 3d$ and $3d \rightarrow 4f$ transitions give access to the magnetic properties of transition metal and rare earth atom containing compounds. The measured dichroic intensities are linked via fundamental sum rules to spin and orbital magnetic moments and their anisotropy, paving the way for the use of XMCD for quantitative magnetometry.

The use of linear polarized X-rays (in *X-ray Magnetic Linear Dichroism* (XMLD)) is suited for the study of antiferromagnetic systems, where there is no net magnetization but still a non-zero value of the local magnetic moment. The linear dichroism, i.e. the difference spectrum between horizontal and vertical polarization of the incident X-rays, allows one to determine the orientation of the antiferromagnetic axis.

The XMC(L)D technique requires the use of synchrotron radiation and of dedicated beam lines with switchable circular (right, left) or linear polarization. The experimental effort is thus considerable. However, XMC(L)D can be used in combination with XPEEM for spectro-microscopic domain imaging on ferromagnetic and antiferromagnetic surfaces, with spatial resolution at the nanoscale. For a comprehensive treatment of nanoscale magnetism and respective characterization techniques we refer to the excellent textbook of Stöhr and Siegmann (2007).

Vibrational properties. Vibrations at surfaces and thin films may be investigated by the absorption of infrared photons (*Reflection Absorption Infrared Spectroscopy* RAIRS) or by the inelastic scattering of photons (*Raman Spectroscopy*) or electrons (*High-Resolution Electron Energy Loss Spectroscopy* HREELS). The inelastic processes in all cases result from vibrational dipolar excitations.

In RAIRS, IR radiation is reflected from a plane surface through a surface layer, in which the reflected light loses intensity due to the excitation of vibrational modes in the surface layer. For sufficient sensitivity, the surface must have high reflectivity, the absorption coefficient must be sufficiently large, p-polarization must be used (i.e. linear polarization in the plane of incidence) and the angle of incidence of the IR radiation must be close to grazing. The electric field component perpendicular to the surface of p-polarized radiation is strong at grazing incidence, allowing for strong coupling to vibrational modes with dipole components perpendicular to the surface. This constitutes the so-called surface selection rule. In addition, in order to be infra-red active, a vibration must support a change of the dynamical dipole moment.

Whereas in RAIRS a photon is resonantly absorbed by the sample exciting transitions between vibrational energy levels, the Raman process describes the inelastic scattering of monochromatic photons (in the visible, near-IR, near-UV range) to excite vibrational, rotational or other low-frequency modes in a system. Conceptually, the Raman effect may be visualized as a transition from a (vibrational) initial state to a virtual excited state, followed by its decay with emission of a photon of lower or higher energy (inelastic scattering). After the scattering event, the sample is in a different vibrational (or rotational) final state. The Raman signal is either shifted to lower energy (down-shifted, Stokes line) or to higher energy (up-shifted, Anti-Stokes line) with respect to the incoming light. The latter is possible if the system is in an excited vibrational state, from which it returns via Raman decay into the vibrational ground state, giving up the energy to the scattered light. The Stokes/Anti-Stokes lines intensity ratio depends thus on the population of vibrational states. For a vibration to be Raman active, a change in the electric dipole polarizability is necessary, and the selection rules for Raman activity are different but complementary to those of IR absorption. However, the cross section for Raman scattering is very weak. Raman spectroscopy has therefore rarely been used for vibrational studies of well-defined oxide film surfaces, due to its low sensitivity, but it is a powerful tool for the examination of high-surface area polycrystalline oxides (oxide powders) and nanoparticle ensembles. It has been shown that Raman peak positions are sensitive to lattice stress/strain and that peak widths may be related to the crystallinity of the material (Riviere and Myhra, 2012).

Low-energy electrons scattered from a surface, thereby exciting surface vibrations, are the basis of HREELS. The long-range Coulomb field of the electrons interacts with surface vibrating dipoles with components perpendicular to the surface, providing dipole selection rules as in RAIRS (Ertl and Küppers, 1985). The dipolar scattering mechanism is operative predominantly around the specular scattering direction. In addition to dipole scattering, another scattering mechanism is operative in HREELS, impact scattering, which arises from short-range interaction of the electrons with the atomic cores. Impact scattering is characterized by a broad angular distribution of the inelastically scattered electrons (in contrast to the specular direction preference of dipole scattering) and follows different selection rules (Ibach and Mills, 2013).

HREELS instrumentation makes monochromatization at low kinetic energies (typically 2–6 eV) of the primary electrons necessary, in addition to a high-resolution electron analyser. Energy resolution of 1 meV can presently be achieved, which necessitates a special treatment of the inner surfaces of the instrument (keeping the work functions stable and reducing secondary electron emission) and a careful shielding from external magnetic fields. For angle-dependent measurements outside the specular direction, as e.g. for measurements of phonon dispersion, a movable electron analyser and higher primary electrons are required, the latter to enable sufficient momentum transfer in the scattering event.

Comparing the HREELS and RAIRS methods for surface vibrational analysis, RAIRS is distinguished by the higher resolution (1–2 cm^{-1}), by a high measurement speed (in particular if using Fourier-Transform methods (Bell, 1984)) and by the possibility of measuring in a non-vacuum environment (which may be of importance in

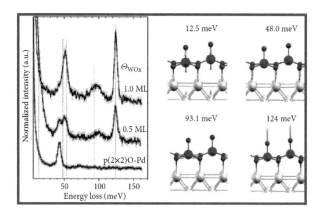

Fig. 3.18 Left panel: HREELS phonon spectra of p(2 × 2)-O (bottom), 0.5 ML WO₃ (middle), and 1.0 ML W-oxide (top) on Pd(100). The vertical dashed lines correspond to the calculated phonon frequencies. Right panel: principal vibrational modes of the 2-D WO₃ structure model on Pd(100). Reprinted from Doudin *et al.* (2016) with permission. Copyright 2016 American Chemical Society.

surface chemistry studies). Conversely, the HREELS resolution is poorer (1 meV = 8 cm^{-1}), but the sensitivity is much higher and a larger spectral range can be assessed (basically the whole IR region, whereas the region < 400 cm^{-1} is unattainable in RAIRS).

In the study of oxide materials, it has been found that HREELS phonon spectroscopy provides a valuable diagnostic means for identifying novel, non-bulk-like surface structures (Surnev *et al.*, 2012). Experimental phonon spectra may provide insight into key elements of the geometric structure. Common structural building blocks of different oxide phases may result in similar phonon losses, rendering vibrational spectra sensitive to the occurrence of common structural elements. More subtle structural details can then be revealed by comparing the experimental spectra with the calculated phonon frequencies of theoretically derived model structures (Surnev *et al.*, 2012). Figure 3.18 shows the HREELS phonon spectra (left panel) of a 2-D WO₃ layer on a Pd(100) surface (Doudin *et al.*, 2016). The right part of Fig. 3.18 illustrates the principal vibrational modes of the 2-D structural model, which consists of a WO₂ plane of (100) type, plus an oxygen layer on top. Each W atom is surrounded by four oxygen atoms in the WO₂ plane and bonded to one oxygen atom on top of it (W=O tungstyl group). The latter is particular to this structure model, and the prominent phonon mode at 124 meV in the experimental spectrum is unequivocal proof of the correctness of the proposed model. It is noted parenthetically that the agreement between calculated and measured phonon frequencies is very good.

Surface chemical properties. Many of the experimental techniques presented previously are suited for the study of surface chemistry. Since at least one reactant of a surface chemical reaction is in an adsorbed state, the investigation of adsorption phenomena is at the heart of surface reaction studies. For example, the reactivity of a reactant molecule may depend on the adsorption site on the surface. The adsorption site of an adsorbate may be directly visualized by scanning probe techniques (STM, AFM), it may be derived from vibrational spectroscopies, and in the case of an ordered adsorbate layer diffraction techniques such as LEED may be successful in its elucidation. Likewise, the molecular orientation and the surface bonding may be determined from UV and X-ray photoelectron spectroscopy, NEXAFS and vibrational spectroscopies, or best by a combination of these techniques. The presence of OH groups is an important feature on oxide surfaces, since OH groups may have a profound influence on the chemical properties of a surface. Their detection is best achieved via their characteristic O-H vibrational modes in IR or HREELS spectra.

A method particularly apt for adsorption and chemical reaction studies at surfaces is (TDS) *Thermal Desorption Spectroscopy* (TDS). In TDS (sometimes also called *Temperature Programmed Desorption* (TPD)) an adsorbate covered surface is heated in a controlled way and the particles (molecules) desorbing into the gas phase are detected as a function of the surface temperature in a mass spectrometer. The heating is often performed at a linear rate, the peak in the desorption spectrum recorded in a UHV system–mass spectrometer signal versus temperature–corresponds to the maximum desorption rate (Rivière, 1990; Woodruff, 2016).

The rate of evolution of adsorbed material from a surface is given by:

$$-\frac{dN}{dt} = k_m N^m \exp\left(\frac{-E_d}{kT}\right) \tag{3.13}$$

with N the surface concentration (particles per unit area), k_m a frequency factor, m the order of the desorption reaction (typically 1 or 2), and E_d the activation energy of desorption. The TDS spectrum gives quantitative information about the surface coverage from the area under the desorption peak (after calibration with the help of other methods such as AES, XPS, LEED). The desorption peak temperature is related to the nature and strength of the adsorbate-substrate bonding (via E_d), though sometimes only in a fingerprinting way. The shape of the desorption spectrum is dependent on the order of the desorption reaction (Woodruff, 2016). Surface reaction products can also be detected and analysed in the mass spectrometer. If the surface reaction is the rate limiting step for the desorption by creating the reaction product at the desorption temperature, *Temperature Programmed Reaction Spectroscopy* (TPRS) can give information on thermal evolution of a surface reaction. The experimental set-up for TDS is relatively simple, requiring only a mass spectrometer in line-of-sight with the sample surface, and TDS is often an additional standard technique in surface science UHV systems. Quantitative information requires however some instrumental precautions to avoid artefacts, such as desorption of gas from surfaces other than the desired one or re-adsorption during the heating step (Rivière, 1990).

The variety and complexity of oxide materials and their surfaces require typically that a combination of different experimental methods is necessary to characterize the particular physical and chemical properties. In ultra-thin films, 2-D oxides, and oxide nanostructures, this complexity is enhanced by the additional degrees of freedom of reduced size and dimensionality. The latter give rise to novel phenomena of entangled structure, composition, and phase behaviour. For a fundamental understanding of oxide nanostructure properties it is therefore necessary to combine the experimental characterization with theoretical modelling approaches. The latter are the subject of the following section.

3.2 Theoretical characterization

3.2.1 Electronic structure methods

Oxides display a wide variety of electronic structures, ranging from insulating (e.g. MgO, Al_2O_3), semi-conducting (e.g. ZnO, TiO_2, Fe_2O_3, $SrTiO_3$), metallic (e.g. Fe_3O_4, RuO_2), or even supraconducting (e.g. doped or oxygen deficient $SrTiO_3$) in the bulk. Some transition metal oxides display a metal-insulator transition as a function of temperature, of the Mott–Hubbard type (e.g. Ti_2O_3, V_2O_3) or of a mixed Peierls–Mott–Hubbard type (e.g. VO_2), as well as various magnetic characteristics. A theoretical determination of the electronic structure including an accurate treatment of correlation effects is thus mandatory to account for oxide properties such as magnetism and local moments, band gaps, excitons, collective excitations, metal-insulator transitions and low dimensionality effects in ultra-thin films or nanoparticles (NPs).

In the following, we present a short overview of the quantum theoretical methods available today, in order of increasing complexity, focusing more on the underlying physical concepts or approximations than on the mathematical formalism that can be found in many reviews or books.

Methods based on wave functions. A first set of methods relies on a variational approach in which a given form of the electronic wave functions is assumed. This leads to Schrödinger-like equations that are solved in a self-consistent way or not.

At the lowest level of sophistication, the *Tight-Binding* (TB) and the *semi-empirical Hartree–Fock* (HF) methods reduce the many body problem to the diagonalization of an effective one-electron Hamiltonian matrix. The variational electronic wave functions are expanded on a minimal basis set of atomic or Slater orbitals centred on the atoms and usually restricted to valence electrons. The matrix elements are self-consistently determined or not, depending upon the method.

In the pure *TB method*, the elements of the Hamiltonian matrix are treated as adjustable parameters to be fitted to experimental or first-principles calculation results. To estimate the total energy, an additional short-range repulsion term is added to the electronic contribution. The dependence of all parameters upon interatomic distances is determined so that the bulk equilibrium structure and its elastic constants are well reproduced. The parameters are assumed to be transferable from bulk to surfaces or low-dimensional systems. While successful in the field of semi-conductors and transition metals, the TB method completely disregards the variations of the electrostatic potential with the local environment of the atoms, which is the reason for wrong predictions having been made in the past for some oxide compounds.

The *HF method* is a variational self-consistent method which treats electron-electron interactions at a mean-field level. It approximates the N-body wave function of the system by a single Slater determinant of N spin-orbitals. Only Hartree and exchange interactions are accounted for, the latter resulting from the Pauli principle. The method can be implemented either in its spin restricted form (RHF) for closed shell systems, or in the unrestricted form (UHF) for open-shell or strongly correlated systems. In the first case, the one-electron spin-orbitals are identical for electrons of both spin directions, while UHF can reproduce non-uniform spin densities. The one-electron orbitals, which are determined in the course of the self-consistent solution of the HF equations, are expanded on an over-complete basis set of optimized variational functions and all matrix elements are calculated on this basis.

When treated at a *semi-empirical* level, the HF method can be viewed as a self-consistent TB approach, since it incorporates the self-consistent relationship between charges and electrostatic potentials in the expression of the Hamiltonian matrix elements. The empirical parameters either fix the spatial dependence of the basis set wave functions or are included directly in some Hamiltonian matrix elements (Pople and Segal, 1965; Pople and Segal, 1966; Harrison, 2012; Noguera *et al.*, 2010). Since they rely on an Hartree–Fock approach to which the Koopman's theorem applies, the conduction band minimum (CBM) and valence band maximum (VBM) positions with respect to the vacuum level can provide values for the oxide ionization potential and electron affinity.

When applied to oxides, semi-empirical HF calculations can reproduce the shifts of the atomic orbital effective levels due to the changes of the electrostatic potential in various atomic environments, i.e. the modifications of local electronegativity (Noguera, 1996). This is a very important point, both for obtaining reliable eigenstate spectra, and for properly describing the specific reactivity of atoms in inequivalent environments. They can also account for the local modifications of the charge distribution, and for their influence on the structural degrees of freedom. Last but not least they allow simulating large size systems, as exemplified in Fig. 3.19. Recent developments include the DFTB method (Seifert, 2007) and the order-N PHFAST code (Noguera *et al.*, 2010; Cabailh *et al.*, 2019).

The *ab initio HF method*, in which no parametrization of the basis set functions or matrix elements is made, was initially designed for atomic and molecular systems but has later been implemented for periodic systems, such as bulks and surfaces of crystalline materials (Pisani *et al.*, 1992). It has proved very useful in the description of magnetic insulators (Towler *et al.*, 1994), and of the surface properties of a large number of simple oxide surfaces (Mackrodt, 1988). However, one very serious problem is the unscreened nature of the Coulomb and exchange interactions. The bare value of a typical intraatomic Coulomb integral U is in the range 15–20 eV, while it is known that screening in solids weakens it by more than a factor of 3. A large overestimation of the energy difference between the CBM and the VBM in insulators results. However, there have been quite accurate predictions of gap values or $d \rightarrow d$ excitation energies, based on HF total energy differences.

By definition, correlations are all effects which result from electron-electron interactions beyond HF. Quantum chemistry approaches have long tried to account for them, at various levels of sophistication, along two main lines. In the *Configuration Interaction* (CI) method, the variational wave function is expanded on a basis set of Slater determinants built from one-electron HF eigenfunctions which include ground state and excited state configurations. The CI method would be exact if the basis set were complete. Depending upon the number and nature of configurations kept, some

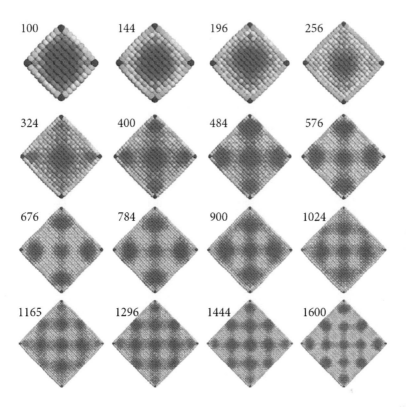

Fig. 3.19 Strain map of metal-supported MgO monolayer islands of increasing numbers of atoms (from 100 to 1600), obtained with the order-N PHFAST code. Expansion and contraction zones are represented in red and green, respectively, thus highlighting the introduction of an increasing number of interfacial dislocations as size grows. For the sake of visibility, all islands are represented with the same area, whatever their atom number. Reprinted from Noguera *et al.* (2010) with permission. Copyright 2010 by the American Physical Society.

excited state properties can be satisfactorily reproduced (Ostlund and Szabo, 1996). Alternatively, *Coupled Cluster (CC)* methods build the variational multi-electron wave function by applying the so-called cluster operator e^T to the HF Slater determinant (Bartlett and Musiał, 2007). T is the sum of single, double, triple, etc., excitation operators. Even when the sum in T is truncated, application of e^T produces an expansion in powers of the selected operators up to infinite order, which is an advantage over perturbation approaches of the CI type. Despite their ability in accounting for correlation effects, CI and CC methods are not used nowadays in the simulation of oxide ultra-thin films or oxide NPs for they can only treat small size systems containing a limited number of electrons.

Methods based on the electronic density. At variance with the previous methods which rely on the use of variational wave functions, another class of electronic structure methods has emerged, based on the electronic density. Most state-of-the art first principles simulations of solids and liquids nowadays rely on the *Density Functional Theory* (DFT) (Kohn and Sherrill, 2014; Engel and Dreizler, 2013) which has impressively improved in the past decades, reaching a level of quasi-chemical accuracy. On the basis of two theorems proved by Hohenberg and Kohn (1964), it gives a prescription for calculating the ground state properties of an assembly of atoms at fixed positions, as functionals of the electronic density $n(\vec{r})$ only. If additionally it is assumed that there exists a system of non-interacting electrons with the same $n(\vec{r})$, the ground state electronic properties of the interacting system can be expressed in terms of the eigenfunctions and eigenvalues of the non-interacting system. The latter are solutions of one-electron self-consistent equations, named Kohn–Sham equations (Kohn and Sham, 1965), which include an external potential V_{ext} due to the ions, a Hartree potential V_H, and a non-local exchange-correlation potential V_{xc}:

$$\left[-\frac{\nabla^2}{2} + V_{\text{ext}} + V_{\text{H}} + V_{xc} \right] \phi_{n\vec{k}}(\vec{r}) = E_{n\vec{k}} \phi_{n\vec{k}}(\vec{r}) \tag{3.14}$$

There exists no exact expression for V_{xc}, but several ansatz are currently used.

The *Local Density Approximation* (LDA) replaces V_{xc}, which is by essence non-local, by a local potential built from the properties of the homogeneous electron gas. The method has been extended to open-shell or magnetic systems, by building an exchange and correlation functional which depends on both the electron density and the spin density. This is the *Local Spin Density Approximation* (LSDA) (Parr and Weitao, 1989; Jones and Gunnarsson, 1989; Nagy, 1998).

The *Generalized Gradient Approximation* (GGA) was conceived to better account for the electronic density inhomogeneous distribution in solids. It makes a semi-local approximation of V_{xc}, in which not only the density but its gradient are locally taken into account (Perdew *et al.*, 1992; Perdew *et al.*, 1996*b*). The successful PBE (Perdew–Burke–Ernzerhof) functional (Perdew *et al.*, 1996*a*) corrects a large part of the systematic overestimation of the LDA cohesive energies. In particular, at surfaces, it usually yields a much better description of chemisorption than the LDA. Nevertheless, non-local long-range forces, such as dispersion forces, are not well accounted for in LDA and GGA, which either ignore them or only cover their short-range exponentially decaying part. There are specific methods for including dispersion forces in the DFT formalism (Dion *et al.*, 2004; Grimme, 2006; Klimeš *et al.*, 2011; Klimeš and Michaelides, 2012). Beyond GGA, the *Meta-GGA DFT* functionals involve the Laplacian of the density or the kinetic density operator, in addition to the density and its gradient (Becke and Johnson, 2006; Tran and Blaha, 2009).

In order to solve the Kohn–Sham equations, an expansion of the one electron wave functions on a basis set is performed. Both localized and plane-wave basis sets are currently used. Localized basis sets have the advantage of their small size. However they are attached to the atomic positions, which yields non-zero Pulay forces in geometry optimization and molecular dynamics (Pulay, 1969). Plane waves, on the other hand, provide a uniform sampling of space, whatever the specific conformation of the system. They are independent of the atomic positions, but they require the use of pseudo-potentials to mimic core electrons and a large number of reciprocal vectors is necessary in standard surface calculations.

The density functional theory is a theory for the ground state. It is not conceived to predict excitation properties. In particular, there is no theoretical justification to identify the Kohn–Sham eigenvalues with quasi-particle energies, although this is currently done. In particular, the Kohn–Sham band gap severely underestimates the energy difference between the highest occupied and lowest unoccupied states measured in direct and inverse photoemission experiments, respectively, which is equal to the ionization potential minus the electron affinity. In addition, DFT-LDA and GGA fail in localizing electrons in highly correlated systems due to their mean-field character and, because of their approximate treatment of exchange, electrons see a part of their own potential (electron self-interaction). To overcome these various limitations, several schemes have been proposed.

For an improved inclusion of exchange, *hybrid exchange-correlation functionals* have been developed, as linear combinations of the HF exact exchange and DFT exchange-correlation functionals. For example, the Becke exchange and the Lee Yang and Parr correlation functionals are mixed in the B3LYP (Becke, 3-parameters, Lee–Yang–Parr) method (Becke, 1993) and the PBE exchange-correlation functional is used in the PBE0 method (Adamo *et al.*, 1999). In the range-separated HSE (Heyd–Scuseria–Ernzerhof) method (Heyd *et al.*, 2003), the short-range part of the Hartree–Fock exchange is kept to correctly describe atomic-like properties, while its long-range part is screened by a mixing with the PBE functional, in order to better account for extended or metallic-like states. The parameters determining the relative weights of the two contributions are obtained by fitting the theoretical predictions to experimental or accurately calculated thermochemical data. However, the use of a single mixing parameter for all materials limits the applicability of the methods. Attempts have been made to determine this parameter for each individual system from its own electronic structure and in particular its dielectric properties (Marques *et al.*, 2011). The addition of an exchange part to the DFT functionals suppresses a part of the electron self-interaction and increases the electronic localization, inducing a higher ionic character of the cation-oxygen bonds, systematic contractions of the lattice parameter, and strengthening of the elastic constants and bulk moduli. Predictions of the properties of transition metal oxides (Guo *et al.*, 2012), of complex solid catalysts (Paier, 2016) and of MgO ultra-thin films grown on Ag(100) (Prada *et al.*, 2016) have, for example, recently been performed, using hybrid functionals.

In the HF mean-field approximation, both the Hartree and exchange energies contain a spurious electron self-interaction which is only cancelled out when the sum of the two contributions is performed. However, in DFT-LDA or GGA, due to the local or semi-local treatment of exchange, this cancellation is incomplete, yielding systematic errors for finite systems and localized states in extended systems. They have been summarized in Perdew and Zunger (1981), which proposes a *Self-Interaction Correction* (SIC) method which replaces the LDA exchange-correlation potential by an orbital-dependent single-particle potential. Within this scheme, band gaps and magnetic moments of transition metal oxides were shown to be better reproduced (Svane and Gunnarsson, 1990).

An improvement may also be obtained by expressing the correlation energy within the *Random Phase Approximation* (RPA) (Hedin, 1965) and adding it to the exact exchange, a procedure which removes the self-interaction problem. The correlation energy, which is a measure of how much the movement of one electron is influenced by the presence of all other electrons, is related to the properties of screening of the system, i.e. to its dielectric function or, in other terms to its polarizability. The polarizability P_0 is a non-local and energy-dependent function which involves virtual transitions between filled $\phi_i(r)$ and empty $\phi_j(r)$ states (for example the Kohn-Sham or hybrid wave functions) associated to energies ϵ_i and ϵ_j. Within the RPA, it reads:

$$P_0(r, r', \omega) = \sum_{i,j} \frac{\phi_i(r)\phi_j^*(r)\phi_i^*(r')\phi_j(r')}{\omega - (\epsilon_i - \epsilon_j)} \tag{3.15}$$

Lattice constants, atomization energies of solids, and adsorption energies on metal surfaces evaluated within RPA turn out to be in very good agreement with experiments (Harl and Kresse, 2009). Moreover, non-local dispersive forces, such as van der Waals forces, are accounted for in this approach. Within RPA, for example, the correct relative stability of TiO_2 rutile and anatase phases is obtained, at variance with standard DFT methods (Cui *et al.*, 2016).

One of the most striking failures of DFT is its inability to correctly describe Mott–Hubbard insulators, i.e. compounds which are insulators, especially at low temperature, while predicted to be metallic by standard band theory. Examples are transition metal monoxides MnO, CoO, NiO, or sesquioxides Ti_2O_3, V_2O_3. In these compounds, the strong electron-electron repulsion partially prevents the electron band delocalization. It forces the electrons to move in a correlated way and localize on atoms. The one-band Hubbard model (Hubbard, 1963; Hubbard, 1964) captures the essence of this competition. Its Hamiltonian contains an electron hopping term t between neighbouring sites $< i, j >$, and a screened on-site repulsion term U between electrons of opposite spins σ:

$$H_{\text{Hub}} = t \sum_{<i,j>,\sigma} (c_{i,\sigma}^\dagger c_{j,\sigma} + h.c.) + U \sum_i n_{i,\uparrow} n_{i,\downarrow} \tag{3.16}$$

where $c_{i,\sigma}^\dagger$, $c_{i,\sigma}$, and $n_{i,\sigma}$ are creation, annihilation, and number operators for electrons of spin σ on site i. To this simple Hubbard Hamiltonian, one may add an exchange term, proportional to a screened exchange on-site interaction J. In the regime $t \gg U$, delocalization effects are dominant and mean-field approximations, such as DFT LDA or GGA, correctly predict the ground state electronic structure. In the opposite limit $t \ll U$, strong correlations are present which require a treatment beyond mean-field.

A multi-band Hubbard model is needed to account for correlations in oxides. Zaanen and Sawatzky have stressed how the strength of U affects the nature of the insulating gap (Zaanen *et al.*, 1985). They have clarified the distinction between charge-transfer oxides and correlated oxides, which relies on the competition between two types of electronic excitations. One is the charge-transfer excitation, well-described by band theory, which is associated to the transfer of an electron from an oxygen anion to a cation X : $X^{n+} + O^{2-} \rightarrow X^{(n-1)+} + O^-$. Its energy is approximately equal to the difference Δ between the oxygen second affinity A_2 and the cation nth ionization potential I_n, both corrected by electrostatic effects (the Madelung potential which involves the surrounding ion and Hartree contributions). The second excitation type is

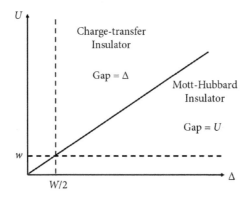

Fig. 3.20 Sawatzky–Zaanen's prediction on the nature of the insulating state as a function of the cation-cation charge-fluctuation energy U and the energy Δ for anion-cation charge-transfer. W and w are the band widths associated with anion-cation and cation-cation electron delocalization, respectively.

a cation charge fluctuation: $X^{n+} + X^{n+} \rightarrow X^{(n-1)+} + X^{(n+1)+}$. Its energy, which corresponds to the U value in the Hubbard model, is equal to the difference between the $(n+1)$th and nth cation ionization energies. Zaanen and Sawatzky have thus proposed the phase diagram reported in Fig. 3.20, which locates the charge-transfer insulators, such as MgO, TiO_2, etc., in the region $U > \Delta$ and the Mott–Hubbard insulators, as Ti_2O_3 or V_2O_3, in the region $U < \Delta$.

The *DFT+U* method incorporates these concepts within the framework of DFT. To the LDA or GGA exchange correlation functional, it adds a Hubbard term for strongly correlated electronic states, such as d or f states, while sp states remain treated in the standard way (Anisimov *et al.*, 1991; Anisimov *et al.*, 1997; Dudarev *et al.*, 1998). This procedure yields a negative energy shift of the occupied bands, and a positive one of the unoccupied bands, thus leading to an enhanced gap width. This jump in energy between occupied and unoccupied states, as well as the correlated jump of potential when the number of electrons goes through an integer value (Perdew *et al.*, 1982), are absent in the LDA-GGA, which thus misses an important contribution to the gap value. The screened Coulomb and exchange parameters U and J may be either taken as adjustable parameters or estimated from first principles (Aryasetiawan *et al.*, 2006). In the former case, the question arises whether their values are transferable from bulk to surfaces or low-dimensional configurations. Recent applications of the GGA+U method were performed in the study of pure and mixed transition metal oxides in the bulk and as honeycomb monolayers (Sadat Nabi and Pentcheva, 2011; Köksal *et al.*, 2018; Le *et al.*, 2018; Goniakowski and Noguera, 2018; Goniakowski and Noguera, 2019)(Fig. 3.21).

The *Dynamical Mean Field Theory* (DMFT) is also a non-perturbative method to treat local electron-electron correlations. Its specificity is to introduce dynamical effects in an approximate way within the LDA+U approach (Georges *et al.*, 1996), as part of the so-called LDA++ method (Lichtenstein and Katsnelson, 1998). It maps the

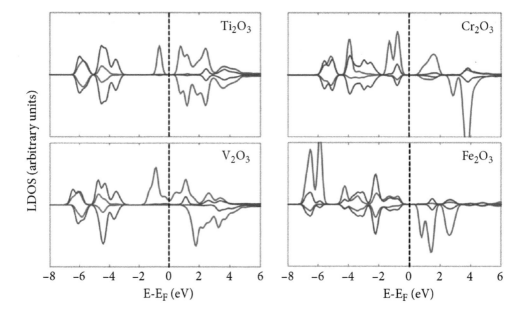

Fig. 3.21 Projected density of states in unsupported honeycomb monolayers of transition metal sesquioxides, obtained within an DFT+U approach. Red and blue curves represent cation and oxygen contributions, respectively. The vertical dashed lines indicate the position of the Fermi level. The top of the VB has a pure cation character in Ti_2O_3 and V_2O_3, showing that these oxides are Mott-Hubbard insulators. At variance, in Cr_2O_3 and Fe_2O_3, oxygen states contribute to the LDOS at the top of the VB, evidencing their mixed charge-transfer/Mott-Hubbard character. Adapted from Goniakowski and Noguera (2019).

crystal Hamiltonian onto a single-site (or small cluster) one with the constrain that the 'impurity' properties (more specifically the impurity Green's function) reproduce the local properties of the whole crystal. This is obtained through the use of an effective time-dependent mean-field potential (self-energy). DMFT can account for a phase transition between a metallic phase and an insulating phase of the Mott-Hubbard type as the strength of correlations increases. Many applications of combined DFT and DMFT methods to strongly correlated bulk oxides, mainly in the perovskite structure, have been performed (e.g. Dang *et al.* (2014)).

Many DFT codes are available, among others ABINIT, CASTEP, CPMD, CRYSTAL, DMol, FHI-AIMS, GAUSSIAN, QUANTUM-ESPRESSO, VASP, WIEN2K, and so on. They all have dedicated websites where instructions to obtain them are given.

Many-body Green's function methods. As long recognized (Hybertsen and Louie, 1986; Godby *et al.*, 1988), the resolution of the Kohn–Sham equations is a practical way of obtaining the DFT ground state properties, but their eigenvalues are not equal to charged excitation energies, nor can their differences yield neutral excitation energies. The *Many-Body Perturbation Theory* (MBPT), based on a Green's function approach, provides a systematic and controlled way to account for multiple particle excitation processes, within a Feynman diagram formalism (Gross and Runge, 1986). Truncated at the first and second levels, it allows describing one-particle and two-particle electronic excitations, using the GW approximation method for the former and the resolution of the Bethe–Salpeter equation for the latter (Onida *et al.*, 2002; Martin *et al.*, 2016).

Direct photoemission experiments probe occupied electronic states, by detecting the electrons which are emitted by photon absorption, leaving a positively charged hole in the system. As its name suggests, inverse photoemission corresponds to the reverse process in which a (negatively charged) electron is added and occupies a previously empty electronic state. In both cases, a charged excitation is produced, which modifies all other electronic states and, itself, is screened by them. This is the basis of the quasi-particle concept, tightly related to that of band structure in a crystalline material. The quasi-particle eigenfunctions $\phi_{n\vec{k}}(\vec{r})$ and eigenenergies $E_{n\vec{k}}$ (n the band index and $\vec{k}$ a point in the Brillouin zone) are solution of an equation of the type:

$$\left[-\frac{\nabla^2}{2} + V_{\text{ext}}(\vec{r}) + V_{\text{H}}(\vec{r})\right]\phi_{n\vec{k}}(\vec{r}) + \int d^3\vec{r}' \, \Sigma(\vec{r},\vec{r}';E_{n\vec{k}})\phi_{n\vec{k}}(\vec{r}') = E_{n\vec{k}}\phi_{n\vec{k}}(\vec{r}) \quad (3.17)$$

in which $V_{\text{ext}}(\vec{r})$ is the ionic potential and $V_{\text{H}}(\vec{r})$ the Hartree potential. Equation (3.17) resembles a Kohn–Sham equation in which V_{xc} is replaced by Σ, the self-energy operator. Σ is a non-local and non-Hermitian quantity. It contains the exchange and correlation contributions and is state $(n\vec{k})$ dependent. Alternatively, rather than solving eqn (3.17) for the quasiparticle eigenfunctions, one can resort to a Green's operator formalism. One introduces the one-particle Green's operator G whose imaginary part is the spectral function (the energy distribution function) and whose poles are the excitation energies. It is related to the independent particle Green's operator $G_0(E) = [E-H_0]^{-1}$ associated to the non-interacting Hamiltonian $H_0 = -\nabla^2/2 + V_{\text{ext}}(\vec{r}) + V_{\text{H}}(\vec{r})$ and Σ by the Dyson equation:

$$G = G_0 + G_0\Sigma G \quad (3.18)$$

The full evaluation of Σ is a very difficult task, because of its non-locality and energy dependence. A possible approximation is the *GW approximation* (Hedin, 1965; Hedin and Lundqvist, 1970; Aryasetiawan and Gunnarsson, 1998), in which a diagrammatic perturbation expansion of the self-energy is constructed and stopped at the first order. As shown in Fig. 3.22, this scheme requires the self-consistent resolution of a set of five coupled equations—the so-called Hedin's equations—which involve G_0, the self-energy Σ, the irreducible one-particle polarizability P which describes screening effects, a vertex operator Γ and the screened Coulomb potential W. In the GW approximation, the vertex operator, which contains the interaction between the electron-hole pairs participating to screening, is assumed equal to identity ($\Gamma = I$, neglect of vertex corrections) and, consequently, the self-energy operator reads $\Sigma = GW$. This result gives its name to the method.

The resolution of these equations may be performed in a self-consistent way or in a perturbative way with respect to $\Sigma - V_{xc}$ (Tarantino *et al.*, 2018). In the second case, carefully chosen eigenstates and eigenenergies (HF, PBE, HSE, depending on the system) have to be used as starting points for the implementation of the method. The GW approximation is very successful to compute band structures, even for correlated

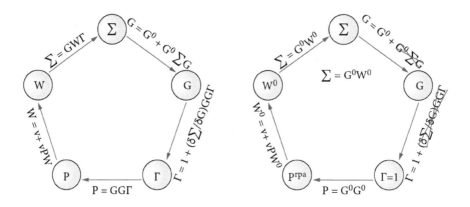

Fig. 3.22 Hedin's pentagon, showing the five steps necessary to calculate the one-particle Green's operator in a self-consistent (left) or non-self-consistent (right) scheme. In the GW approximation, $\Gamma = I$ and the polarizability operator is estimated within the RPA: $P = G_0 G_0$. Within the non-self-consistent scheme (right panel), G is approximated by the non-interacting G_0 operator and the self-energy evaluation is stopped after the first cycle, which yields $\Sigma = G_0 W_0$.

oxides (Rödl *et al.*, 2009) but is nowadays also used for nanostructures, clusters, and molecules (Martin *et al.*, 2016; Reining, 2018). In many cases, the GW approach corrects a large part of the PBE band-gap error. However, although many improvements have been proposed for an efficient calculation of the dielectric response function P from first principles (Hybertsen and Louie, 1988; Reining *et al.*, 1997), still this method is computationally heavy, both in time and memory.

The GW approximation may be regarded as a Hartree–Fock theory with a frequency- and orbital-dependent screened Coulomb interaction. Similarly, the LDA+U method represents a static approximation to the GW method, in which the non-local and energy-dependent self-energy is approximated by a frequency-independent but non-local screened Coulomb potential.

At variance with photoemission processes, optical absorption and electron energy loss processes (EELS) create an electron-hole pair, i.e. a neutral excitation, with the electron in the conduction band and the hole in the valence band. These two particles of opposite charge interact, sometimes giving rise to bound states (excitonic states) with energies in the band gap. An approach of the GW-type is insufficient to describe such a process which requires including the interaction between the two quasiparticles. One has to resort to a two-particle formalism and solve the so-called *Bethe–Salpeter Equation* (BSE) (Hanke and Sham, 1979; Onida *et al.*, 2002).

Within the MBPT, the GW one-particle Green's function is obtained thanks to the neglect of the so-called vertex corrections ($\Gamma = I$). However, Γ contains electron-hole interactions and depends on the two-particle Green's function G_2. This means that, through it, the one- and two-particle Green's functions are coupled. Neglecting the coupling to higher order particle Green's functions leads to the BSE, an equation of the Dyson-type for the two-particle Green's operator G_2. Written in terms of the two-particle correlation operator $L = GG - G_2$, it reads:

$$L = GG + GGKL \qquad (3.19)$$

In this equation, K is the two-particle interaction kernel, which includes contributions from the unscreened exchange term and the screened Coulomb interaction between the two particles.

In typical BSE implementations, DFT simulations provide the independent one-particle Green's operator. Then the GW one-particle Green's operator G is calculated and used to solve the Bethe-Salpeter equation. This is a computationally heavy procedure which, to our knowledge, has not been applied to large systems such as supported oxide ultra-thin films or oxide NPs. We will not further develop the method but, instead, refer the reader to the excellent book by Martin, Reining and Ceperley (Martin *et al.*, 2016) for a thorough review.

Actually, the EELS or optical spectra are obtained from contractions of the two-particle Green's function, which means that a part of the information derived from the Bethe–Salpeter approach is useless. A more efficient method for the direct calculation of absorption spectra is the *Time-Dependent Density Functional Theory* (TDDFT). It relies on a theorem which establishes a one-to-one correspondence between the external time-dependent potential and the density, thus playing the same role as the Hohenberg–Kohn theorem in static density-functional theory (Runge and Gross, 1984). It is possible to generalize the Kohn–Sham scheme to time-dependent situations and, from it, derive the response functions which describe the neutral excitations of the system. The main ingredient is the time-dependent exchange-correlation potential $V_{xc}[n](r,t)$, which depends on the density at all points in space and at all past times. If V_{xc} were known, TDDFT would be an exact theory for the description of neutral excitations. Several approximations have been proposed (Gross and Burke, 2006; Marques *et al.*, 2006; Botti *et al.*, 2007; Martin *et al.*, 2016), such as the adiabatic TDDFT (Furche and Ahlrichs, 2002) in which, at each time t, the exchange-correlation functional of the ground state calculations is used, but applied to the instantaneous density $n(r,t)$. Information about electronic excited states may be obtained from TDDFT through the linear response theory formalism. The linear response χ of the system is related to the independent electron response χ_0 by a Dyson-type equation:

$$\chi = \chi_0 + \chi_0 f_{xc} \chi \qquad (3.20)$$

Setting $f_{xc} = 0$ in eqn (3.20) corresponds to the random-phase approximation. The time-dependent local-density approximation (TDLDA) is the simplest approximation to f_{xc} as the functional derivative of V_{xc}^{LDA} with respect to density. However, TDLDA fails in the calculation of absorption spectra of solids due to the missing long range part of the Coulomb interaction in f_{xc}. The construction of an exchange-correlation kernel with the proper long-range behaviour has been the subject of numerous works, as reviewed in (Botti *et al.*, 2007; Marques *et al.*, 2006; Martin *et al.*, 2016). Due to the relative ease of its implementation, a growing number of applications of TDDFT is presently being performed.

3.2.2 Total energy approaches

Electronic structure calculations provide the total energy of an assembly of atoms and, thanks to the Hellmann–Feyman theorem the forces acting on the atoms. From them, it is possible to search the atomic arrangement of lowest energy and explore the potential energy surface (PES). However, depending on the degree of refinement of the method, the computational cost to make simulations of large size system or of long time processes may become prohibitive. Empirical PESs, fitted on experimental or first principles results, are thus very useful in this context.

Historically, the cohesion of solids has long been modelled by force field methods with empirical potentials, whose analytical form was chosen in agreement with physical considerations. In these models, the formation energy E with respect to the non-interacting atoms or ions, is expanded into a sum of pair, triplet, and higher-order terms:

$$E = \frac{1}{2} \sum_{i \neq j} E_{ij} + \frac{1}{2} \sum_{i \neq j \neq k} E_{ijk} + \dots \tag{3.21}$$

Restricting the summation to the E_{ij} terms yields the so-called pair-potential approximation.

In oxides, electrostatic interactions and ionic charges play a major role in the cohesion (Catlow and Stoneham, 1983). This was recognized in the *Born models* (Tosi, 1964) in which the E_{ij} terms between charges Q_i and Q_j located at distance r_{ij} include the Coulomb charge-charge interaction, a short-range repulsion term, and a van der Waals contribution:

$$E_{ij} = \frac{Q_i Q_j}{r_{ij}} + B_{ij} e^{-\frac{r_{ij}}{\rho_{ij}}} - \frac{C_{ij}}{r_{ij}^6} \tag{3.22}$$

The parameters B_{ij}, ρ_{ij}, and C_{ij} entering E_{ij} are fitted either to experimentally measured characteristics or to those calculated by means of first principles techniques. In the simplest approaches, ions are treated as point species, with a net charge independent of their environment.

Many body interactions, included in the higher order terms in eqn (3.21), are necessary to obtain a correct description of more covalent systems. Their contribution depends in a more complex way on the environment of the atoms, such as bond-angle terms or coordination numbers. The former, of key importance around tetrahedrally-coordinated atoms or in organic molecules, are accounted for in the Stillinger and Weber (1985) or Tersoff (1988) potentials, for example. Covalent contributions are not proportional to the number of neighbours but rather vary with its square root (Noguera, 1996), as in metals where such a dependence has been derived from the second moment approximation to TB models (Finnis, 2003).

Because they are not computationally expensive, classical methods may treat large size systems or allow molecular dynamics (MD) simulations over long times. They have been widely used in the past, for example to predict inorganic crystal structures, relaxation effects at oxide surfaces, but also lattice dynamics or point defect properties (Woodley *et al.*, 1999; Mackrodt *et al.*, 1987). Even now, they remain useful to understand complex mineral bulk or surface structures (Geysermans and Noguera, 2009), to derive structural phase diagrams of mixed oxides (Purton *et al.*, 2006), or to make a first screening of low-energy configurations of oxide NPs (Catlow *et al.*, 2010). However, because they include no information on the electronic structure, their parameters, determined for the bulk atomic environment, are not necessarily transferable to surfaces or nano-objects.

For example, when low-coordinated atoms are present, there is a need to account for their polarization by the non-vanishing electric fields to which they are submitted, and which are generally absent in high-symmetry crystalline environments. The *shell model* (Dick Jr and Overhauser, 1958) was designed to account for this effect, in an approximate way. It represents ions by a core and a massless shell, of charges Z_c and Z_s, respectively ($Z_c + Z_s$ then equals the total ionic charge). Cores and shells are coupled by harmonic springs and may move with respect to each other, thus creating an electric dipole which mimics the electronic polarization. This correction to the Born's model is useful at surfaces, but it is also suggested that it partly corrects the neglect of covalent interactions.

The development of concepts such as local electronegativity or site-specific acid/base character (Noguera, 1996), has led to the recognition that, even in highly ionic materials, the electronic structure is not frozen but may vary as a function of the local environment of the atoms. Methods intermediate between fully classical and fully quantum methods have emerged, such as those based on the *electronegativity equalization principle* (Streitz and Mintmire, 1994; York and Yang, 1996; Hallil *et al.*, 2006), which account for environment-dependent ionic charges. They are very efficient to address issues in which large inhomogeneous systems are involved. They have, for example, been applied in the MD simulation of the oxidation of an aluminium NP, in which interactions between more than eight hundred thousand atoms were accounted for (Fig. 3.23) (Campbell *et al.*, 1999).

Most classical models do not permit bond rearrangements in the course of a chemical reaction or a MD simulation. At variance, the reactive force field *ReaxFF* (Van Duin *et al.*, 2003), associated to the electronegativity equalization method, takes care of how the strength of a chemical bond depends on the local environment (coordination numbers, angles, and bond lengths). These features allow dynamical breaking and forming of bonds. It has for example been recently applied to describe stoichiometric and partially reduced ceria bulk, surfaces, and NPs (Broqvist *et al.*, 2015).

Several numerical codes are based on such classical total energy approaches, among which are GULP, LAMMPS, DLPoly, METADISE, and others. They allow calculating crystal, molecule, surface, cluster, defect properties by energy minimization or MD simulations. They generally include data bases for the parameters of a wide variety of classical potentials applied to a large number of compounds, as well as fitting algorithms to determine new ones.

Another conceptually different approach to the determination of PESs, is based on a direct fitting of electronic structure data employing mathematical flexible functions, with no relation to the underlying physics. Machine-learning methods (Khorshidi and Peterson, 2016), whether based on neural networks (Behler, 2011; Kolsbjerg *et al.*, 2018) or not, may be used to map the atomic interaction energies and forces from DFT to predefined functional expressions. There is thus no need to first make an analysis of the bond character and energy contributions as in classical force-field methods.

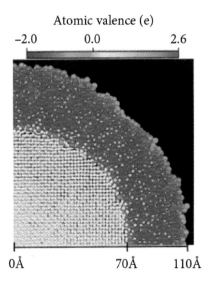

Atomic valence (e)

-2.0   0.0   2.6

0Å   70Å   110Å

Fig. 3.23 Snapshot of a small slice $(115 * 115 * 8 \text{ Å}^3)$ of an Al nanocluster, showing its oxidation state after 466 ps of simulation time. The larger spheres correspond to oxygen and smaller spheres to aluminium; colour represents the sign and magnitude of the charge on an atom. The central part of the particle is still metallic with aluminium atoms in a zero-valence state. When moving towards the outer rim of the particle, oxygen and aluminium atoms appear charged with -2 and $+3$ valence, respectively, which reveals that oxidation has occurred. Reprinted from Campbell *et al.* (1999) with permission. Copyright 1999 by the American Physical Society.

With unprecedented speed, they can predict reasonably good DFT energies and forces without needing an actual DFT calculation (Botu *et al.*, 2016; Meldgaard *et al.*, 2018).

3.2.3 Structural determination

The prediction of oxide ultra-thin film or NP structures is a very tough task because they nearly always strongly depart from bulk-cut pieces and involve large numbers of atoms. Indeed, when a few monolayer thick, oxide ultra-thin films usually display large complex unit cells and exotic stoichiometries. Similarly, below some critical size, a so-called non-scalable regime exists in which NP structures have nothing to do with the most stable bulk structure, and as size increases, the number of metastable states (i.e. of local minima in the free energy configurational space) exponentially grows. This applies to both NPs and ultra-thin films, but the same is also true for surface reconstructions. The probability is thus very low that starting from a bulk-cut configuration and applying local optimization techniques lead to the structural ground state. Nearly always, the procedure ends up trapped in a local minimum. Aside from chemical intuition, physical principles for generating promising starting configurations are thus of crucial importance. These configurations may then be refined by local or global minimization techniques.

Finding reasonable starting configurations. Whatever minimization technique is used, restricting oneself to a single starting point is most often insufficient and one has to generate a full set of starting configurations. Moreover, in ionic compounds like oxides, the latter cannot be produced by randomly positioning atoms, due to the strong electrostatic repulsion which takes place between atoms of similar type. Anions tend to be surrounded by cations and vice versa, so that thinking in terms of 'building blocks' or of 'polyhedra' around cations is the right way to start with. Comprehensive rules to build them have been developed a long time ago by Pauling (1929), which are known as the five Pauling's rules:

- 'A coordinated polyhedron of anions is formed about each cation, the cation-anion distance being determined by the radius sum (the sum of the r_+ and r_- cation and anion ionic radii) and the coordination number Z of the cation by the radius ratio.' The radius ratio r_+/r_- suggests which polyhedron is the most stable. Geometric considerations allow defining the minimum ratio above which a given polyhedron is the most likely (0.225 for tetrahedra ($Z = 4$), 0.414 for octahedra ($Z = 6$), and 0.732 for cubes ($Z = 8$)).
- 'In a stable coordination structure the electric charge of each anion tends to compensate the strength of the electrostatic valence bonds z/v (z the electric charge and v the valence) reaching to it from the cations at the centers of the polyhedral of which it forms a corner.'
- 'The presence of shared edges, and particularly shared faces, in a coordinated structure decreases its stability; this effect is large for cations with large valence and small coordination number, and is especially large in case the radius ratio approaches the lower limit of stability of the polyhedron.'
- 'In a crystal containing different cations those with large valence and small coordination number tend not to share polyhedron elements with each other.'
- 'The number of essentially different kinds of constituents in a crystal tends to be small.'

Pauling's rules have been very successful in rationalizing or predicting complex bulk structures in the fields of inorganic chemistry and mineralogy (Bragg, 1937). More recently, a careful analysis of a bunch of $SrTiO_3$ and other perovskite surface reconstructions has shown that, in all cases, these rules were also well obeyed at surfaces (Andersen *et al.*, 2018). It is our opinion that they may represent a useful tool to exclude unreasonable configurations at the start of local or global optimization procedures.

Local optimization. Starting from an assumed structure, relaxation of the structural degrees of freedom may be performed using standard minimization techniques, such as the steepest descent, the conjugate gradients or the Broyden's method (Press *et al.*, 1992). These algorithms have to be coupled to an evaluation of total energies and forces, which may be performed using advanced electronic structure methods or empirical ones (Sections 3.2.1 and 3.2.2). They all lead to the local minimum which is the closest to the starting point.

If a large set of starting structures is generated, it may be computationally demanding to relax all of them with a first principles approach. A more approximate energy model — pair potential or tight-binding— may allow to perform a pre-screening of the lowest energy configurations, which will be subsequently refined with a first-principles approach. It often turns out that the ordering of low-energy configurations obtained in the second step is different from that resulting from the pre-screening. This is especially true when many metastable states exist close in energy. This is why a substantial set of configurations has to be kept after the first step. As a whole, applying this strategy is simple but there is no certainty that the global minimum has been found.

Global optimization. Global optimization techniques are designed to counter the deficiencies of local optimization. They all aim at exploring the topography of the energy hypersurface, but according to how they do it, they can be divided into two families.

In the first family, only few starting points are chosen, which are continuously modified in the course of a *Monte Carlo simulation* (Binder, 1986), in which random atomic displacements, angle modifications, and cell variations (in periodic systems) are performed. The exploration of the energy landscape is driven by the rate of acceptance of the moves, which is determined by a Metropolis algorithm at finite temperature. The higher the simulated temperature, the smoother the energy landscape appears, thus allowing to overcome the barriers between the local minima and explore a more extended part of the configurational space. Gradually lowering the temperature in a so-called *simulating annealing* procedure leads to low-energy minima.

However, as described, this method is not computationally very efficient. Various algorithms have been developed to distort the energy hypersurface in the course of the simulation in order to smooth it and reduce the number of energy minima. The most successful is the *Basin-Hopping* algorithm which has the advantage of not changing the global minimum nor the relative energies of the local minima. The exploration of the energy landscape is performed by successive Monte Carlo steps and at each point the energy is assigned to that of the closest local minimum obtained by local geometry optimization. The energy landscape is thus transformed into a set of staircases with plateaus corresponding to all the attraction basins (Wales and Doye, 1997). The Monte Carlo-Basin-Hopping technique has, for example, been recently applied to the search for low-energy isomers of hydroxylated silica clusters (Cuko *et al.*, 2017).

Another family of global optimization methods relies on *evolutionary algorithms* to explore the configurational space (Johnston, 2003). These methods mimic some aspects of biological evolution: a population of initial structures evolves to low energy by mating (creation of an 'off-spring' structure from two or more 'parents') and mutation (creation of the 'off-spring' from a single 'parent'). Starting from a population of initial structures, mating or mutation are performed and the new configurations so-produced are relaxed by a local optimization method at a chosen level of theory (empirical potentials, DFT, hybrid functionals, etc.). An algorithm for the survival of the fittest is implemented, for example it is kept if its energy is lower than at least one of the parents and discarded otherwise. If it is kept, the parent with the highest energy is suppressed so as to keep the population constant. The procedure is continued for many steps, ensuring a truly global search. Various processes other than matings and mutations may be introduced to make the procedure even more efficient (Lyakhov *et al.*, 2013).

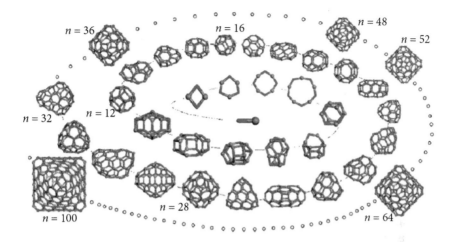

Fig. 3.24 Ground state structure of $(ZnO)_n$ clusters for n = 1 to 32 and extended to larger sized octahedral bubbles. Grey represents zinc atoms and red oxygen atoms. Reprinted from Catlow *et al.* (2010). Copyright 2010 by the Royal Society of Chemistry. Permission conveyed through Copyright Clearance Center, Inc.

In the field of oxides, evolutionary algorithms have been used to determine the structural phase diagram of complex bulk oxide structures (Glass *et al.*, 2006; Wu *et al.*, 2014; Van den Bossche *et al.*, 2018), the surface reconstructions of oxide surfaces such as MgO(111), TiO_2(110), or SiO_2 (Zhu *et al.*, 2013a; Wang *et al.*, 2014; Feya *et al.*, 2018), the structure of metallic or molecular adsorbed species on oxide surfaces (Sierka *et al.*, 2007; Vilhelmsen and Hammer, 2014; Kolsbjerg *et al.*, 2018), or low-energy isomers of oxide NPs, such as ZnO, SiO_2, or MgO (Catlow *et al.*, 2010; Farrow *et al.*, 2014; Bhattacharya *et al.*, 2014; Neogi and Chaudhury, 2014). As an example, Fig. 3.24 gives an overview of the ground state structure of ZnO NPs of increasing sizes obtained with a combination of methods, including evolutionary ones.

3.3 Summary

A bunch of experimental and theoretical methods are nowadays available to explore the structural, electronic or magnetic properties of oxide thin films and nanoparticles, as well as their chemical composition. Nevertheless, these are complex materials for which a single technique is often insufficient to reach a conclusive answer. Combining several experimental techniques with numerical simulations often prove necessary to solve the structure of films with large cell reconstructions or Moiré patterns. Developments of theoretical techniques to treat large size systems or detailed spectroscopic features are still ongoing.

4
Oxide thin film properties

All metal surfaces in contact with our atmosphere are covered with a dry or hydrated oxide film. In some cases, like amorphous alumina films grown on aluminium objects or sacrificial chromium oxide films which form on inox surfaces, the oxide films prevent further oxidation and provide a good protection against corrosion, while, on other surfaces, like pure iron ones, oxidation keeps propagating towards the bulk of the material, with detrimental consequences on its integrity. For a long time, only the existence of oxide films, their thickness, and their porosity were examined, but not really their size-dependent physico-chemical properties. It is the growing demand of miniaturization in electronic devices and the development of nanotechnologies which have placed ultra-thin films — i.e. with thickness of the order of one or several nanometres — to the forefront and that precise protocols to control their thickness, crystallinity, and chemical composition were established. This has allowed the emergence of a new field of fundamental research on oxide ultra-thin films. Pioneer works have been accomplished in René Franchy's (Franchy, 2000), Wayne Goodman's (Street *et al.*, 1997), and Hajo Freund's groups (Freund and Pacchioni, 2008). The initial aim was the use of STM/STS on insulating compounds, which was impossible on bulk samples due to their lack of conductivity. In contrast, electrons can tunnel into a metal substrate through a thin insulating layer, thus opening the way to studies of the local structural and electronic characteristics of oxide surfaces.

However, oxide thin films quickly proved to be much more than mere models or substitutes for oxide surfaces, in several respects. Beyond the presence of under-coordinated atoms around which local structural distortions and modifications of the electronic structure take place, their finite thickness is responsible for electronic confinement effects, and sometimes for a whole change of the film characteristics. In the few nanometre thickness range, due to the enhanced ratio of surface to bulk atoms, crystalline polymorphs which are unstable or only metastable in the bulk under the same thermodynamic conditions may become the thin film ground states. Additionally, thin oxide film properties are strongly influenced by the substrate on which they are grown, since fabrication by exfoliation is more the exception than the rule. Their atomic structure has to adjust to the stress exerted by the substrate, to the presence of interfacial dislocations, to the possible mixing with substrate atoms, etc., and their electronic structure may be modified by interfacial electron transfers with consequences on their chemical reactivity.

Oxide Thin Films and Nanostructures. Falko P. Netzer and Claudine Noguera, Oxford University Press (2021). © Falko P. Netzer and Claudine Noguera. DOI: 10.1093/oso/9780198834618.003.0004

The present chapter focuses on these various physico-chemical properties, and, in particular on their thickness dependence. It aims at bridging the gap between the now well established knowledge acquired on semi-infinite oxide surfaces (Noguera, 1996; Goniakowski and Noguera, 2016)—surfaces with a semi-infinite bulk below—, and the peculiarities of truly two dimensional (2-D) oxides which are the subject of Chapter 5.

4.1 Structural effects

The structural characteristics which differentiate oxide ultra-thin films from their bulk or infinite surface counterparts originate from thickness effects as well as from their interaction with the substrate on which they are grown. Due to their small number of layers, they contain a large proportion of under-coordinated atoms, around which relaxation effects take place or which may induce a transition towards a state which is metastable in the bulk. Phonon modes and ferroelectric distortions are also affected by thickness effects. Especially when it is strong, the interaction between a thin film and its substrate is always a source of strain, leading to the formation of a network of interfacial dislocations which require several layers to vanish.

4.1.1 Surface orientation and termination–Polarity

As for semi-infinite surfaces obtained by cleaving or cutting a bulk material, the stability of an oxide thin film primarily depends on its orientation and termination, which determine the density of broken bonds and the environment of the surface atoms.

Compared to metals or tetrahedral semiconductors, oxides present an impressive variety of crystallographic structures—rock-salt, rutile, corundum, perovskite, for the simplest ones—which is reflected in the atom organization at their surfaces. Except in very rare cases (e.g. rock-salt (100)), several terminations usually exist for a given surface orientation. Moreover, by tuning the experimental conditions, such as temperature or partial gas pressures (oxygen and water in particular), surfaces of different compositions may be stabilized.

Among this diversity, a classification exists, which is based on electrostatic arguments (Tasker, 1979). Three types of surfaces are differentiated, according to whether the layers are charged and the structural repeat unit bears a dipole moment (Fig. 4.1). Type 1 surfaces (non-polar) are usually the most stable, with no charge in their layers nor dipole moment in their repeat units, as opposed to Type 3 surfaces (polar surfaces) which experience a strong electrostatic instability due to the presence of a dipole moment in the repeat unit. Type 2 surfaces involve charged layers and repeat units without dipole moment. However, they may be non-polar or polar depending on their termination. More precisely, one defines a polar orientation/termination, or equivalently a non-charge-neutral surface, when starting from vacuum there is no dipole moment in the repeat unit, or equivalently, when starting from a choice of bulk repeat unit without dipole moment—which is most of the time feasible—, the remaining layers close to the surface globally bear a non-zero charge.

The specificity of semi-infinite polar surfaces comes from the existence of a macroscopic polarization (dipole moment per unit volume) perpendicular to the surface. For a long time, they were believed not to exist because an infinite surface energy results from this polarization. Indeed, as shown by a simple capacitor model of alternating

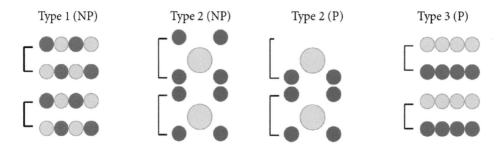

Fig. 4.1 Classification of compound surfaces (profile views) according to Tasker (1979). NP and P refer to non-polar and polar terminations, respectively. At Type 1 surfaces, the layers are neutral, and the repeat unit bears no dipole moment. In more complex crystallographic structures, the layers are charged. If, starting from the surface, the repeat unit bears no dipole moment, as schematised in Type 2 (NP) surfaces, the surfaces are non-polar and generally stable. If not, they share the polar character with Type 3 surfaces, at which both the layer charge and the dipole moment borne by the repeat unit are non-zero. These 'polar' surfaces require specific compensation mechanisms to be stable.

layers with charge densities $\pm\sigma$ (Fig. 4.2), the electrostatic potential varies monotonically across the system, inducing surface instability and leading to the so-called 'polar catastrophe' at infinite thickness. However, in this limit, compensating charge densities $\Delta\sigma_\infty$ on the outer terminations, equal to:

$$\Delta\sigma_\infty = \sigma \frac{R_1}{(R_1 + R_2)} \tag{4.1}$$

in the configuration of Fig. 4.2 or more generally (repeat unit of length a containing m layers of charge density σ_i located at positions z_i):

$$\Delta\sigma_\infty = \frac{\sum_i^m z_i\sigma_i}{a} \tag{4.2}$$

may stabilize polar surfaces by creating the required depolarization field, leading to a charge neutral surface. In non-centro-symmetric compounds such as ZnO along the polar (0001) orientation, the asymmetry of the electron density around the nuclei produces an additional contribution to the macroscopic polarization $\Delta\sigma_\infty$ (Stengel and Vanderbilt, 2009; Goniakowski *et al.*, 2007*a*). Many polar oxide surfaces have been prepared and studied in the past, such as MgO(111), ZnO(0001), SrTiO$_3$(110) or (111), Fe$_3$O$_4$(100) ... and this field remains very active (Noguera, 2000; Goniakowski *et al.*, 2007*a*).

   In thin films, the question of whether polarity is still relevant needs to be addressed, since, due to their finite thickness, there is no actual divergence of the electrostatic potential (Noguera and Goniakowski, 2012), but the answer differs in the large and small thickness regimes. In the former, little difference is expected with respect to semi-infinite surfaces. As reviewed in Noguera (2000) and Goniakowski *et al.* (2007*a*), a modification of the surface region composition by an adequate density of charged

(a)   (b)

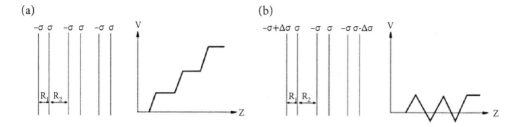

Fig. 4.2 (a) Capacitor model of a polar surface with alternating layers of charge density $\pm\sigma$ and spatial variation of the electrostatic potential V. (b) same model, but compensating charges $\Delta\sigma_\infty = \sigma R_1/(R_1 + R_2)$ are added on the outer layers, suppressing the monotonic variation of the electrostatic potential.

vacancies—whether disordered or forming facets or large cell reconstructions—, or the adsorption of charged foreign species—in particular protons or hydroxyl groups resulting from the dissociation of water molecules—, are the most common sources of compensating charges. A deep modification of the surface electronic structure may also provide them (see Section 4.3.2). These processes lead to original structural or electronic properties of semi-infinite surfaces.

They all may be met in polar thin films. However, in the ultra-small thickness regime, because of the structural flexibility, new ways to lower their energy exist in contrast to semi-infinite surfaces or thick films. One is the so-called uncompensated polarity in which, below some critical thickness, no charge compensation is required, see Section 4.3.2. Another one is a transition to a different structural ground state having a non-polar termination, see Section 4.1.3.

4.1.2   Intrinsic structural effects

At non-polar oxide thin film surfaces, the balance between short-range repulsive forces and the prevailing attractive interatomic forces—electrostatic in highly ionic compounds, electron delocalization in metals, or mixed ionic and covalent in oxides— is broken and atomic relaxation takes place. In thin films, this leads to variations of interplane spacings and also impacts the in-plane lattice parameters to relax the surface stress. The resulting distortions depend on the prevailing ionic or covalent character of the oxide.

As a general statement, in oxides of strong ionic character, a contraction of bondlengths occurs around under-coordinated atoms. In the simplest Born model of cohesion, the equilibrium bond-lengths R_0 around atoms of coordination Z result from the competition between long-range charge-charge interactions (α the Madelung constant, R the first neighbour distance) and short-range repulsion terms (A their amplitude and m their exponent). Locally, the energy E can be written as:

$$E = -\frac{\alpha Q^2}{R} + \frac{ZA}{R^m} \tag{4.3}$$

and its minimization with respect to R yields the value of R_0, equal to:

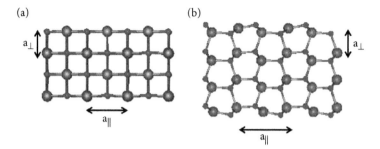

(a) (b)

Fig. 4.3 (a) Structural relaxation via bond contraction as exemplified in a rock-salt (100) thin film. The outermost interlayer distance $a_\perp$ is expanded while the in-plane lattice parameter $a_\parallel$ is shortened with respect to the bulk values. The small film rumpling in which the most polarizable surface atoms (generally oxygen atoms) are pushed outwards is not visible at this scale. (b) Relaxation via bond rotation in a wurtzite ($10\bar{1}0$) thin film. At the surface, a layer flattening takes place, which leads to an increase of $a_\parallel$. Anions and cations are represented by small and large spheres, respectively.

$$R_0 \propto \left(\frac{Z}{\alpha}\right)^{1/(m-1)} \tag{4.4}$$

R_0 is an increasing function of Z since α grows less quickly than Z (due to the repulsion between second neighbours). Bond-lengths thus get shorter around under-coordinated atoms.

The extension of this simple model to non-polar surfaces predicts a shortening of the outermost inter-plane distance, the lateral lattice parameter being fixed by the underlying bulk. In contrast, in free-standing thin films, bond contraction mainly occurs within the surface layer and yields a decrease of the in-plane lattice parameters $a_\parallel$, which is stronger and stronger as the film becomes thinner. It is accompanied by a tetragonal-type distortion characterized by an expansion of the inter-plane distances $a_\perp$ so as to roughly preserve the local atomic volume (Fig. 4.3, left). Such a behaviour has been found in free-standing MgO(001) thin films by DFT simulations (Fig. 4.4, left) and in RHEED experiments on Ag-supported MgO(100) films (Kiguchi *et al.*, 2002)—a case of weak oxide-support interaction (Fig. 4.4, right).

The same prediction of bond contraction applies to (elemental) metals in which the attractive interaction is due to the electron delocalization, responsible for the formation of metallic bands. However, in covalent oxides such as ZnO, other degrees of freedom exist which cost less energy than bond contraction. A back-bond rehybridization around under-coordinated atoms transforms the bulk-like sp_3 bonding into a surface-like sp_2 one, as at surfaces of elemental semiconductors (Chadi, 1979). This is the so-called rotational relaxation mechanism. It induces a film flattening and an increase of the in-plane lattice parameters (Fig. 4.3, right), as well as contraction of the structure in the perpendicular direction to preserve the atomic volumes. In ZnO (0001) thin films, such an expansion of the lateral lattice parameters has indeed been demonstrated by DFT simulations, as well as the flattening of the surface layers revealed by an increase in the average surface bond angle (Sponza *et al.*, 2016).

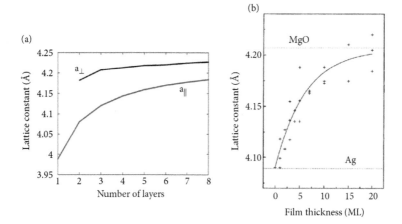

Fig. 4.4 Left: DFT variations of the lateral $a_\parallel$ (red curve) and average vertical $a_\perp$ (black curve) lattice constants of free-standing MgO thin films as a function of film thickness. Both $a_\parallel$ and $a_\perp$ increase with the number of layers, as a result of the increased mean atom coordination in the films. $a_\perp$ is larger than $a_\parallel$, revealing the tetragonal distortion of the film. Courtesy J. Goniakowski. Right: lateral lattice constant $a_\parallel$ of Ag-supported MgO thin films measured by RHEED, as a function of film thickness. The values of the bulk Ag and MgO lattice parameters are indicated with dotted lines. Reprinted from Kiguchi *et al.* (2002). Copyright 2002, with permission from Elsevier.

Many reconstructions have been observed at oxide polar or non-polar surfaces. However, at variance with metals or semi-conductors, they are not intrinsic but rather due to vacancy ordering created under specific preparation conditions or as a way to compensate polarity (see Section 4.5). Similar trends are expected in ultra-thin films.

4.1.3   Polymorphism–Global structural change

In the small thickness regime, oxide films do not necessarily adopt the bulk crystalline structure which is the most stable under the same thermodynamic conditions. A global structural change may occur beyond mere bond relaxations around under-coordinated atoms or changes of in-plane lattice parameter. Many oxides possess several bulk poly-morphs. This is for example the case for ZnO, which, beside the wurtzite ground state at ambient conditions, displays several metastable structures, such as the zinc-blend, the h-BN, the body-centred tetragonal, the sodalite or the cubane structures. Al_2O_3 possesses a similarly rich phase diagram including α, β, γ, δ, θ, and κ phases in which the Al local structure includes a variable ratio of octahedral-to-tetrahedral cation sites (Levin and Brandon, 1998).

   If a global transformation toward a metastable bulk structure helps expose more compact surfaces of lower energy, such transformation may be thermodynamically favoured below a critical thickness. The latter is determined by the competition be-tween the cost of bulk energy, which scales as the film thickness and the gain of surface energy. Examples of such a scenario can be found in all types of materials, from metals

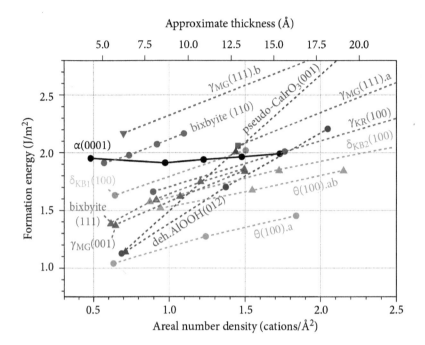

Fig. 4.5 Variations of DFT formation energies per formula unit with respect to bulk α-Al_2O_3 of thin films obtained by cutting various bulk polymorphs, as a function of film thickness (or equivalently Al atom density per unit area). In this plot, only the curve corresponding to the α-Al_2O_3 structure becomes flat as thickness increases. For the other polymorphs, the slopes of the curves scale with the difference in their bulk energy with respect to the α phase. Note, at the smallest thickness, the stability of films cut from unexpected bulk polymorphs, such as dehydrated $AlOOH(012)$ or pseudo-$CaIrO_3$ (001). The intersections of the lowest energy curves mark the critical thicknesses for structural phase transitions. The bulk α-Al_2O_3 phase is predicted to become the film ground state above ≈ 30 Å thickness. Lines are added to guide the eye. Reprinted from Van den Bossche *et al.* (2020).

to semiconductors or insulators. It has, for example, been evidenced theoretically in alumina ultra-thin films (Clausen and Hren, 1984; Aykol and Persson, 2018; Van den Bossche *et al.*, 2020) as illustrated in Fig. 4.5. In the thickness range 1 to 4 MLs, films with the $\theta(100)$, $\gamma(001)$, $\delta(100)$ structures have lower formation energies than those cut out of the α-(0001) phase, consistently with their lower surface energies (0.95, 1.07, 1.39, and 2.01 J/m^2 in DFT, respectively). The transition to the bulk stable α phase is predicted to occur at roughly 30 Å thickness (Van den Bossche *et al.*, 2020).

In the special case of polar orientations, a structural transformation may also be driven by the gain of energy obtained by reducing or even suppressing the surface polarity. The same competition between bulk and surface energy terms determines the critical thickness below which the film structure is globally different from the bulk structure. For example, theoretical and experimental works agree in finding that, at low thickness, a variety of binary oxide polar films, along the (111) or (0001) orientation,

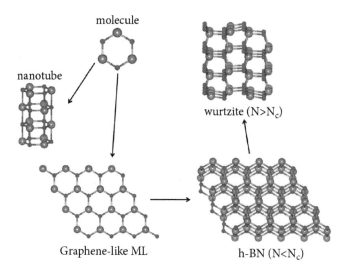

molecule

nanotube

wurtzite (N>N$_c$)

Graphene-like ML

h-BN (N<N$_c$)

Fig. 4.6 Bulk and low-dimensional ZnO objects. In contrast to the local Zn environment in the stable wurtzite bulk phase, a $(ZnO)_3$ molecule adopts a flat six-member ring structure. ZnO nanotubes, ZnO(0001) monolayer, and thin ZnO(0001) films with a thickness below a critical value N_c display this building unit in their most stable configurations derived from the hexagonal BN structure.

adopt a hexagonal boron nitride (h-BN) structure, which consists of a stacking of graphene-like, atomically flat six-member rings with an equal number of cations and oxygen atoms. While the h-BN structure is only metastable in the corresponding bulk oxides, non-polar h-BN(0001) films are favoured over polar rock-salt (111), as in MgO (Goniakowski *et al.*, 2004; Kiguchi *et al.*, 2003), or wurtzite (0001) ones (like ZnO (Claeyssens *et al.*, 2005; Tusche *et al.*, 2007; Morgan, 2009) and other wurtzite compounds (Freeman *et al.*, 2006)) below a critical film thickness (Fig. 4.6).

4.1.4 Substrate-induced structural effects

Aside from oxide nanosheets obtained by direct synthesis or exfoliation of lamellar compounds, most oxide thin films are grown on substrates. As recalled in the Introduction, because the initial focus was to perform STM experiments on oxide surfaces, throughout nearly one decade, metallic substrates were chosen to grow crystalline oxide thin films. A direct oxidation of the metal was initially tried but it led to bad crystallinity of the oxide films due to the usually large lattice mismatch between a metal and its own oxide. Better-quality films were obtained by growth on other metals, with the various methods described in Chapter 2. In the last 10–15 years, growth of oxide films on oxide substrates has also been successfully accomplished.

In order to obtain good crystalline quality, several conditions need to be fulfilled (Nilius, 2009). First, there should be a good symmetry matching of the two lattices. (001)-oriented cubic films, like rock-salt oxides or perovskites, are more easily obtained on a metal substrate of cubic symmetry, like Ag(100), Pd(100), Mo(100), or on rock-salt or perovskite (001) oxide substrates. In contrast, substrate surfaces of hexagonal symmetry, like Pt(111), Ag(111), or Al_2O_3(0001) are well suited to grow oxide films of hexagonal symmetry, like (111)-oriented rock-salt oxides, (0001) wurtzite or corundum oxides.

However, the symmetry condition is generally insufficient, since lattice distortions occur at the interface when the mismatch between the two lattices is important. The lattice mismatch is defined as the relative difference between the film and substrate lattice parameters $\delta = (a_{ox} - a_{sub})/a_{sub}$. It may be characterized by two or one values according to whether the two vectors of the unit cell differ or not. As a rule of thumb, good crystallinity is expected if δ is less than $\approx 5\%$. This is for example the case for MgO(100)/Ag(100) ($\delta = 3\%$), but there are a number of exceptions. When δ is large, good coïncidence cells may still be obtained by rotation of one lattice with respect to the other. At this point, it is worth recalling that the intrinsic lattice parameter of a film depends on its thickness (see preceding section) and usually increases with thickness. An evaluation of the lattice mismatch from the oxide bulk lattice parameters is thus subject to errors, in particular for the formation of the first few atomic layers.

The competition between the film-substrate interaction and the elastic energy associated to the interfacial distortions controls the consequences of the lattice mismatch on the film structure. The former favours a commensurate phase, in which the film adopts a pseudomorphic structure with the periodicity of the substrate and can even display high-energy surfaces. When the latter prevails, misfit dislocations set in to relax the interfacial strain, with a density which grows with the misfit. Zones of good matching alternate with zones of poor matching, with a periodicity $A = a_{sub}(1 + \delta)/\delta$ which is the smallest lattice parameter in common between the substrate and the overlayer, and which increases as δ becomes smaller. For example, a 5% misfit induces a coïncidence pattern with a lattice parameter of the order of twenty times that of the substrate. In STM images, the misfit dislocation network is revealed by a Moiré pattern, as shown in Fig. 4.7, and in X-ray or LEED diffraction by the presence of extra satellites. This scenario represents the two-dimensional extension of the original Frenkel–Kontorova model, which had been devised to describe the structure of one-dimensional elastic chains on a substrate (see Appendix B).

There are many examples of interfacial misfit dislocation networks observed at oxide-substrate interfaces. Among them, pioneering work considered the FeO(111) film grown on Pt(111) (Galloway *et al.*, 1996; Ritter *et al.*, 1998; Giordano *et al.*, 2007). At the monolayer limit, a hexagonal close-packed structure was observed, with a periodicity $(\sqrt{84} \times \sqrt{84})$R10.9°. With increasing coverage, four coincidence structures with slightly different lateral lattice constants and rotation misfit angles against the platinum substrate were observed. Another well-studied system is MgO/Mo(001) characterized by a misfit of -5.4% and a coïncidence cell lattice parameter $A = 55$ Å. While in-plane modulations in the lattice parameter are most pronounced in the MgO interface layer, they also remain detectable in subsequent planes. With increasing film thickness, the region of minimal Mg-O bond lengths expands and finally develops into a square zone in the third MgO plane. Additionally, during the relaxation of the film, a surface deformation becomes evident. LEED spots display a fourfold splitting along the <100> MgO directions that depends linearly on the scattering vector (Benedetti

(a)   (b)

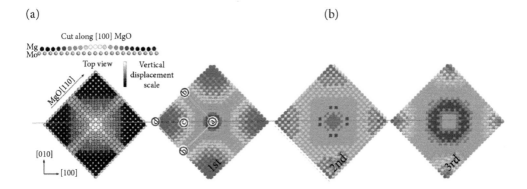

Fig. 4.7 (a) Structural model of the MgO/Mo(001) interface, including a cross section along the MgO [100] direction and a top view of the MgO superstructure cell. The model demonstrates the interrelation between the interfacial dislocation grey lines and the surface corrugation. Reprinted from Benedetti *et al.* (2008), with permission. Copyright 2008 American Chemical Society. (b) Colour-coded 2-D maps of the in-plane parameter in the first three MgO planes from left to right. Long and short O-Mg distances are depicted with red and blue colours, respectively. Reprinted from Benedetti *et al.* (2013), with permission. Copyright © 2013 Wiley-VCH Verlag GmbH and Co. KGaA, Weinheim.

et al., 2008; Benedetti *et al.*, 2013). Moiré patterns may also form at oxide-oxide interfaces, as confirmed by recent studies of $SrTiO_3$ films grown on MgO(100) (Zhu *et al.*, 2013*b*; Dholabhai *et al.*, 2014).

When the lattice modulation is large, a fracture of the film may occur, leading to configurations in which the planes are tilted, thus forming a mosaic structure. This scenario was observed in NiO films on Ag(100) (Wollschläger *et al.*, 2001). While the 3 MLs film is perfectly pseudomorphic on Ag(001), above 5 MLs its strain starts being released. In SPA-LEED diagrams additional streaky satellites along the <100> directions are observed, together with weaker satellites with a fourfold symmetry. The behaviour of the satellites positions, which move toward the central spot both for increasing film thickness and decreasing scattering phase, allowed ascribing them to scattering from mosaics with a slightly tilted angle with respect to the (001) surface, angle which decreases with increasing film thickness. As a general statement, as thickness increases, there is a progressive evolution from a regime determined by metal/oxide interactions to the bulk-like characteristics.

4.2 Vibrational and ferroelectric properties

As semi-infinite surfaces, thin films possess specific phonon modes determined by quantum confinement at the interface with their support and/or at their free surface(s). In particular, the macroscopic optical Fuchs–Kliewer modes display strong variations with the film thickness, which have been evidenced both by experiments and simulations. Confinement also affects some structural phase transitions, such as the paraelectric to ferroelectric transition, which are driven by the condensation of soft phonons. The second part of this section summarizes how ferroelectric characteristics—transition temperature and macroscopic polarization—vary with the film thickness, surface characteristics and substrate interaction.

4.2.1 Vibrational properties

Semi-infinite surfaces and thin films possess vibration modes classified as microscopic or macroscopic according to their penetration into the inner layers. Microscopic modes mostly affect a few inter-plane spacings. Wallis (1964) and Lucas (1968) were the first authors to predict their existence at semi-infinite surfaces. They showed that these modes, characterized by displacement fields perpendicular to the surface in the former case and parallel to the surface in the latter, are generally located in the gaps of the bulk phonon spectrum. However, at some reciprocal space vectors, they may become degenerate with the bulk modes, thus transforming into surface resonances. In thin films, microscopic modes are present at the free film surfaces and at the interface with the substrate. Due to their very short penetration depth, their frequencies may not vary with the film thickness. Their description requires a precise account of the atomic structure and force constants, which makes them sensitive to stress, relaxation, or reconstruction effects at surfaces and interfaces.

The macroscopic modes—acoustic Rayleigh mode or optic Fuchs–Kliewer (FK) mode—have an attenuation length which varies as the inverse of the normal component of their wave vector. They penetrate deeply inside the film. On this length scale, the precise atomic structure is unimportant and the elasticity theory of continuous media or the dielectric theory may be used, which means that their characteristics only depend on the bulk vibrational properties and on the dielectric constant of the external medium. The first mention of surface phonons is due to Lord Rayleigh (1885) who, relying upon the elasticity theory, predicted the existence of a surface acoustic mode with a sound velocity lower than in the bulk. From a similar matching procedure at the surface, Fuchs and Kliewer have predicted the existence of macroscopic surface optic modes in ionic crystals, with a frequency intermediate between the transverse optic and longitudinal optic frequencies.

A simplified derivation of their result makes use of the following expression of the bulk dielectric constant $\epsilon_1(\omega)$ as a function of frequency ω, in the phonon frequency range (ω_{TO} and ω_{LO} the frequency of transverse and longitudinal optic modes, ϵ_0 and ϵ_∞ the zero and infinite frequency dielectric constants):

$$\epsilon_1(\omega) = \epsilon_\infty + (\epsilon_0 - \epsilon_\infty)\frac{\omega_{TO}^2}{(\omega_{TO}^2 - \omega^2)} \tag{4.5}$$

or, according to the Lyddane–Sachs–Teller relation, as:

$$\epsilon_1(\omega) = \epsilon_\infty \frac{\omega_{LO}^2 - \omega^2}{\omega_{TO}^2 - \omega^2} \tag{4.6}$$

The matching procedure at a free surface yields a new phonon mode, whose frequency ω_s given by the roots of the equation $1 + \epsilon_1(\omega_s) = 0$ is intermediate between ω_{TO} and ω_{LO}:

$$\omega_s = \omega_{TO}\sqrt{\frac{\epsilon_0 + 1}{\epsilon_\infty + 1}} \tag{4.7}$$

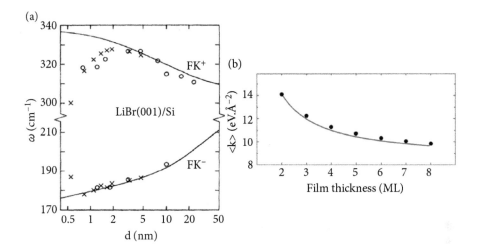

Fig. 4.8 (a) Positions of the FK loss peaks in the HREELS spectra of LiBr/Si(100) epitaxial layers against film thickness (open circles: experimental results from Gao *et al.* (1993); full curves: prediction from the dielectric approximation). Reprinted from Senet *et al.* (1995) with permission. Copyright 1995 by the American Physical Society. (b) DFT average stiffness $< k >$ of free-standing MgO(001) films of increasing number of layers. The red curve shows the results of a simple energy expression which sums local elastic interactions around atoms of various coordination, and dots are the result of explicit ab initio simulations. Reprinted from Cabailh *et al.* (2019) with permission. Copyright 2019 by the American Physical Society.

A similar procedure can be used to predict the values of the FK mode frequencies in thin films of thickness d, deposited on a substrate of dielectric constant ϵ_2 assumed to be constant (Froitzheim *et al.*, 1975; Lambin *et al.*, 1991). The dielectric constant of the thin film then reads as a function of wave vector Q and frequency ω:

$$\epsilon(Q,\omega) = \frac{(\epsilon_1 + \epsilon_2)(\epsilon_1 - 1) - (\epsilon_1 - \epsilon_2)(\epsilon_1 + 1)e^{-2Qd}}{(\epsilon_1 + \epsilon_2)(\epsilon_1 + 1) - (\epsilon_1 - \epsilon_2)(\epsilon_1 - 1)e^{-2Qd}} \qquad (4.8)$$

The poles in this expression correspond to the macroscopic FK eigenmodes of the system. There are two branches of the FK modes that can be considered as surface and interface modes, respectively, and whose frequencies $\omega_\pm$ are solutions of:

$$\epsilon_1(\omega_\pm) = -\tanh^{\pm 1}\left(\frac{Qd \pm X}{2}\right) \quad \text{with} \quad X = \sinh^{-1}\left(\frac{\epsilon_2 - 1}{\epsilon_2 + 1}\sinh(Qd)\right) \qquad (4.9)$$

Figure 4.8a shows the variations of these two modes as a function of the thickness of an LiBr film deposited on Si(100) (Senet *et al.*, 1995). Beyond $d \approx 3$ nm, the experimentally measured surface mode frequency ω_+ decreases as d becomes larger, while the interface one ω_- increases (Gao *et al.*, 1993), in good agreement with the dielectric approach. However, when d becomes smaller than 3 nm, the ω_+ branch starts to deviate significantly from the predictions of electrostatics. This comes from

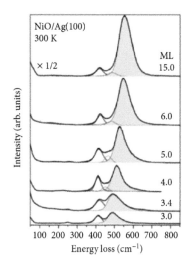

Fig. 4.9 HREELS spectra of NiO(100) films with different thicknesses (indicated in the figure), as grown on Ag(100) at 300 K. The FK phonon mode is shaded. Reprinted from Kostov *et al.* (2016) with permission. Copyright 2016 by the American Physical Society.

the formation of optical longitudinal standing-waves confined in the layer which are not correctly described by the dielectric model (Rücker *et al.*, 1992).

Vibrational characteristics of several thin oxide films have been investigated by high-resolution electron energy loss spectroscopy. At the ML limit, the MnO vibrational spectrum displays a strong and narrow phonon at 368 cm^{-1}, while for larger d values, a Wallis and a FK mode (382 and 547 cm^{-1}, respectively) are present. The former has constant intensity and frequency, while the latter has an increasing intensity and a decreasing frequency with coverage, as predicted by the dielectric theory (Sachert *et al.*, 2010). In NiO/Ag(100) thin films, the FK phonon intensity shows a strong thickness dependence (see Fig. 4.9). The experimental data present an excellent agreement with the predictions of the dielectric theory for film thicknesses above 15 ML. In contrast, a strong FK phonon softening is observed for thin films below 5 ML that cannot be explained by the dielectric theory nor by the formation of phonon standing waves. This softening is attributed to the presence of surface stress, which results from the lattice mismatch between NiO and Ag. (Kostov *et al.*, 2016). In 20–30 ML thin MgO films, surface FK and Wallis modes at 677 cm^{-1} and 524 cm^{-1} have been observed. In thinner films, standing wave optical phonons confined in the MgO layer are present, whose frequencies depend strongly on the film thickness. At the monolayer limit, a single intense loss peak remains at 427 cm^{-1} (Savio *et al.*, 2003). The thickness dependence of Wallis and FK phonons has also been identified in thin BaO layers on Pt(001) (Goian *et al.*, 2018).

These observations show that in ultra-thin films, not only is the atomic structure modified, but also the interatomic forces, so that their elastic properties differ from their bulk analogues. This is well-recognized in nanowires, for example ZnO nanowires, in which the large surface-to-volume ratio results in strongly size-dependent stiffening

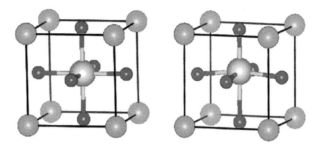

Fig. 4.10 Crystal structure of perovskite ferroelectric $BaTiO_3$. Left: high temperature, para-electric cubic phase. Right: low temperature, ferroelectric tetragonal phase showing the up--polarization variant. Barium, titanium and oxygen atoms are represented by green, blue and red balls, respectively.

(Chen *et al.*, 2006). In MgO nanoislands and ultra-thin films, the stiffness increase at small size has recently been demonstrated both experimentally and theoretically, as shown in Fig. 4.8(b) (Cabailh *et al.*, 2019). It results from the increasing weight of five-fold coordinated atoms at the surface of the film with respect to the six-fold coordinated ones in the film centre. It is accompanied by a decrease of the film lattice parameter, as shown in Fig. 4.4. More generally, this has led to the idea that it is possible, by changing the dimensions of a nano-object, to fine tune its physical and chemical properties via strain engineering (Li *et al.*, 2014a).

4.2.2   Critical thickness for ferroelectricity

Ferroelectric materials exhibit a spontaneous electric polarization that can be switched under an applied electric field (Ahn *et al.*, 2004; Osada and Sasaki, 2012; Lichtensteiger *et al.*, 2012). This switching process is associated with a hysteresis loop and there exists a critical temperature T_C above which a phase transition to a high-symmetry unpolarized phase takes place. Ferroelectricity is produced by spontaneous atomic off-centre displacements in the unit cell, leading to non-centrosymmetric structures. It is closely linked to piezoelectricity and thermoelectricity (Walia *et al.*, 2013). The resulting dipole moments of the unit cells create a macroscopic polarization, which can vary over many orders of magnitude in different materials (Lichtensteiger *et al.*, 2012). Many ferroelectric oxides are ABO_3 perovskites—like $BaTiO_3$, $PbTiO_3$, $Pb(Zr,Ti)O_3$, $LiNbO_3$, etc.—but there also exist ferroelectrics of more complex structure derived from the perovskite structure, such as $SrBi_2Ta_2O_9$, $Bi_4Ti_3O_{12}$, Bi_3TiNbO_9 and Bi_2WO_6 which crystallize in the Aurivillius phase, or $RbBiNb_2O_7$ and $CsBiNb_2O_7$ in a Dion–Jacobson phase (Benedek *et al.*, 2015). Ferroelectric materials are employed in many applications, ranging from pyroelectric infra-red detectors, ultrasound transducers, surface acoustic wave devices, ultrasonic micro-motors, to ferroelectric tunnel junctions, and memristor devices. With the advent of miniaturization, ferroelectric thin films have found applications in non-volatile random access memories, high-density data storage devices, sensors and actuators, and in tuneable microwave circuits (Setter *et al.*, 2006).

Most ferroelectric phase transitions in oxides are displacive, which means that they result from the shift of some ions away from high-symmetry positions. In ABO_3

perovskites, the phase transition can be visualized by the motion of the B cation with respect to the centre of the octahedral oxygen cage, with a concomitant small deformation of the unit cell due to this polar atomic displacement—see Fig. 4.10. This is the case for the prototypical ferroelectric $BaTiO_3$. In $PbTiO_3$, the same displacement takes place, but the polarization is enforced by simultaneous moves of the Pb atoms from the corners of the unit cell.

In ultra-thin films, short- and long-range interactions responsible for the phase transition are modified, as compared to the bulk, due to the importance of boundary conditions at surfaces and interfaces. On general grounds, in confined materials of reduced size, a decrease of the critical temperatures for phase transition is expected whenever the relevant correlation length becomes of the order of the size of the object. These intrinsic size effects have long been thought to degrade ferroelectricity, leading to the idea that it should disappear below some critical thickness, a conjecture that was supported by earlier experimental observations. However, the sensitivity to parasitic extrinsic effects due to poor sample quality (defects, impurities, grain boundaries) also increases upon size reduction and the existence of a 'dead layer', i.e. a physically and chemically distinct layer with smaller permittivity close to the ferroelectric-electrode interfaces, has now mostly been ascribed to extrinsic effects induced by the fabrication process.

Various experimental evidences in a number of oxide systems through piezo-electric response AFM measurements have proved that ferroelectricity actually only vanishes in thin films at extremely low thicknesses. Typical recorded critical thicknesses are of the order of 3–10 unit-cells and sometimes ferroelectricity remains down to 1 unit-cell. This is true in $BaTiO_3$, whether free-standing (Li *et al.*, 2015) or grown on various substrates (Choi *et al.*, 2004; Tenne *et al.*, 2009), in $PbTiO_3$ (Streiffer *et al.*, 2002; Fong *et al.*, 2004), in $Pb(Zr,Ti)O_3$ (Tybell *et al.*, 1999; Nagarajan *et al.*, 2006; Gao *et al.*, 2017) and in $BiFeO_3$ (Chu *et al.*, 2007; Maksymovych *et al.*, 2012). Results from all these studies are qualitatively consistent, indicating that although the transition temperature is markedly reduced in the thinnest films, epitaxial films still display a ferroelectric behaviour down to a thickness of just a few unit cells.

The macroscopic polarization $\vec{P}$ in thin film is very sensitive to surface and electro-static effects. Surface effects have been theoretically addressed at free (001) surfaces of $BaTiO_3$ and $PbTiO_3$, assuming that the spontaneous polarization lies parallel to the surface. The deviations of the ferroelectric distortions from the bulk value were found to be confined to the first few atomic surface layers. In the case of perpendicular polarization, (001) oriented films were found to display a significant enhancement of the polarization at the surface (Ghosez and Rabe, 2000; Meyer and Vanderbilt, 2001). Actually, the ferroelectric state with perpendicular polarization, which is often encountered in applications, presents many analogies with polarity. To maintain a vanishing internal electric field which stabilizes the ferroelectric state, surface charges $\sigma_{pol} = \vec{P}.\hat{n}$ ($\hat{n}$ the unitary outward vector normal to the interface) need to be present on the surface, equivalent to the compensating charges in polar films. They may be provided either intrinsically by metallization of the surfaces (Watanabe *et al.*, 2001) or domain formation (Lichtensteiger *et al.*, 2012), or by several types of extrinsic processes, such as non-stoichiometric reconstructions (Gao *et al.*, 2017), impurities (Fong *et al.*, 2006),

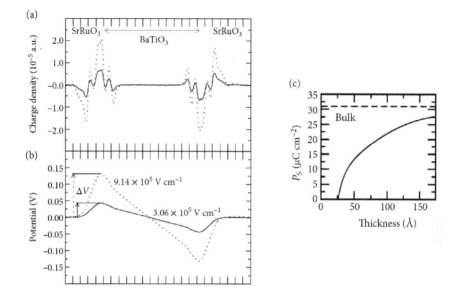

Fig. 4.11 Electrostatic properties of a $BaTiO_3$ capacitor sandwiched between metallic $SrRuO_3$ electrodes under short-circuit conditions: change in (a) the macroscopic total charge density and (b) the electrostatic potential with respect to the paraelectric phase, for two frozen polarization. The thickness of the $BaTiO_3$ layer corresponds to 6 unit cells. The magnitude of the electric field for each configuration is indicated. ΔV is the potential drop across the interface. (c) Evolution of the spontaneous polarization P with thickness, deduced from an energy minimization. Reprinted from Junquera and Ghosez (2003) with permission from Nature, Copyright 2003.

and so on. They all act in a way very similar to compensation mechanisms in polar thin films (see Sections 4.1.1 and 4.3.2).

In devices, the ferroelectric films are sandwiched between metallic electrodes, which are able to provide interfacial charges to screen the internal electric field, but generally a full screening is never attained. The remaining electric field induces a decrease of the film polarization at all thicknesses and a suppression of the ferroelectric ground state below a critical value, as shown in Fig. 4.11. The thin film behaviour is thus driven not only by intrinsic confinement effects, but also by long-range surface/interface interactions. In addition, the polarization may be affected by short-range interface effects, such as local charge redistribution due to the difference in the oxide-substrate work functions (Sai *et al.*, 2005) or the formation of interfacial bonds (Stengel *et al.*, 2009).

There are concomitant evidences from both experiments and theory that strain effects have a strong impact on the ferroelectric state. Huge T_c variations between ≈ 40 K and more than 900 K in the thickness range 2–10 nm have for example been recorded in $BaTiO_3$ thin films grown on $SrTiO_3$ by Raman spectroscopy, see Fig. 4.12(a). The highest value exceeds the bulk T_c by more than a factor 2 (Tenne *et al.*, 2009). Similarly, a T_c rise of nearly 500°C and an increase of at least 250% of the remanent polarization have been found in $BaTiO_3$ films deposited on $GdScO_3$ or

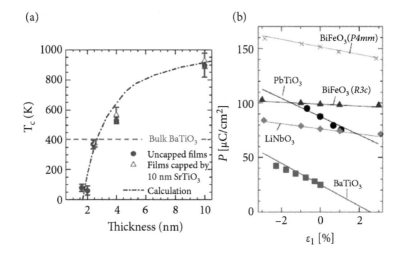

Fig. 4.12 (a) Variation of the ferroelectric critical temperature of $BaTiO_3$ thin films grown on $SrTiO_3$ as a function of thickness, obtained from Raman spectroscopy measurements. Reprinted from Tenne *et al.* (2009) with permission. Copyright 2009 by the American Physical Society. (b) Calculated evolution of the spontaneous polarization of various perovskite ferroelectrics as a function of strain. Reprinted from Ederer and Spaldin (2005) with permission. Copyright 2005 by the American Physical Society.

$DyScO_3$, compared to bulk $BaTiO_3$ (Choi *et al.*, 2004). It has even been demonstrated that room temperature ferroelectricity is induced in a $SrTiO_3$ thin film—normally not a ferroelectric—by strain effects due to its interaction with a $DyScO_3$ substrate (Haeni *et al.*, 2004). These experimental findings are in line with results from first principles simulations in which the strain effect is represented by a mere change of the lateral lattice parameter (Ederer and Spaldin, 2005) (Fig. 4.12(b)) or is fully taken into account in simulated heterostructures, such as $BaTiO_3$ thin films sandwiched between $SrRuO_3$ metallic electrodes grown on a $SrTiO_3$ substrate (Junquera and Ghosez, 2003).

4.3 Electronic structure

In oxides, structural and electronic degrees of freedom are strongly coupled. The electronic structure of thin films thus presents specific features associated to the relaxation effects and distortions present in the film. Some are intrinsic—low coordination of the atoms, quantum confinement—while others are substrate-induced.

4.3.1 Intrinsic electronic properties

By comparison with the bulk, three main effects impact the intrinsic electronic structure of thin films (Fig. 4.13). The first one is a splitting and/or a shift of the atomic orbital energies ϵ from which the electronic structure is built (Fig. 4.13(a,b)).

On the one hand, compared to the bulk, the change of symmetry of the local atomic environment modifies the p or d multiplet splitting. This particularity concerns the Mott–Hubbard insulators in which, due to the cationic d character of both the VB

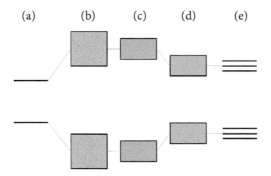

Fig. 4.13 Schematic representation of the various effects which impact an oxide thin film electronic structure: (a) Relevant atomic orbitals forming the VB and CB. (b) Band formation in the bulk. The positions of the top of the VB and the bottom of the conduction band define the HOMO-LUMO band gap G. (c) Local band narrowing at under-coordinated atoms. (d) Shift of the bands due to variations of the local Madelung potential. (e) Quantization of states driven by quantum confinement.

and CB edges, the d-d electronic excitation energies significantly depend on the cation local environment. For example, compared to the bulk, at the NiO(001) surface (Freitag *et al.*, 1993) and in thin NiO films deposited on MgO (Guo *et al.*, 1998), new transitions have been evidenced by HREELS at energies lower than the band gap, and attributed to surface Ni $d - d$ transitions. In the thin film, they gradually shift to the values observed for single crystal NiO(100) with increasing thickness. The lower symmetry of the crystal field at surfaces also impacts the electronic structure of charge transfer insulators with a pronounced covalent character, in which back-bond rehybridization may change the surface bonding from sp_3 to sp_2 and induce the presence of dangling bond states with energies in the gap region.

On the other hand, in (highly or partially) ionic oxides, energetic and electronic properties are mainly controlled by electrostatic interactions. The strong electrostatic (Madelung) potential V_i exerted on each ion i by its environment shifts the effective atomic orbital energy ϵ_i (the diagonal terms of the Hamiltonian projected on an atomic orbital basis set) with respect to its value ϵ_i^0 in the isolated neutral atom, approximately as $\epsilon_i = \epsilon_i^0 - V_i - U_i Q_i$ (Noguera, 1996). This Hartree-type expression involves a term associated to the ionic charge Q_i and the on-site electron-electron repulsion U_i, and a Madelung contribution V_i. The two corrections to ϵ_i^0 are usually in competition— for example $U_i Q_i < 0$ and $V_i > 0$ on anions, and the opposite for cations—and their balance depends on the value of the optical dielectric constant which controls the screening properties, but the Madelung term is often dominant. In thin films, under-coordinated atoms are subject to a reduced Madelung potential which brings the oxygen and cation effective levels closer to each other and, in charge transfer insulators, induces surface states in the gap region (Fig. 4.13(d)).

As a whole, whatever the case—Mott–Hubbard or charge transfer oxide, covalent or ionic oxide—-, the above effects, consequence of bond breaking at the film surface, induce states in the HOMO-LUMO gap and thus reduce its effective width. Conversely, structural relaxation, whether by means of bond-length contraction or orbital rehy-bridization, causes an opposite trend on the effective levels and tends to reopen the gap.

The second effect induced by bond breaking is a narrowing of the local band widths (Fig. 4.13(c)). Its importance roughly scales as the square root of the atomic coordina-tion number (Noguera, 1996). In Mott-Hubbard insulators, the decrease of the number of cation-cation second neighbours locally reduces the widths of both the CB and the VB. In charge transfer insulators, the local widths of the CB and VB rather depend on the number of cation-cation and oxygen-oxygen second neighbours, respectively. The band narrowing, on its own, tends to increase the local band gap, but the accom-panying bond contraction acts in the opposite direction both as regards the band and the gap widths.

Finally, quantum confinement, i.e. the quantification of electronic states propagat-ing in the direction perpendicular to the film surface, may also soundly modify its elec-tronic structure. In particular, the discretization of bands associated to perpendicular wave vectors in the direct vicinity of the VB maximum and the CB minimum produces an overall increase ΔG of the gap (Fig. 4.13(e)). In semiconductor nanocrystallites, ΔG is of the order of $\hbar^2 \pi^2 / 2\mu R^2$, with R the effective radius of the nanoparticle and μ the reduced electron-hole effective mass. A similar expression holds in thin films where R is replaced by the film thickness. Moreover, since the effective mass is determined by the band curvatures at the top of the VB and the bottom of the CB, it is smaller in simple metal sp charge transfer oxides than in transition metal Mott–Hubbard ones and should thus lead to larger quantum confinement effects in the former. Quan-tum confinement effects have been observed in various oxide thin films, such as ZnO (Coli and Bajaj, 2001; Bowen *et al.*, 2008; Li *et al.*, 2013), Cu_2O (Poulopoulos *et al.*, 2011), Fe_2O_3 (Fondell *et al.*, 2014; Garoufalis *et al.*, 2017), NiO (Garoufalis *et al.*, 2018), BeO (Shahrokhi and Leonard, 2016) or EuO (Prinz *et al.*, 2016), or perovskites (Zhong *et al.*, 2013; Choi *et al.*, 2015). When d electrons contribute to the band edge states, the simple model of quantum confinement based on quasi-free electronic waves can no longer been used and the more localized nature of the electronic states has to be taken into account. The observed increase in the absorption transition energy in all these oxides (an example is given in Fig. 4.14) is assigned to both the quantization of the wave functions (quasiparticle effect) and the increase in the exciton energies (electron-hole interaction) when the radius of the latter becomes of the order of the film thickness.

In ultra-thin films, these various effects induce positive or negative variations of the HOMO-LUMO gap G. Their competition is driven by the nature of the oxide (charge-transfer or Mott–Hubbard), its degree of ionicity, the nature (bond rotation or contraction) and the strength of relaxation effects, and the film thickness. However, the electronic effects due to atomic under-coordination are essentially local, so that a gap opening at the film surfaces does not alter the gap of the entire film, which is determined by the fully coordinated atoms in the film core. Conversely, quantum confinement effect is global and the induced gap opening concerns the entire film.

Several experimental works have confirmed that gap variations occur as a function of the film thickness. At non-polar surfaces of ionic oxides, a gap narrowing is confirmed

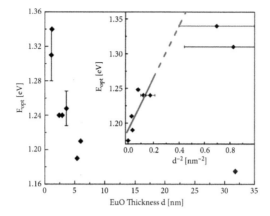

Fig. 4.14 Dependence of EuO optical band gap on film thickness (d between 1.1 nm and 32 nm). Error bars indicate representative statistical errors. In the inset, the data are plotted against the d^{-2} law which accounts for quantum confinement in a simple potential well model. Error bars indicate a monolayer (0.26 nm) deviation of the thickness. The line represents a linear fit to the data. Reprinted from Prinz *et al.* (2016) with the permission of AIP Publishing.

by experimental results or more advanced theoretical simulations in MgO(001) (Didier and Jupille, 1994; Goniakowski and Noguera, 1995), CaO(001) (Protheroe *et al.*, 1983) or NiO(001) (Dudarev *et al.*, 1997), while at the (0001) surface of Al$_2$O$_3$, the huge inward relaxation of the surface Al atoms counterbalances the electrostatic effect and leads to an increase of the gap width (Gautier-Soyer *et al.*, 1996).

It is sufficient to grow three (001) layers of MgO on a metal like silver to see converged properties in terms of band gap, as accurate experimental measurements by STS (Schintke *et al.*, 2001), and first principles simulations (Giordano *et al.*, 2003) have confirmed. In NiO(001) thin films grown on Au(001), the measurement of valence band spectra indicates metallicity below 3 MLs, due to the metal induced gap states at the interface. At larger thicknesses, for which the influence of the substrate becomes negligible, electron energy-loss analysis reveals that the band gap decreases from ≈ 3 eV at 10 MLs to ≈ 1.8 eV at 5 MLs (Visikovskiy *et al.*, 2013).

4.3.2 Specificities of polar orientations in thin films

In thick stoichiometric polar films, the stacking of (alternatively) charged atomic layers $\pm\sigma$, such as met e.g. in (111)-oriented rock-salt films, triggers a monotonic increase of the macroscopic electrostatic potential V across the film, which progressively shifts the local band structure and brings the VB maximum of one film termination close to the CB minimum of the other one (Fig. 4.15(a,b)). Assuming a frozen electronic configuration, the overall potential jump ΔV would scale (linearly) with the film thickness. However, as soon as it becomes of the order of the surface gap, a band overlap takes place allowing an electron transfer from the local VB on one side of the film to the local CB on the other side. A reduction of the surface charge density occurs which

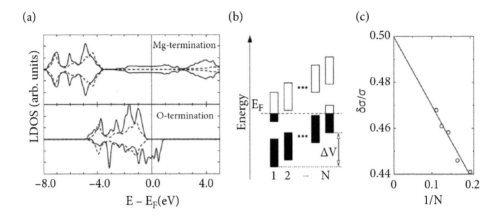

Fig. 4.15 (a) Spin resolved LDOS on the magnesium (top) and oxygen (bottom) terminations of a MgO(111) polar film, highlighting the metallization of both surfaces and the spin polarization at the oxygen termination. (b) Sketch of the local band position and filling across a polar thick film. (c) Thickness dependence of $\Delta\sigma/\sigma$, deduced from a Bader analysis, in thick MgO(111) films, evidencing its $1/N$ asymptotic behaviour, consistent with eqn. 4.10. Reprinted from Noguera and Goniakowski (2008) with permission. Copyright © 2008 IOP Publishing Ltd. All rights reserved.

provides the compensating charges $\Delta\sigma$, necessary to heal polarity. It fixes ΔV at a value close to the band gap and is associated to a metallization of the terminations (a 2-D electron gas), sometimes called 'polar catastrophe', electronic reconstruction, or 'Zener breakdown', as shown in Fig. 4.15.

The existence of such a 2-D electron/hole gas (2DEG/2DHG) had long been predicted by atomistic simulations (Tsukada and Hoshino, 1982; Pojani *et al.*, 1997; Noguera, 2000) but rarely confirmed experimentally. The likely reason is that this mechanism of compensation by electronic excitation across the band gap costs a great deal of energy, especially in large gap oxides, so that the surface energy remains high even after compensation. However, recently the observation of a 2DEG at the interface between SrTiO$_3$ and LaAlO$_3$(100) (a polar orientation in the latter) has generated much excitement in the physics community (see Section 4.3.4). 2DEGs have also been observed at (001) non-polar (Santander-Syro *et al.*, 2011) and (110) and (111) polar (Rödel *et al.*, 2014) SrTiO$_3$ surfaces, but in both cases, they were due to the presence of oxygen vacancies (see Section 4.5.1), and not to polarity compensation. Similarly, 2DEGs have been evidenced at hydrogenated surfaces of ZnO(10$\bar{1}$0) (Wang *et al.*, 2005), and SrTiO$_3$(100) (D'Angelo *et al.*, 2012) with electrons coming from the ionization of the adsorbed hydrogen atoms, with no relationship to polarity.

In the large thickness regime, little difference is expected between polar films and semi-infinite polar surfaces, except for a thickness dependence of the compensating charge density $\Delta\sigma(N)$ coming from surface metallization. It has been shown that, in the configuration of Fig. 4.2, it depends on the number of repeat units N, the HOMO-LUMO energy difference G, and the optic dielectric constant ϵ_∞ (Noguera

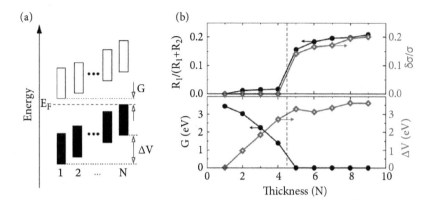

Fig. 4.16 (a) Sketch of the shift of the local electronic bands across a polar uncompensated film for $N < N_c$. Note that a gap subsists between the top of the VB on the right termination and the bottom of the CB on the left one. (b) Thickness dependence of unsupported MgO (111) film properties in the metastable zinc blend phase. Top panel $R_1/(R_1 + R_2)$ (filled symbols) and excess Bader charge densities $\Delta\sigma/\sigma$ (open symbols). Bottom panel gap G (filled symbols) and total voltage ΔV jump (open symbols). Reprinted from Goniakowski *et al.* (2007*b*) with permission. Copyright 2007 by the American Physical Society.

and Goniakowski, 2008), according to:

$$\Delta\sigma(N) = \Delta\sigma_\infty - \frac{G\epsilon_\infty}{4\pi N(R_1 + R_2)} \qquad (4.10)$$

The $1/N$ dependence of $\Delta\sigma(N)$, deduced from a Bader analysis in thick free-standing MgO(111) films is illustrated in Fig. 4.15(c) (Noguera and Goniakowski, 2008).

However, below a critical thickness, the shift ΔV of the electrostatic potential may remain smaller than the width of the band gap, producing a specific 'uncompensated polar', or 'subcritical' regime (Goniakowski *et al.*, 2007*b*). In this regime, the total dipole moment of the film grows linearly with its thickness, while its band gap simultaneously decreases, as shown in Fig. 4.16(a). The electronic structure remains insulating up to a critical thickness N_c at which ΔV produces an overlap between the VB and CB at opposite sides of the film. This transition is associated with an insulator-to-metal transition and a strong discontinuity of surface charges, which prevent a further increase of the film dipole moment and of the electrostatic potential jump ΔV. In this scenario, it manifests itself by the appearance of a 2DEG at the film surface. Beyond N_c (Fig. 4.16(b)), the excess surface charge density converges towards the value expected at semi-infinite polar surfaces. The film flexibility at small thickness may further help reducing the repeat unit dipole moment below N_c and increasing the critical thickness thanks to a strong structural distortion of the film. This scenario has been proposed in a study of the MgO(111) film phase diagram (Goniakowski *et al.*, 2007*b*) and invoked to explain the metal-insulator transition observed at the interface between SrTiO$_3$ and LaAlO$_3$ thin films (Thiel *et al.*, 2006). It is also relevant for understanding the electronic structure of thin kaolinite particles (see Chapter 7) (Hu and

Michaelides, 2010), of (0001) interfaces between two wurtzite materials (Goniakowski and Noguera, 2014), and of some ultra-thin Al_2O_3 film (Van den Bossche *et al.*, 2020).

4.3.3   Electronic structure at oxide/metal interfaces

When thin oxide films are grown on a substrate, their intrinsic electronic properties, as described in the previous sections, remain unchanged only in the case of weak film-substrate interaction. Otherwise, the stress exerted by the substrate distorts the films in a region close to the interface (see Section 4.1.2), leading to modifications of the bonding network and oxygen-cation hybridization. Moreover, interfacial electron transfers may take place, induced by direct bond formation across the interface or by the offset between the substrate and the film band structures. All these effects have more and more spectacular consequences when the oxide thickness decreases down to 1 ML, as described in Chapter 5. The present section focuses on the case of metallic substrates, while the following one will deal with oxide substrates.

The main impact of a metallic substrate on the electronic structure of an oxide thin film is determined by the band offset between the two materials. At metal-*sp* semiconductor interfaces, this has been conceptualized by the Metal Induced Gap States (MIGS) theory (Tersoff, 1984; Monch, 1990; Bordier and Noguera, 1991). The MIGS are the tails of the metal wave functions at energies in the gap of the semiconductor. They have an evanescent character with penetration lengths λ usually of the order of a few lattice spacings, but which depend on the proximity of their energy to the gap edges, as shown in Fig. 4.17. Close to the band edges, λ is large, and when the energy reaches the top of the VB or the bottom of the conduction band, the MIGS gradually transform into propagating waves. In contrast, at the gap centre, the damping length is the smallest and varies inversely with the gap width. The portion of space close to the interface on the semiconductor side thus acquires a metallic character due to the MIGS penetration.

In the semiconducting half space, the MIGS formation is correlated to a decrease of the number of valence and conduction band states and displays a smooth DOS in the band gap energy range. The semiconductor remains neutral if the MIGS are filled up to some energy called the zero-charge point E_{ZCP}, often assumed to be close to mid-gap. However, in most cases, the metal Fermi level E_F does not coincide with E_{ZCP}. In metals of large work functions, $E_F < E_{ZCP}$ implies an electron transfer from the semiconductor to the metal, while the transfer is in the opposite direction for metals with smaller work functions ($E_F > E_{ZCP}$). In all cases, the interfacial dipole thus created bends the bands on both sides of the interface, as represented in Fig. 4.17(a). In a first approximation, the interfacial dipole and the bending potential $-eV(z)$ are proportional to $E_F - E_{ZCP}$.

Such a description also applies to interfaces between a semiconducting/insulating oxide thin film and its metallic substrate. For example, larger charge transfers have been reported between CeO_2 films and Cu(111), Ag(111), or Au(111) compared to Pt(111), which can be explained by considering the larger work function of the latter (Luches *et al.*, 2015). However, for films a few monolayer thick, the decay length of the MIGS may exceed the film thickness, thus strongly modifying not only the interfacial LDOS but the whole electronic structure of the film. Additionally, in most cases, the

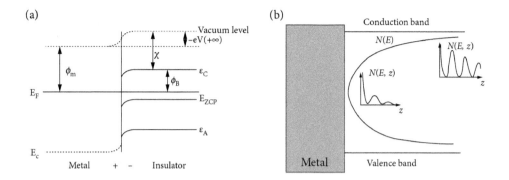

Fig. 4.17 (a) Sketch of the band bending at a metal-oxide interface. The metal is on the left and the oxide on the right. ϵ_A and ϵ_C refer to the top of the valence band and bottom of the conduction band of the oxide, respectively. Φ_m and Φ_B are the metal work function and Schottky barrier height, respectively, and $-eV(z)$ the potential which bends the bands. (b) Schematic representation of the MIGS density of state in the oxide band gap and the MIGS wave function at the gap centre or close to the band edges. Reprinted from Bordier and Noguera (1991) with permission. Copyright 1991 by the American Physical Society.

nearly free electron picture valid for small band gap *sp* semiconductors is not accurate enough for describing the more complex interfacial electronic structure. On the one hand, for metals with a high oxygen affinity, states due to chemical bonding between the metal and the oxygen atoms of the oxide film may be present in the LDOS. The electron redistribution due to the filling of those states located below E_F participate to the formation of the interfacial dipole. Such a complexity has been evidenced in simulations of metal/MgO interfaces as a function of the metal electronegativity χ (Goniakowski and Noguera, 2004). Metals with small χ transfer electrons towards MgO, whereas metals with higher χ receive electrons from MgO. This is consistent both with the conventional MIGS model, in which the charge transfer is driven by the difference between $E_F - E_{ZCP}$, and with the formation of bonds across the interface and the resulting electron transfer.

On the other hand, in Mott–Hubbard oxides with a small gap, the band shift at the interface may push a narrow filled d or f state from below to above E_F or vice versa, thus enhancing the interfacial charge transfer and leading to a change (increase or decrease) of oxidation state of the cations close to the interface. Such a scenario is consistent with the observation of reduced Ce^{3+} entities at the $CeO_2/Pt(111)$ interface (Luches *et al.*, 2015) and with the DFT prediction of the oxidation of the Ti and V cations in Ti_2O_3, V_2O_3, or mixed oxide MLs deposited on Au(111) (Goniakowski and Noguera, 2019) (see Chapter 5).

Finally, at the interface, the electronic density in the metal half space gets polarized by the electrostatic field exerted by the oxide ionic charges, an effect which does not significantly exist at metal/*sp* semiconductor interfaces. By emptying the space below the interfacial oxygen atoms and by populating those under interfacial cations, it creates image charges in the metal. Very early, it has been realized that the two major parameters which determine the insulating behaviour of oxides, namely the U on-site

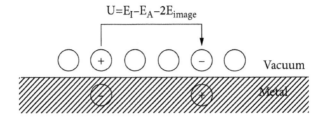

$$U=E_I-E_A-2E_{image}$$

Vacuum

Fig. 4.18 Combined electron removal and addition process for a thin film close to a metal surface, defining the on-site Coulomb interaction U for an open-shell system. During this process opposite image charges seem to appear below the surface, thereby reducing the ionization potential (E_I) and increasing the electron affinity (E_A) each by the amount E_{image}, and thus decreasing U by $2E_{image}$. Reprinted from Altieri *et al.* (1999) with permission. Copyright 1999 by the American Physical Society.

Coulomb repulsion and the anion-to-cation charge-transfer excitation Δ (see Chapter 3, Section 3.2.1) (Zaanen *et al.*, 1985) are modified in thin films deposited on metallic substrates due to the presence of the image charges in the metal (Duffy and Stoneham, 1983; Altieri *et al.*, 1999). This screening process is schematized in Fig. 4.18 for the U term which corresponds to an on-site charge fluctuation, and is qualitatively equal to the difference between the cation ionization potential E_I and electron affinity E_A. The interaction of the virtually excited electron with the image charge E_{image} reduces E_I and increases E_A, thus decreasing U by $2E_{image}$. The same conclusion is reached for the charge-transfer energy Δ which depends on the difference between the oxygen ionization energy and the cation electron affinity.

In thin films, E_{image} is larger when the charge fluctuations take place closer to the interface, due to its $1/d$ dependence. In average, the reduction of U and Δ is thus expected to be larger in thinner films. This trend has been verified by spectroscopic experiments on MgO/Ag(001) films (Altieri *et al.*, 1999). By recording the oxygen $1s$ and Mg $1s$ and $2p$ core level shifts and the changes in the Mg $KL_{23}L_{23}$ Auger peak position as a function of the oxide film thickness, the authors have determined the values of the extra-atomic relaxation energy and highlighted a large reduction of both parameters between 20 MLs and 1 ML films, as given in Table 4.1. The influence of the substrate is still noticeable relatively far from the interface in 10 ML films, which results from the long range character of Coulomb interactions. Similar results have been obtained experimentally on NiO(001) films grown on Ag(001) (Yang *et al.*, 2013)

Table 4.1 Thickness dependence of the intra-atomic electron-electron Coulomb repulsion U and charge transfer energy Δ corresponding to a transition O2p $\rightarrow$ Mg3s, as determined from the change of positions of the Mg and oxygen $1s$ core levels and the Mg $KL_{23}L_{23}$ Auger peak in MgO(100) films on Ag(100). All values are referred to the bulk ones, assumed equal to the 20 ML results (Altieri *et al.*, 1999).

	1 ML	2.7 MLs	10 MLs	20 MLs
$\delta U(\mathrm{Mg}2p)$	−1.8	−0.8	−0.4	0
$\delta\Delta(\mathrm{O}2p \rightarrow \mathrm{Mg}3s)$	−2.5	−1.2	−0.5	0

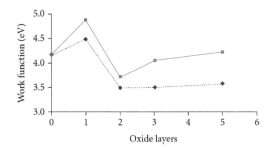

Fig. 4.19 DFT+U/PBE predictions of the evolution of the Ag(100) work function as a function of the thickness of a deposited NiO film, assuming either a rumpled (square symbols) or a flat (diamond symbols) oxide ($U = 5.3$ eV, except for the Ni atoms at the interface for which $U = 4$ eV). Reprinted from Sementa *et al.* (2012) with permission. Copyright 2012 by the American Chemical Society.

and by DFT simulations (Barcaro *et al.*, 2010). They all suggest large variations of the magnetic properties of oxide thin films interacting with a substrate, as will be indeed shown in Section 4.4.

The U parameter may also vary with the strain resulting from the lattice mismatch, but in a way which, according to RPA calculations in the series LaTiO$_3$, LaVO$_3$ and LaCrO$_3$ (Kim *et al.*, 2018), depends on the occupation of the d shell and their degree of localization (d^1, d^2 and d^3, respectively).

The deposition of oxides on a metal surface is a long-recognized way to modify the metal work function. As described in more details in Chapter 5, it results from the cooperative or competitive contributions of three dipoles formed at the interface, due to the compression of the metallic electrons which spill out of the surface, to the interfacial charge transfer and the film rumpling. Among those, it is mainly the third one which depends on the oxide thickness, leading to a variation of the total work function as a function of thickness, as exemplified in Fig. 4.19 in the case of NiO/Ag(001).

Before closing this section, the stabilizing role played by a metallic substrate and the concomitant modification of the electronic structure of a polar film are addressed (Goniakowski *et al.*, 2010). Indeed, a very specific electron transfer occurs at the interface, which has nothing to do with the previous considerations. It is similar to the one which exists between a semi-infinite polar oxide surface and an adsorbed metal layer (Goniakowski and Noguera, 1999; Goniakowski and Noguera, 2002). The substrate transfers electrons to the film when it is in contact with the negatively charged termination or receives electrons in the opposite case. This transfer is of the order of the compensating charge $\Delta\sigma$, so that the interfacial ions recover a valence close to their bulk value. In the substrate, due to metallic screening, the charge is mostly localized in the vicinity of the interface. The electronic structure on the other side of the film, at the interface with vacuum, on the other hand, is little affected by the presence of the support. A metal substrate provides a large energy stabilization because the electron excitation occurs between the metal Fermi level E_F and one of the oxide film gap edges, not across the oxide gap. The associated gain of energy is reflected in a strong interfacial adhesion. This process was invoked to support the observation of a structurally and chemically abrupt Cu/MgO(111) interface (Muller *et al.*, 1998; Imhoff *et al.*, 1999).

4.3.4 Electronic structure at oxide/oxide interfaces

While for many years thin oxide films have been grown on elemental metals or alloys—with the exception of SiO_2 layers in electronic devices—, a burst of interest for oxide/oxide interfaces arose after the discovery of a 2DEG at the interface between $LaAlO_3$(001) and $SrTiO_3$ (001). The unexpected metallicity at the interface between two insulating materials paved the way towards a search of new states of matter between materials which do not display them in the bulk. Although the occurrence of metallization at a polar surface was already well-known in the oxide surface community, and although the physics of charge transfer at oxide/oxide interfaces is not very different from what has been previously described, the variety of new phenomena occurring at the contact between two oxides, especially transition metal oxides (TMOs), is much larger. This is due to the diversity of electronic and magnetic ground states triggered by the localized character of the correlated d electrons, and/or by doping. As a consequence, in the last fifteen years, a wealth of results has been gathered on the so-called 'emergent phenomena' at oxide/oxide interfaces, including electron reconstruction, interfacial doping/charge transfer, orbital polarization effects, magnetism, and ferroelectricity, which are associated to the coupling between charge, spin, and orbital degrees of freedom, as schematized in Fig. 4.20. Most studies have considered heterostructure configurations, and very few have focused on the thickness dependence of the physical effects. However, since they are usually strongly localized within a few unit cells of the interface for structural or electronic reasons (small Thomas–Fermi lengths), the insight gained in these works is fully relevant to the physics of ultrathin oxide films deposited or grown on an oxide substrate. In the following, we will mainly focus on electronic properties and present a few selected examples, sending the reader to the recent reviews by Mannhart and Schlom (2010), Hwang *et al.* (2012) or Chakhalian *et al.* (2014) for a more extensive overview.

The physical process at the root of many interface properties is the interfacial charge transfer, which depends on the cation electronegativity and valence and on the polar character of the interface (Chen and Millis, 2017). With regard to the latter, interfaces may be classified in three families. There are the non-polar/non-polar interfaces such as $SrTiO_3$/MgO(001), $SrMO_3$/SrM'O_3(001), or M_2O_3/M'$_2O_3$(0001), the polar/non-polar interfaces such as $LaAlO_3$/SrTiO$_3$(001) or $LaMO_3$/SrMO$_3$(001) (M = Ti or V), and the polar/polar interfaces. Among the latter, one can distinguish (Goniakowski and Noguera, 2014) those displaying a polarization discontinuity, e.g. ZnO(0001)/MgO(111) or ZnO/(Zn,Mg)O(0001), from those with no polarization discontinuity, e.g. $LaMO_3$/LaM'O_3(001). Although, in the latter, the layer piling is polar, the absence of polarization discontinuity gives the interface a non-polar character with no need of compensating charge.

At non-polar interfaces—truly non-polar layer staking in both oxides or polar/polar interfaces with no polarization discontinuity—, a charge transfer generally exists, which

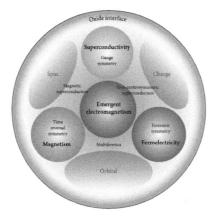

Fig. 4.20 Schematic diagram showing the symmetries and degrees of freedom of correlated electrons that can be engineered at oxide interfaces. Inversion symmetry is broken by the structure itself, and the two-dimensionality usually enhances electron correlation effects. Reprinted from Hwang *et al.* (2012) with permission by Nature. Copyright 2012.

may be partial or complete depending on the band-offset between the two oxide band structures, the nature of the states at the band edges, and the bond formation at the interface. The dipolar potential resulting from this charge transfer bends the bands in both oxides. To foresee its direction and magnitude, arguments based on band alignment in the bulk are approximate but useful guides (Zhong and Hansmann, 2017). At the $SrTiO_3/MgO(001)$ interface, for example, a small electron transfer occurs from MgO to $SrTiO_3$, which is mainly determined by the formation of interfacial bonds, and does not modify the insulating character of the overall system (Casek *et al.*, 2004). Electrons are also transferred at the $M_2O_3/M'_2O_3(0001)$ interfaces (M, M' = Ti, V, Cr, Fe) (Le *et al.*, 2019), from the lighter cations to the heavier one, with a magnitude which depends on the nature of the states at the band edges. In particular, due to the Mott–Hubbard character of Ti_2O_3 and V_2O_3 oxides with narrow d states at the band edges, the band bending is able to completely deplete or fill a d state at Ti_2O_3/V_2O_3 and Ti_2O_3/Fe_2O_3 interfaces, resulting in a change of oxidation states of the interfacial cations, similar to a redox reaction. This substantial cation-to-cation charge transfer is traced back to the overlap of the Ti_2O_3 valence band maximum of purely d character with the conduction band minima of V_2O_3 and Fe_2O_3. It is consistent with the evidence of a Ti 4+ oxidation state in mixed $TiVO_3$ and $TiFeO_3$ oxides (Le *et al.*, 2018).

Interfaces between perovskites have been the subject of many detailed studies. Along the (001) orientation, $AMO_3/AM'O_3$ heterostructures have been synthesized, which display a continuity of $MO_6/M'O_6$ octahedra along the c axis. When the A cation is divalent, the layer stacking is non-polar, while when it is trivalent, the layer stacking is polar but there is no polarization discontinuity at the interfaces. The electron transfer at the interface may be zero as in $SrTiO_3/SrVO_3$, partial as in $SrVO_3/SrMnO_3$, or complete as in $LaTiO_3/LaFeO_3$ (Kleibeuker *et al.*, 2014) (see Fig. 4.21). This is consistent with the valence and conduction band offsets of the two oxides and the electronic structure known in the corresponding double perovskites $A_2MM'O_6$

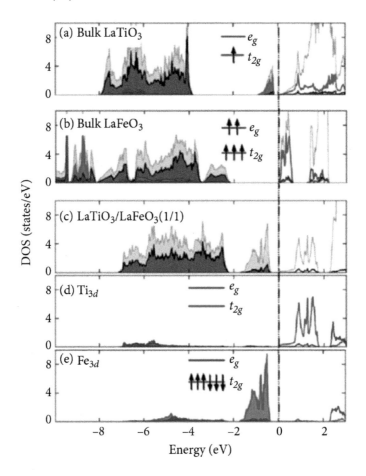

Fig. 4.21 Atomic and orbital projected DOS of (a) bulk $LaTiO_3$, (b) bulk $LaFeO_3$, and (c)–(e) a (1:1) $LaTiO_3$:$LaFeO_3$ superlattice. Total DOSs are marked in grey, O p states in black, Fe and Ti t_{2g} states in red, and Fe and Ti e_g states in blue. The Fermi level E_F is indicated by the dotted line. The absence of Ti d states below E_F and of Fe t_{2g} states above E_F point to the transfer of formally one electron from Ti to Fe in the heterostructure, and thus to their 4+ and 2+ oxidation states, respectively. Reprinted from Kleibeuker *et al.* (2014) with permission. Copyright 2014 by the American Physical Society.

which are ordered 50%:50% mixed oxides (Chen and Millis, 2017). The redox process which drives Ti and Fe cations to +4 and +2 oxidation states, respectively, has the same origin as that which takes place at the (0001) interface between the corundum oxides Ti_2O_3 and Fe_2O_3.

The historical example of a polar/non-polar interface is $LaAlO_3$/$SrTiO_3$(001). $LaAlO_3$ repeat units bear a dipole moment which requires a compensating charge density $\pm 1/2$ electron per formula unit, while $SrTiO_3$(001) is a non-polar surface. As a result, a 2DEG is formed at stoichiometric interfaces by overlap of the (shifted) $LaAlO_3$ VB with the $SrTiO_3$ CB. However, this overlap can only take place above a

LaAlO$_3$ thickness of 4 MLs. For thinner films, as mentioned and explained previously (see Section 4.3.2), an uncompensated (sub-critical) insulating regime exists (Hwang *et al.*, 2012). Similar results are obtained at the LaVO$_3$/SrTiO$_3$ interface with a critical thickness of 5 MLs (Hotta *et al.*, 2007). This description applies to ideal interfaces and does not take into account the possible existence of cation intermixing, vacancies or point defects present at real interfaces.

Finally, the interface between ZnO(0001) and mixed Zn$_{1-x}$Mg$_x$O(0001) provides an example of a polar-polar interface with a small polarization discontinuity. As long as the Mg content x of the mixed oxide remains small (typically less than 0.2–0.3), Zn$_{1-x}$Mg$_x$O retains the insulating wurzite structure of ZnO, but with a gap which increases with x and with an internal parameter u of the wurtzite structure slightly different from that of ZnO. Both the lack of inversion symmetry of the structure, responsible for an electronic contribution to the polarization and the u discontinuity induce a polarization discontinuity. However, it is weak, as well as the required compensating charge $\Delta\sigma$. In thick heterostructures, 2DEG is formed at the interface which has been revealed by the observation of Shubnikov–de Haas oscillations and Quantum Hall Effect (Tsukazaki *et al.*, 2007). In thin heterostructures, one could expect an uncompensated regime, due to the small value of the polarization discontinuity, as in the AlN/GaN superlattice illustrated in Goniakowski and Noguera (2014).

The generation of a charge transfer at the interface has various consequences, among which is the so-called orbital reconstruction associated to the difference between the crystal field splitting of the d orbitals at the interface and in the bulk. For example, at the LaAlO$_3$/SrTiO$_3$(001) interface, it was shown by X-ray absorption spectroscopy that the degeneracy of the Ti t_{2g} states is removed and the $3d_{xy}$ levels become the first available states for conducting electrons (Salluzzo *et al.*, 2009). The interface thus represents a system with a purely 2-D partially filled single band in which correlation effects are expected to be strongly enhanced. At the same interface, below 0.2 K, the 2DEG exhibits superconductivity (Reyren *et al.*, 2007). Spin reconstruction may also be induced by the charge transfer. For example, a ferromagnetic spin polarization was evidenced at the (001) non-polar/non-polar interface between the antiferromagnetic insulator CaMnO$_3$ and the paramagnetic metal CaRuO$_3$, where the ferromagnetic double-exchange between Mn spins is enhanced by the interfacial charge transfer, at the expense of antiferromagnetic super-exchange interactions (Takahashi *et al.*, 2001).

4.4 Magnetic properties

Transition metal oxides display a variety of magnetic characteristics, which are even enhanced in thin films due finite size effect and interaction with the substrate.

Among the antiferromagnetic (AF) oxides of the first transition series—MnO, FeO, CoO, and NiO—NiO is the compound whose magnetic properties have been the most studied. The first systematic variation of the Néel temperature T_N was mentioned by Alders *et al.* (1998) in a linear polarized X-ray absorption study of thin NiO(100) films grown on MgO(100) (Fig. 4.22). The authors determined T_N values of 295 K, 430 K and 470 K for NiO thicknesses of 5, 10 and 20 MLs, respectively, much lower than the NiO bulk Néel temperature ($T_N = 530$ K). This behaviour was reproduced by Monte Carlo simulations of an Ising model with parameters relevant to NiO which showed

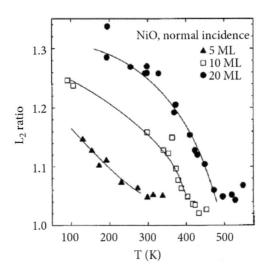

Fig. 4.22 Ratio of the two peaks in the Ni L_2 X-ray absorption spectrum of NiO (100) films of nominal thickness of 5, 10, and 20 MLs taken at normal incidence. This ratio is proportional to the short-range spin-spin correlation function. T_N values of 295 K, 430 K and 470 K were inferred from these curves, thus highlighting the strong thickness dependence of the Néel temperature. Reprinted from Alders *et al.* (1998) with permission. Copyright 1998 by the American Physical Society.

that the decrease in T_N occurs when the spin-spin correlation length becomes of the order of the film thickness (Noguera and Mackrodt, 2006). A decrease of the magnetic transition temperature at low thickness is quite a general trend and has been observed in other oxide thin films, whether antiferromagnetic as MnO/Ag(100) (Kundu *et al.*, 2018) or ferromagnetic as EuO/SiO_2 (Müller *et al.*, 2009), or $SrRuO_3/SrTiO_3$ (Si *et al.*, 2015; Ishigami *et al.*, 2015). In the latter case, the magnetic moments decrease at low thickness, a metal-insulator transition occurs at 4–5 MLs, and ferromagnetism disappears at 2 ML, replaced by an antiferromagnetic ground state (Toyota *et al.*, 2005; Chang *et al.*, 2009; Xia *et al.*, 2009).

By comparing the magnetic properties of NiO films on an insulating (MgO) or metallic substrate (Ag), Altieri et al. have shown that the Néel temperature is enhanced on the metallic substrate compared to the insulating one (390 K versus less than 40 K, at 3 ML thickness) (Altieri *et al.*, 2009). This enhancement was ascribed to the difference in polarizability of the two substrates. At an interface, the super-exchange interaction J_{se} is modified, via its dependence on the charge transfer energy Δ and the on-site Hubbard U parameter of the oxide (t the anion-cation transfer integral):

$$J_{se} = -\frac{2t^4}{\Delta^2}\left(\frac{1}{\Delta} + \frac{1}{U}\right) \tag{4.11}$$

These two electronic excitations have reduced energies in the vicinity of a highly polarizable medium (Duffy and Stoneham, 1983) (see Section 3.3) and experiments on NiO/Ag(100) (Altieri *et al.*, 1999) confirmed that this image charge screening effect could yield a 50% reduction of U and Δ, and thus a substantial enhancement of J_{se} and T_N. A similar trend was evidenced in MnO(100) thin films deposited on Ag(100), in which, as thickness increases, the Néel temperature saturates at a value which exceeds the bulk T_N by 23 K (Kundu *et al.*, 2018). Comparison of the Curie temperatures of ferromagnetic EuO when deposited on Mg or on MgO also evidences a larger Curie temperature in the former (Altendorf *et al.*, 2018).

The magnetic anisotropy in oxides is mainly determined by the magneto-crystalline anisotropy and the dipolar interactions. The former comes from a coupling of the spins to the crystallographic axis via spin orbit interactions. It vanishes for MnO and NiO and for all ions having a singlet ground state, or a filled t_{2g} ground state and partially occupied e_g states, since the spin-orbit interaction does not couple different e_g states. As a consequence, magneto-crystalline anisotropy is the main driving force for orienting the spins only in CoO and FeO in the first TMO series. In MnO and NiO, it is the dipole-dipole interactions which are dominant and which orientate the spins along the (111) and equivalent directions. This degeneracy is lifted in thin films due to the breakdown of translational symmetry in the direction perpendicular to the films, which favours in-plane domains (shape anisotropy phenomenon).

Interfacial strain as well as thickness effects influence the spin orientation in oxide thin films. In the case of compressive strain, they both act towards a stabilization of in-plane spin orientations, while for tensile strain, they act in opposite ways, leading to either in-plane or out-of-plane orientation (Finazzi and Altieri, 2003). As an illustration of these concepts, Altieri et al. showed that the Ni L_2 edge X-ray magnetic linear dichroïsm is reversed when NiO is grown on MgO or Ag, upon which the oxide film is tetragonally compressed or stretched, respectively (Altieri *et al.*, 2003). By contrast, by recording the polarization dependent Co L_{23} XAS spectra, Csiszar et al. found that it was opposite for CoO/MnO(100) or CoO/Ag(100). In that case, the magnetocrystalline anisotropy is dominant within the CoO films. The easy axis is perpendicular to the film on MnO (tensile strain) and parallel on Ag (compressive strain) (Csiszar *et al.*, 2005). Similar types of arguments were put forward by Ingle and Elfimov (2008), who predicted that it would be possible to increase the Curie temperature of EuO, provided it is submitted to a sufficiently strong biaxial strain. Unfortunately, experiments performed on a Ni(100) substrate evidenced a rapid relaxation of strain as the film thickness increases, leaving little hope of a substantial increase in the Curie temperature (Förster *et al.*, 2011).

Proximity effects between two magnetic materials modify both the magnetic transition temperatures and the spin orientation (Manna and Yusuf, 2014). Li et al. have determined these quantities at interfaces between antiferromagnetic NiO and CoO (Li *et al.*, 2016), which, in the bulk have Néel temperatures equal to 520 K and 291 K, respectively. NiO(100) films of 2 and 12 nm thickness have been sandwiched between an MgO(100) substrate and a CoO(100) capping layer. The strain induced by the MgO substrate on the two oxides is compressive for CoO/MgO and tensile for NiO/MgO, which is responsible for an easy axis parallel to the CoO/MgO interface in CoO, while in NiO it is perpendicular to the NiO/MgO interface. When NiO is sandwiched between MgO and CoO, it adopts an easy axis parallel to the interface via its interaction with CoO at low thickness, while at larger thickness the spin directions rotate in the

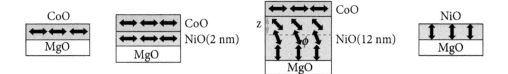

Fig. 4.23 Schematics of AFM spin configurations in CoO (left panel) and NiO (right panel) films deposited on MgO, and in NiO(100) films of 2 and 12 nm thickness sandwiched between an MgO(100) substrate and a CoO(100) capping layer (central panels). Arrows represent the AFM spin alignments. ϕ is the spin canting angle with respect to the film surface and z is the depth from the CoO/NiO interface. Adapted from Li *et al.* (2016).

film from perpendicular at the MgO interface to a finite canting angle ϕ at the CoO interface (Fig. 4.23). The CoO Néel temperature is concomitantly increased by proximity with NiO. This is in agreement with previous results on CoO/NiO superlattices in which the transition temperature of the Co spins of CoO was found to be increased by $\approx$ 80 K (Borchers *et al.*, 1993). In the reversed system NiO/CoO/MgO(100) (Zhu *et al.*, 2014), the NiO spins undergo a spin reorientation transition from in-plane to out-of-plane at a critical thickness which increases with the CoO thickness, while the CoO Néel temperature is greatly enhanced by the proximity with NiO

The interest in magnetic order of low-dimensional oxide systems has been fuelled by the advent of spintronics. Spintronics attemps to overcome the limitations of conventional electronics, based on charge transport, by adding new functionalities by exploring the spin degrees of freedom for information transport (Wolf *et al.*, 2001). The birth date of spintronics can be tied down to 1988, when A. Fert (Baibich *et al.*, 1988) and P. Grünberg (Binasch *et al.*, 1989) discovered the giant magnetoresistance (GMR) effect in metallic multilayers. The most basic two-terminal spintronic device in a 2-D configuration, the magnetic tunnel junction (MTJ), involves an oxide layer as insulating barrier between two ferromagnetic electrodes. Electron tunnelling is a quantum mechanical process, whereby electrons flow from one metallic electrode, across a thin insulating barrier layer, into another metallic electrode. According to the Wentzel–Kramers–Brillouin (WKB) approximation, the tunnelling current is determined by the product of the DOS of the electrodes, the thermal occupation probability of the states involved, and by the transmission coefficient; the latter is the square of the tunnelling matrix element, which involves an exponential dependence on the barrier thickness and the square root of the barrier height. Here lies the basis for the interest in 2-D oxides as barrier layers for MTJs, because they generate uniform barrier thickness facilitating an all-area tunnelling across the junction.

When ferromagnetic metals are used as electrodes, novel physical effects emerge due to their non-equivalent LDOS for spin-up and spin-down states. Assuming that spin is conserved during the tunnel process, the total tunnelling current is the sum of the currents for spin-up ($\uparrow$) and spin-down ($\downarrow$) tunnelling. Accordingly, the conductance G of a MTJ (i.e. the derivative of the current with respect to voltage) in the parallel (P) and anti-parallel (AP) configurations of the electrode magnetizations is proportional to the product of the DOS of the electrodes (Bibes *et al.*, 2011):

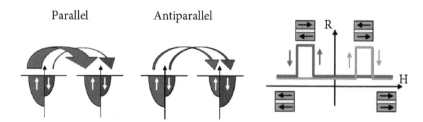

Parallel   **Antiparallel**

Fig. 4.24 Schematic representation of the tunnel current in a MTJ with two identical posi-tively spin-polarized electrodes. In the P state (left), a strong current is carried by the spin-up channel, resulting in a large total current and low resistance. In the AP state, the current is weak in both spin channels, resulting in a weak total current and high resistance. The junction shows two resistance states as the magnetic field is swept, as shown on the right hand side. Reprinted from Bibes *et al.* (2011) by permission of the publisher Taylor and Francis Ltd (www.tandfonline.com).

$$G_P \propto G_{\uparrow\uparrow} + G_{\downarrow\downarrow} \propto N_{1\uparrow}N_{2\uparrow} + N_{1\downarrow}N_{2\downarrow}$$
$$G_{AP} \propto G_{\uparrow\downarrow} + G_{\downarrow\uparrow} \propto N_{1\uparrow}N_{2\downarrow} + N_{1\downarrow}N_{2\uparrow} \tag{4.12}$$

$N_{1,2}$ are the DOS of the electrodes at E_F for majority ($\uparrow$) and minority ($\downarrow$) electrons. The tunnel resistance is different in the P and AP configurations, which defines the tunnel magnetoresistance (TMR) effect (Julliere, 1975):

$$TMR = \frac{R_{AP} - R_P}{R_P} = \frac{G_P - G_{AP}}{G_{AP}} = \frac{2P^1_{spin}P^2_{spin}}{1 - P^1_{spin}P^2_{spin}} \tag{4.13}$$

with P_{spin} the spin polarization of the electrodes: $P^i_{spin} = (N_{i\uparrow} - N_{i\downarrow})/(N_{i\uparrow} + N_{i\downarrow})$

Figure 4.24 gives a schematic illustration of the tunnel process between two iden-tical spin-polarized electrodes. Consideration of the previous formulas leads to the conclusion that electrode materials with high P_{spin} will yield high TMR values. Mate-rials with very high spin polarization (up to 100 percent) are the half-metals: they are metallic for one spin direction (albeit with a low DOS at E_F), but insulating for the other spin direction. Figure 4.25 compares the DOS of a normal metal, a ferromag-netic metal, and a half-metal (from left to right). Many half-metals are oxides, such as CrO_2, Fe_3O_4, mixed-valence manganites, or double perovskites. However, few of them actually display the predicted good MTJ performance (Bibes *et al.*, 2011).

In the WKB approximation, the material of the barrier layer is neglected, just the width and height of the barrier play a role. However, for a real crystalline barrier, the structure and physical properties of the barrier material and of the electrodes and their interfaces influence the tunnelling matrix element and thus the transmission coefficient. The symmetry matching of the complex band structure of electrode and barrier materials is important and has to be considered, leading to a symmetry filter-ing due to wave function matching and the different decays of evanescent waves. For specific combinations, e.g. epitaxial Fe/MgO/Fe MTJ, very large TMR ratios have been reported (Bibes *et al.*, 2011). The use of ferro- or ferrimagnetic insulating ma-terials as tunnel barriers can filter electrons coming from an unpolarized electrode,

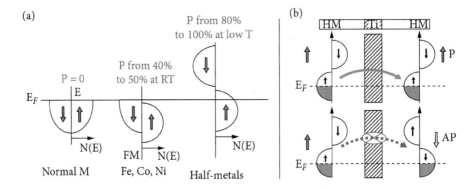

Fig. 4.25 (a) Schematic illustration of the density of states of a normal metal, a ferromagnetic metal, and a half-metal (from left to right). The spin polarization P ranges from 0% to 100% as indicated. Examples of half metals are $La_{0.7}Sr_{0.3}MnO_3$ ($T_c = 365$ K), CrO_2 ($T_c = 398$ K), Sr_2FeMoO_6 ($T_c = 420$ K), or Fe_3O_4 ($T_c = 860$ K) (b) An MTJ with two half-metallic electrodes (the direction of the external field is either up or down). Reprinted from Haghiri-Gosnet *et al.* (2004) with permission. Copyright © 2004 Wiley-VCH Verlag GmbH and Co. KGaA, Weinheim..

according to their spin. This can be employed in spin-filter junctions. The bottom of the conduction band in the barrier material lies at different energies for spin-up and spin-down electrons because of the exchange splitting: electrons will transfer differently depending on their spin, and the current emerging from the barrier is spin polarized. In order to measure the spin-filtering efficiency, a reference layer acting as a spin detector has to be added, in the form of a ferromagnetic (half-metallic) counter electrode. Depending on the orientation of the magnetization of the barrier and the

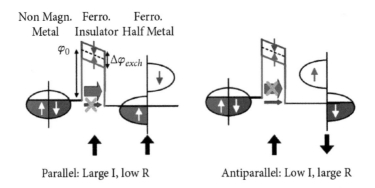

Fig. 4.26 Tunnelling process through a spin filtering tunnel barrier: the vertical black arrows indicate the direction of the magnetization in the magnetic barrier and the ferromagnetic (half-metallic) counter electrode. The red and blue vertical arrows indicate the spin direction and the horizontal one the tunnel current. Reprinted from Bibes *et al.* (2011) with permission of the publisher Taylor and Francis Ltd (www.tandfonline.com).

counter electrode, the detected current through the junction will be small or large, as displayed in Fig. 4.26. Ferromagnetic insulating oxides are scarce, and they typically feature low Curie temperatures T_C, since there are no carriers to mediate the exchange interaction. The dominant exchange interaction is then superexchange via the oxygen ions, which according to the Goodenough–Kanamori–Anderson rules (Goodenough, 1963) cannot be ferromagnetic *and* strong. In compounds with more than one type of magnetic sites, stronger anti-ferromagnetic superexchange can generate ferrimagnetic ordering, which can lead to higher T_Cs. EuO, $BiMnO_3$, and spinel-type oxides ($NiFe_2O_4$) have been employed as ferromagnetic semiconducting (insulating) barrier materials in spin-filtering tunnel junctions (Bibes *et al.*, 2011). Current applications of TMR include magnetic read heads, magnetic sensors, and non-volatile magnetic memory devices (see the two resistance states of the MTJ in Fig. 4.24).

4.5 Non-stoichiometric oxide films

The most common non-stoichiometric defects in oxide thin films, as well as at oxide surfaces, are oxygen vacancies. While much knowledge has been acquired on how they locally modify the oxide atomic and electronic structures, the presence of the substrate may substantially modify their localization, charge, and energy of formation. Doping is another means of creating non-stoichiometric defects, purposely used in the semiconductor community to adjust conductivity. The presence of dopants in oxides and more specifically in supported oxide films leads to localized states in the oxide gap and strong electron redistributions. The analysis of charge transfers between oxygen vacancies or dopants and substrates will thus be at the heart of this section.

4.5.1 Oxygen non-stoichiometry

Oxygen non-stoichiometry covers both oxygen deficiency and oxygen excess. Under specific reaction conditions, at high oxygen pressures, the latter may occur in the film top layers, leading to 'abnormal' stoichiometries, particularly striking at the extreme 2-D limit. Some examples will be given in Chapter 5. Here, we will focus on the characteristics of oxygen vacancies (V_O), commonly encountered as a result of thin film preparation, especially under ultra-high vacuum conditions. Before this fact was fully recognized, it was claimed that oxide surfaces featured many surface states in their gap. However, the states that had been observed were not intrinsic surface states, but rather oxygen vacancy states.

When a neutral oxygen atom is missing, two electrons are left behind, whose localization depends on the nature of the oxide. In non-reducible oxides, the electrons are trapped at the vacancy site, forming a so-called F^0 centre and resulting in a strongly localized state in the oxide gap (for a review on F centre, see e.g. Pacchioni and Freund (2012)). This scenario applies to rock-salt oxides such as MgO, CaO, etc., but also to transition metal oxides of stoichiometry MO (MnO, FeO, VO, etc.) in which the cations do not readily adopt a +1 oxidation state. In contrast, in reducible transition metal oxides with multivalent cations, such as TiO_2, Fe_2O_3, CeO_2, or various perovskites, the excess electrons are redistributed on the neighbouring cations, with a concomitant decrease of their oxidation state.

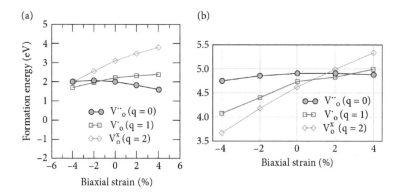

Fig. 4.27 Variations of the formation energy (in eV) of an oxygen vacancy in CaMnO$_3$ (a) and MnO (b) in the neutral ($q = 0$) and charged states ($q = +1$ or $+2$) as a function of a biaxial strain (in %). Reprinted from Aschauer and Spaldin (2016) with the permission of AIP Publishing.

The reduction of the neighbouring cations induces an increase of their ionic radii and a local expansion of the lattice. The strain exerted by the substrate, due to the lattice mismatch, opposes or reinforces this effect with implications on the vacancy formation energy E_{vac} (Aschauer and Spaldin, 2016). For example, DFT simulations of CaMnO$_3$ under biaxial strain have shown that tensile strains lower E_{vac} (Fig. 4.27(a)). The effect of strain to promote the formation of oxygen vacancies has also been invoked at the CeO$_2$/Rh(111) interface (Castellarin-Cudia *et al.*, 2004). In contrast, there are only local distortions of the lattice around neutral F centres, and the variations of E_{vac} upon strain are weaker (Fig. 4.27(b)).

The oxygen vacancy formation energy is not only dependent on the strain exerted by the substrate, but also on the localization of the vacancy in the thin film. As a general statement, oxygen removal is easier in the oxide layers in direct contact with a metal substrate (for a review, see Ruiz Puigdollers *et al.* (2017)). Various arguments have been invoked, such as the distortion which takes place around the vacancy upon charge transfer, the accommodation of the excess electrons by the MIGS in the interfacial oxide layer (Tamura *et al.*, 2009), the hybridization between the vacancy state and the substrate, or the direct electron transfer to the metal Fermi level when the latter is just below the vacancy state (Giordano *et al.*, 2003; Yang *et al.*, 2007; Prada *et al.*, 2011). The favourable vacancy formation at the interface results in an increase of the film adhesion and in an increase of the oxide reducibility compared to the bulk.

The transfer of electrons to a metal support may also change the charge of the vacancy. A detailed theoretical study of the electronic structure of an oxygen-deficient ultra-thin MgO film on Pd(001) has highlighted the disappearance of the doubly occupied vacancy state, typical of neutral oxygen vacancies at the MgO(001) surface, and the transfer of approximately one electron from MgO to the substrate resulting in a F$^+$ charge state of the vacancy. This transfer starts due to the higher position of the vacancy state with respect to the metal Fermi level and continues until the two levels become aligned (Giordano *et al.*, 2003). A similar process has been identified in MgO films on an Ag(001) substrate (Ling *et al.*, 2013). The electron transfer depends on the relative positions of the vacancy state and the Fermi level of the system and is thus highly dependent on the metal substrate. Around charged vacancies, the lattice distortions are much stronger, with a relaxation of the neighbouring oxygen atoms towards the vacancy site and of the cations away from it. The presence of oxygen vacancies decreases the metal work function, but the strength of this effect depends on the charge state and the position of the vacancy.

Before concluding, let us recall that many reconstructions exist at oxide non-polar surfaces, driven by oxygen vacancy ordering under specific preparation conditions. For example, one of the richest diagrams is displayed by the under-stoichiometric α-alumina (0001) surface. After annealing at increasing temperatures, a succession of reconstruction patterns has been observed: (1×1), $(\sqrt{3} \times \sqrt{3})R30°$, $(2\sqrt{3} \times 2\sqrt{3})R30°$, $(3\sqrt{3} \times 3\sqrt{3})R30°$ and $(\sqrt{31} \times \sqrt{31})R \pm 9°$ (Gautier *et al.*, 1991). Each of them corresponds to a different oxygen content in the outermost layers, the density of vacancy increasing with the annealing temperature. Non-stoichiometric reconstructions have also been observed on $TiO_2(100)$ (Bowker, 2006) and many other non-polar surfaces. They are also frequent at polar surfaces. In that case, the vacancies provide the compensating charges, and in many cases, they form ordered structures. For example, on $SrTiO_3(110)$, numerous reconstructions have been observed (Enterkin *et al.*, 2010), and it is recognized that in most cases, the mechanism of stabilization of polar orientations by non-stoichiometric reconstructions is the most efficient. Similar trends are expected in ultra-thin films.

4.5.2 Doped oxide thin films

Most natural oxides have low levels of impurities that occupy interstitial or substitutional sites. By the defect interaction with light, initially transparent oxides may acquire a colour. Some, like ruby and sapphire (chromium-doped α-Al_2O_3) become valuable gemstones, renowned in jewellery. More generally, dopants modify the electronic structure of the host oxide and locally its atomic structure, a pivotal effect in the newly developed all-oxide electronics. Additionally, in case of segregation at the surface, they may impact its catalytic activity. While these various effects in bulk oxides have long been reviewed, much less is known about doping in ultra-thin oxide films (Nilius, 2015).

Intentional doping of oxide ultra-thin film may be performed following two pathways, as schematized in Fig. 4.28. One method consists in co-depositing metal and dopant atoms on the substrate from the gas phase, with the flexibility to choose the doping atoms and their concentration. Another technique, based on high-temperature annealing, makes use of the diffusion of atoms from the support into the film where they substitute oxide cations.

For a given oxide, one identifies two classes of dopants, those forming cations of higher oxidation state than the genuine oxide cations (for example, Cr, Mo, or Eu in MgO or CaO) and those forming cations of lower valencies (e.g. Li, Na, or K in the same oxides). Due to the charge unbalance, compensating charges are expected, which usually are cationic vacancies in the first case and donor-type defects in the latter,

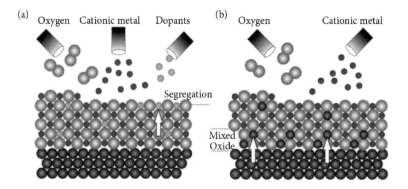

Fig. 4.28 (a) Preparation of doped oxide films via co-deposition of metal and impurity atoms from the gas phase. In some cases, a dopant segregation may occur and lead to the formation of a surface ternary oxide. (b) Preparation via self-doping from the support at elevated temperature. In addition to doping in the whole film, a new ternary phase may form at the interface. Reprinted from Nilius (2015) with permission. Copyright © 2015 IOP Publishing Ltd. All rights reserved.

most often oxygen vacancies. These compensating entities annihilate the electronic impact of doping, contrary to the case of classical semiconductors, for which doping has a direct consequence on conductivity. However, in ultra-thin films deposited on metal substrate, a third compensation mechanism exists via electron transfer to or from the substrate.

An illustrative example of the first family is provided by the doping of MgO thin films by chromium atoms, investigated by means of scanning tunnelling microscopy, X-ray photoelectron spectroscopy, and diffraction. The Cr impurities adopt a 3+ oxidation state, compensated by Mg vacancies. At low doping level, they dissolve in MgO, while for the highest doping levels, they segregate at the surface forming a ternary oxide of stoichiometry close to $MgCr_2O_4$ (Benedetti *et al.*, 2015; Stavale *et al.*, 2012).

In this same family, self-doping by interdiffusion of atoms from the substrate has been obtained in CaO films grown on Mo(001) by high-temperature annealing (Shao *et al.*, 2011*a*). Due to the high valency of the Mo ions in the CaO lattice, the most abundant defect types detected by STM at the film surface are cation vacancies, with a concentration which increases with the doping level (Cui *et al.*, 2014). Close to the interface, the Mo ions possess another way to stabilize their preferred 3+ oxidation state, by donating electrons to the substrate. This is revealed by shifts of the CaO core levels and of the CB onset detected by STM conductance spectroscopy in sufficiently thin films, which are both induced by the dipole associated to the interfacial electron transfer (see Fig. 4.29). Moreover, due to the substantial lattice mismatch with the support, a new interfacial phase of rock-salt structure and stoichiometry Ca_3MoO_4 develops in which Mo atoms replace 25% of the Ca ions. This phase has a negligible lattice mismatch with the Mo(001), enabling the growth of extended, defect-free oxide islands. Such an intermixing does not occur in MgO films grown on Mo(001), likely due to the smaller lattice mismatch between the two systems (Shao *et al.*, 2011*b*).

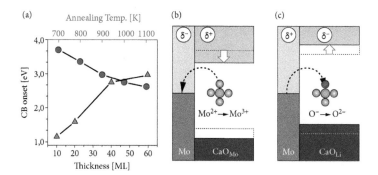

Fig. 4.29 (a) CB onset in Mo doped CaO, measured by STM conductance spectroscopy as a function of temperature (red) and oxide thickness (black). (b) and (c) Schematic relation between the band shifts and charge-transfer processes at the interface between the Mo support and CaO films doped with high-valence Mo (b), and low-valence Li dopants (c). A lowering of the CB onset occurs in the first case, which increases with the doping level (higher annealing temperature in panel a) but tends to vanish at the surface of thicker films. In contrast, the CB onset occurs at higher energies in Li-doped CaO films. Reprinted from Nilius (2015) with permission. Copyright © 2015 IOP Publishing Ltd. All rights reserved.

In view of the known application of Li-doped MgO as a catalyst for the oxidative coupling of methane, thin MgO films grown on Mo(001) have been doped by lithium atoms. This system is representative of the class of low-valence dopants. Charge compensation involves the creation of holes in the MgO valence band, which destabilize oxygen atoms and induce their desorption. As a result, as shown both theoretically and experimentally, the actual compensating charges are charged oxygen vacancies which form F^{2+} centres by transferring their two electrons to adjacent Li ions (Myrach *et al.*, 2010; Richter *et al.*, 2015; Nilius, 2015). In this respect, STM luminescence spectroscopy proved to be an invaluable method of study, allowing the gain of topographic and optical information from the same surface area, and sensitive to both the dopant and the compensating defect electronic states. Prada et al. have theoretically analysed a different mechanism of compensation which does not require the formation of oxygen vacancies. They showed that instead of the creation of a hole in the MgO VB to stabilize the Li ions in their +1 oxidation state, the needed electron may be provided by transfer from the metallic substrate, thus restoring the oxygen −2 formal charges (Fig. 4.29(c)). The resulting interfacial dipole enhances the work function of the whole system (Prada *et al.*, 2012).

4.5.3 Consequences of charge transfer on adsorption and adhesion

As recently reviewed (Honkala, 2014; Nilius, 2015; Pacchioni and Freund, 2018), the charge redistribution processes associated to the presence of oxygen vacancies or dopants in thin oxide films have a strong impact on the charge state of adsorbates, whether atoms, molecules, or metallic clusters, with important consequences on chemical reactivity. Quite inert non-reducible oxides may become active thanks to these processes. For example, low-valence doping which favours the formation of oxygen

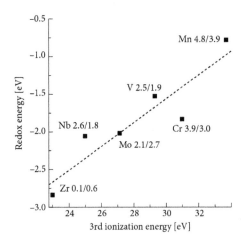

Fig. 4.30 Redox energy associated to the reaction Au/CaO + M^{2+}CaO $\longrightarrow$ Au$^-$/CaO + M^{3+}CaO, plotted as a function of the measured 3rd ionization energy of the various M dopants in the CaO(100) matrix. The two numbers given after the name of the element M refer to the magnetic moments of M^{2+} and M^{3+}, respectively. Reprinted from Andersin *et al.* (2013) with permission. Copyright © 2013 Wiley-VCH Verlag GmbH and Co. KGaA, Weinheim.

vacancies and the desorption of oxygen atoms may promote oxidative reactions. In contrast, high-valence doping which is associated to electron donation may modify the charge of the adsorbates, as shown in the few illustrative examples which follow.

While high-valence dopants located at the interface with the film support are prone to transfer their excess electron to the metallic support, at the film surface, the transfer more easily occurs towards an adsorbate, especially when it is assisted by strong local deformation of the lattice (polaronic distortion). STM measurements have highlighted this process by revealing the formation of superoxo groups O$_2^-$ after adsorption of oxygen molecules at the surface Mo-doped CaO films grown on an Mo substrate (Cui *et al.*, 2015*a*). Similarly, simulations of CO$_2$ adsorption on ultra-thin Al-doped MgO films have found that carboxylate groups CO$_2^-$ are more stable than carbonate groups CO$_3^{2-}$ (Tosoni *et al.*, 2015). Simulations of adsorbed Au atoms on a CaO surface doped with various high-valence cations M have shown that the Au atoms become negatively charged and that their adsorption energy scales with the redox process in the solid phase: Au/CaO+ M^{2+}CaO $\longrightarrow$ Au$^-$/CaO+M^{3+}CaO, itself function of the third ionization potential of the M cation (see Fig. 4.30) (Andersin *et al.*, 2013).

Such charging processes may also modify the shape of adsorbed clusters, as illustrated by studies of gold islands on doped oxides. At the surface of Mo-doped CaO films, a strong electrostatic interaction between the island and the CaO surface results from an electron transfer to the Au island, which consequently tends to maximize its contact with the support. Contrary to the gold 3-D shape in the absence of doping, flat Au islands, one monolayer thick, are observed on the doped substrate, as shown in Figs. 4.31(a,b) (Shao *et al.*, 2011*c*). A similar effect had been found on ultra-thin (undoped) MgO films grown on Ag(001), based on a direct electron transfer from the

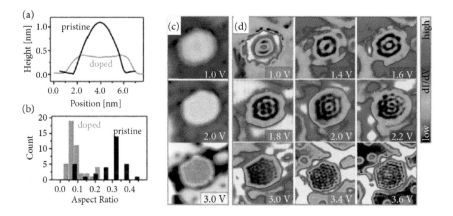

Fig. 4.31 Left: characteristics of a 0.7 ML Au deposit on pristine and doped CaO films. a) height profiles; b) Histogram of particle aspect ratios. Reprinted from Shao *et al.* (2011*c*) with permission. Copyright © 2011 Wiley-VCH Verlag GmbH and Co. KGaA, Weinheim. Right: c) topographic and d) conductance STM maps of an Au island on MgO/Ag(001) taken at different bias voltages (image size 11×11 nm^2). Note the bias-dependent evolution of the standing-wave pattern inside the island. Reprinted from Cui *et al.* (2015*b*) with permission. Copyright 2015 by the American Physical Society.

metal support to the adsorbate (see Chapter 5). In both cases, the strong confinement in the islands gives rise to quantized electronic waves at energies close to the Fermi level revealed by standing-wave patterns in STM conductance images. Their bias dependence indicates a linear electronic dispersion relation, whose slope and energy can be traced back to the Au bulk band structure close to the L point of the Brillouin zone (Cui *et al.*, 2015*b*).

4.6   Summary

Ultra-thin films grown on metal or oxide substrates feature a range of novel properties that their bulk analogues do not possess, and which can be tuned by an adequate choice of substrate, orientation and thickness. Similarly to the well-known saying that, in clusters, 'each atom counts', one can assert that in ultra-thin films, each layer counts. Unexpected crystalline structures, new vibrational modes, critical thickness for metal-insulator transitions, critical temperatures for ferroelectricity or magnetism, and stabilization of new magnetic ground state, all these findings are strongly dependent on the film orientation and thickness. Thanks to the advances in the preparation of ultra-thin film of controlled composition and thickness, it is now possible to engineer artificial systems with the desired properties.

Manipulating the thin film composition by intentional doping or creation of oxygen vacancies is a lever to activate inert non-reducible oxides or modify the reducibility of reducible ones. The resulting adsorbate charging allows tailoring acid-base properties to be used in chemical or catalytic reactions.

The choice of the substrate on which oxide thin films are grown is of paramount importance, not only because it determines the structural quality of the films, but also because it may have an active impact on the nature of interfacial electronic states. Examples have been given of strong modifications of cation valencies at the interface, of 2DEG at the interface between two insulting oxides or of a ferromagnetic layer at the interface between a paramagnetic and an antiferromagnetic oxide. Intermixing at the interface or (purposely created or not) interdiffusion may also generate thin ternary compounds at the interface with the support with possibly interesting properties.

New properties, found in films several ML-thick, become even more varied and more spectacular in the extreme 2-D limit, when the film thickness is reduced to one structural unit, as described in the next Chapter.

5
Two-dimensional oxides

The discovery of graphene by Geim and Novoselov in 2004 (Novoselov *et al.*, 2004) has prompted a veritable hype in the search for 2-D materials, and many novel 2-D materials beyond graphene have been established in the last decade (Butler *et al.*, 2013), amongst them 2-D oxide systems. The demonstrated possibilities of easy manipulation of graphene have opened up new opportunities for the application of 2-D materials, and their novel properties and functionalities in sympathy with the general trend towards miniaturization in nanotechnology have been transformative in creating new research areas. The increased significance of 2-D oxides in research and technology has also been enabled by recent progress in the preparation and growth techniques, by the availability of nanoscale experimental characterization techniques, and by the advancements in theoretical modelling tools.

But what is actually meant by a 2-D oxide? This is a tricky question, because a precise definition does not exist and the term 2-D is used in the literature from truly 2-D systems such as graphene, one atomic layer thick, to several nanometre thick layered systems. Here, we will adopt a definition that appears suitable for oxide materials: a 2-D oxide is a single layer of one bulk unit-cell thick or thinner, typically less, or of the order of 1 nm. The transition from the scalable to the non-scalable regime of properties in size reduction, discussed in Chapter 6, is also an important criterion, but it depends on the characteristic length scale of the property and may be material dependent. True 2-D effects may therefore not necessarily be associated with the advent of the non-scalable regime. We will therefore employ the previously mentioned definition as a guideline in this chapter, but in a pragmatic and somewhat flexible way.

The so-called 'nanosheets' are a particular nanoscale oxide morphology, which are often referred to as 2-D oxides in the literature. In some cases, they do concur with the one-unit-cell thickness definition, but mostly they consist of layered structures with a variable number of 2-D oxide layers in a stacked arrangement. The latter is the result of their mostly non-specific preparation procedures. An important characteristic of nanosheets is their free-standing nature, which does not require a substrate for stability. Since oxide nanosheets have found a variety of important applications in optical, solar energy, and energy storage devices (Tan *et al.*, 2017) and their fabrication has seen a burst of new methodologies in recent years (Osada and Sasaki, 2009; Ma and Sasaki, 2014), mostly of a low-cost wet chemical nature, they will be included in the discussion of this chapter, even if often not strictly 2-D in our sense of definition.

Oxide Thin Films and Nanostructures. Falko P. Netzer and Claudine Noguera, Oxford University Press (2021). © Falko P. Netzer and Claudine Noguera. DOI: 10.1093/oso/9780198834618.003.0005

Raman spectroscopy has been developed as a useful tool to characterize the number of layers and the crystalline nature of oxide nanosheets (Tan *et al.*, 2017). This is based on the fact that some phonon frequencies of 2-D oxide layers are sensitive to the coupling between layers, even the weak coupling of layers in typical van der Waals solids, and the respective Raman bands show frequency shifts as a function of layer numbers and sheets thickness (Kalantar-zadeh *et al.*, 2010; Balendhran *et al.*, 2013*c*).

It is useful to separate the 2-D oxides into two categories: unsupported and supported layers. While in the unsupported systems dimensionality reduction is at the root of their novel properties, the presence of the support creates an interface which may be responsible for new structures and stoichiometries via electronic and chemical interactions. The role of the substrate in ultra-thin films and nanoparticles is treated more explicitly in Chapter 4, but it has to be emphasized here that substrate interface effects are most relevant in the 2-D case. The 2-D character of the materials discussed in this chapter has a number of consequences in terms of particular synthesis methods, structural, electronic, and chemical properties, which will be discussed in the following. An interesting experimental consequence of the 2-D state is that all states of the system are *accessible* to experimental analysis, e.g. in scanning probe techniques, because there is no bulk phase. In the case of supported 2-D oxides, however, the influence of the support has to be taken into account.

Confinement in the lateral dimension of 2-D layers results in the formation of nanoribbons (NRs), which may be regarded as being of intermediate dimensionality between 2-D and 1-D. Due to the limited width, the relative importance of undercoordinated edge atoms introduces new properties and functionalities (Yagmurcukardes *et al.*, 2016). This has been demonstrated for graphene NRs, which can be prepared with different width and edge geometries, yielding width and edge-shape dependent electronic and magnetic properties, functionalities, and quantum transport. In contrast to the edge states in graphene NR, which are non-polar with non-bonding π states at the low-coordinated edge C atoms, oxide NRs may be polar or non-polar depending on their structure and the orientation of their edges. At the time of writing, most oxide NR studies are still of a theoretical nature, but it can be expected that experimental work in this field will soon follow.

5.1 Fabrication–synthesis

It is instructive and convenient to separate the methodologies of 2-D oxide preparation into top-down and bottom-up methods. The top-down methods have a particular relevance for 2-D materials, since the mechanical exfoliation method for the preparation of graphene layers from bulk graphite (the so-called *scotch tape* method) has been used successfully in the original work of Geim and Novoselov (Novoselov *et al.*, 2004) to generate single-atom thick layers of graphene. The top-down methods require the existence of a layered bulk structure, from which individual 2-D layers can be separated by delamination or exfoliation. The delamination/exfoliation of layers can be achieved by micromechanical cleavage (*scotch tape*), by ultrasonic treatment in a solvent (sonication), by intercalation of bulky ions or molecules which weaken the out-of-plane interactions between the individual layers, or by other wet chemical means. In fact, many delamination techniques to obtain 2-D oxide materials have been reported

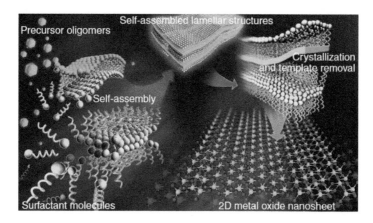

Fig. 5.1 Schematic drawing of the self-assembly of 2-D metal oxide nanosheets. Reprinted from Sun *et al.* (2014) by permission from Springer-Nature. Copyright 2014.

during the last decade (Tan *et al.*, 2017; Osada and Sasaki, 2009; Kalantar-Zadeh *et al.*, 2010; Kalantar-zadeh *et al.*, 2016; Rui *et al.*, 2013; Hanlon *et al.*, 2014; Ma and Sasaki, 2014). Most techniques, however, refer to the fabrication of oxide nanosheets, in which the single-layer thickness criterion is not always strictly fulfilled. In some cases, where layered forms of bulk oxides do not exist, exfoliation of related compounds (e.g. chlorides, hydroxides, hydrated oxide phases) has been successfully achieved, followed by the transformation of the 2-D sheets into oxides by thermal oxidation treatment (Kalantar-zadeh *et al.*, 2010; Feng *et al.*, 2014; Zhao *et al.*, 2016).

The bottom-up methods include the vapour phase deposition techniques (as described for nanostructure and thin film preparation in Chapter 2: PVD, MBE, PLD, CVD) and a variety of special wet-chemical synthesis methods. To give a flavour of the latter, a few representative and original examples are mentioned in the following. Diverse oxide nanosheets (e.g. TiO_2, ZnO, Co_3O_4) have been prepared by solution assembly of oxide precursor oligomers into lamellar structures with surfactant molecules, followed by a hydrothermal reaction to induce oxide crystallization into 2-D nanosheets with atomic thickness (Sun *et al.*, 2014). Figure 5.1 gives a schematic representation of the molecular assembly of 2-D oxide structures from liquid solutions, where metal oxide precursor molecules are assembled into lamellar structures followed by crystallization into 2-D oxide nanosheets (Sun *et al.*, 2014).

A template assisted evaporation-induced self-assembly process with nitrate precursor compounds adsorbed on graphene oxide as templates in solution, followed by a heat treatment, has been used to fabricate a number of transition metal monoxide (e.g. NiO, ZnO, CuO) and dioxide (e.g. TiO_2, ZrO_2, CeO_2) nanosheets and also mixed metal oxide structures (Lei *et al.*, 2018). A salt-templated synthesis method, using water-soluble salt crystals as templates for epitaxial growth of oxide precursor compounds, has been reported as a successful strategy to generate hexagonal MoO_3, MnO, and WO_3 nanosheets (Xiao *et al.*, 2016). In the molten-salt method, cation intercalated 2-D Mn oxide and W oxide nanosheets have been reported to form by reaction of oxide precursor compounds (e.g. ammonium tungsten hydrate) with a molten salt (e.g. NaNO$_3$); it has been conjectured that the 2-D material grows in the molten salt by self-assembly of MO$_x$ seeds in kinetically stabilized structures, but the growth mechanism remains somewhat obscure (Hu *et al.*, 2017). A particularly simple method has been described to grow MoO$_3$ nanosheets; the method has been termed 'van der Waals epitaxy' by the authors, and is essentially a vapour deposition method without strict evaporation control (Molina-Mendoza *et al.*, 2016). A Mo metal foil, covered by a thin oxide film (muscovite mica), is heated under atmospheric conditions, and MoO$_3$ from the oxidizing Mo surface is transferred to the mica substrate via sublimation, thus creating a thin layer of MoO$_3$. The substrate is then removed and quenched at room temperature; the 2-D MoO$_3$ crystals can then be transferred to another desired substrate, using a wet polymer stamp procedure. The thickness of the material is however not completely homogeneous, ranging between 1.4 nm (1 bulk bilayer unit-cell thick) and 4 nm.

A problem with many wet-chemical synthesis methods of oxide nanosheets is the lack of morphological control (thickness, lateral size). Although the latter is apparently not strictly necessary for some optoelectronic and electrochemical applications, it does constitute a bottleneck for the use of oxide nanosheets in high-end nanotechnology devices, where identical nanostructures with exact thickness and size control are necessary (Osada and Sasaki, 2009). Accurate control of 2-D oxide growth is possible in vapour-phase deposition techniques, where single-layer structures can be fabricated with atomic precision, albeit at the expense of experimental complexity and expenditure. Also, scalability is another problem with many fabrication methods.

The growth of free-standing 2-D nanocrystals of non-layered materials is a very challenging endeavour in general, because selective growth along one nano-facet is required and the 2-D layers may be unstable. One way out is to grow the 2-D layers on a substrate, which may provide mechanical stability to the 2-D structure. In addition to providing stability, the substrate may add new design capabilities to generate non-bulk-like materials, with novel stoichiometries and structures, i.e. the substrate may play an active role in creating new non-native phases (Surnev *et al.*, 2012; Wu *et al.*, 2009). The substrate may also introduce lattice strain in the epitaxially grown layer, which generates new functionalities in the system. The removal of 2-D materials from the growth substrate is not trivial and often impossible, in particular if the substrate, in addition to providing mechanical stability, is also necessary for the chemical stability. Some 2-D materials, however, could be successfully removed from their growth substrate, e.g. graphene grown by CVD on metal surfaces (Li *et al.*, 2009*b*; Reina *et al.*, 2008; Bae *et al.*, 2010), or 2-D SiO$_2$ grown on a Ru(0001) single-crystal surface (Büchner *et al.*, 2016).

For fundamental scientific studies of the physical and chemical properties of 2-D oxides, the vacuum-based deposition methods provide the highest precision in fabrication and materials quality. This may also apply for the use in high-end device fabrication technologies. However, for a number of optoelectronic, photochemical and electrochemical applications, the cheap chemical preparation methods have a large potential and useful device prototypes have been demonstrated (Alsaif *et al.*, 2014*a*; Kalantar-zadeh *et al.*, 2016; Tan *et al.*, 2017; Hu *et al.*, 2017).

5.2 Geometric structure aspects

5.2.1 Free-standing 2-D layers

The 2-D materials obtained by delamination of layered bulk solids retain their principal layer geometry after exfoliation. The strong covalent bonding within the 2-D layers provides a robust geometric intralayer stability, which is hardly influenced by the weak interlayer van der Waals coupling in the bulk phase. Therefore, most studies in the literature addressing exfoliated 2-D oxide systems have devoted little attention to structural modification. However, some structural relaxations may still be expected due to the fact that a large proportion of atoms, if not all, are at the surface of the 2-D sheets. Indeed, a careful TEM study of delaminated titania $Ti_{0.85}O_2$ monolayer sheets has revealed small ($<5\%$) but significant distortions of the TiO_6 octahedra within the layers as a result of Ti vacancies and surface boundary effects (Wang *et al.*, 2010*b*). The structures of a number of 2-D oxide systems, prepared by a wet-chemical self-assembly method of precursor molecules (TiO_2, ZnO, Co_3O_4, WO_3), have been analyzed by DFT calculations (Sun *et al.*, 2014); comparing the structures and charge density distributions of the self-standing 2-D layers with corresponding planes in the bulk solids, it was concluded that the 2-D sheets present some lattice distortions accompanied by a partial redistribution of electron density, but new structural phases have not been proposed.

In contrast, in some systems, significant hybridization of the bond character and correspondingly new structures have been predicted (see Chapter 4, Section 4.1.3). A prototypical example is ZnO, which transforms from a wurtzite structure to a planar graphene-like structure (actually a BN-like structure, with Zn and O replacing B and N) in the transition from 3-D to 2-D. The planar hexagonal ZnO phase has been experimentally confirmed for ZnO monolayer phases on Ag(111) (Tusche *et al.*, 2007) and Pd(111) substrates (Weirum *et al.*, 2010). The same structure has been derived from theoretical considerations of polarity compensation for a MgO(111) monolayer (Goniakowski *et al.*, 2004), where the transition is from a rock-salt (111)-type bilayer structure to the hexagonal planar BN-like structure. Non-native oxide phases in nanostructured systems, such as hexagonal MnO, CoO, and ZnO have been fabricated by kinetic reaction control in the synthesis paths (sometimes with the help of a substrate) (Nam *et al.*, 2012; Nam *et al.*, 2017). The flexibility of 2-D lattices at the nanoscale facilitates the transition from cubic to hexagonal; moreover, the large surface energy contribution to the total energy also plays an important role for the stabilization of these metastable phases. It has been predicted that, in 2-D, the wurtzite MnO transforms spontaneously into a BN-like structure featuring half-metallic ferromagnetism as a consequence of small hole doping, which would open up a way toward 2-D magnetic oxide materials with prospective spintronics applications (Kan *et al.*, 2013). Experimental realization of a 2-D hexagonal MnO phase is, however, not yet available.

The properties of hypothetical 2-D hexagonal MO_2 phases (M = transition metal) in honeycomb-type lattice geometries have been investigated in a comprehensive DFT study (Ataca *et al.*, 2012). As in MS_2 compounds, two hexagonal structure types may exist: the H structure (honeycomb) and the T structure (centred honeycomb), which are depicted in Fig. 5.2. The MO_2 sheets consist of a trilayer, a positively charged

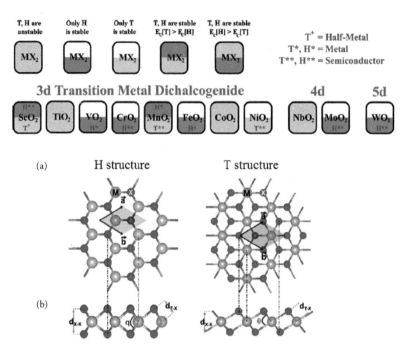

Fig. 5.2 Stability and atomic structure of 2-D MO_2 layers in the H and T structures. (a) Top and (b) side views showing the primitive unit-cell and internal structure parameters (M: grey; O: red). The resulting structures may be half-metallic (+), metallic (*), or semiconducting (**). Reprinted from Ataca *et al.* (2012) with permission. Copyright 2012 American Chemical Society.

transition metal atom layer sandwiched between two negatively charged oxygen layers. In the H structure, the two oxygen layers are in registry, whereas in the T structure they are shifted with respect to each other by a fraction of a unit-cell. Figure 5.2 also gives the stabilities of the different (hypothetical) MO_2 compounds in terms of single layer H and/or T structures. These 2-D MO_2 systems have not yet been realized experimentally in free-standing layer form, but denser trilayer structures in hexagonal close packed arrangements have been observed in metal-supported oxide systems (see below).

Similarly, a thorough DFT study of the structural, electronic, and magnetic properties of 1 unit-cell thick, free-standing M_2O_3 (0001) layers (M = Al or a 3d transition metal) has been performed (Köksal *et al.*, 2018; Goniakowski and Noguera, 2018). While crystallizing in the bulk in the corundum structure where anions and cations are not coplanar, at the 2-D limit, the layers are flat and display a fully coordinated network of twelve member rings with an equal number of oxygen and metal atoms. The same is true for unsupported mixed $MM'O_3$ layers (Goniakowski and Noguera, 2018).

5.2.2 Metal-supported 2-D layers: general properties

The use of a substrate opens up new opportunities for the fabrication and provides additional degrees of freedom in the design capabilities of 2-D oxide systems. The substrate can be a support for mechanical stability, an atomic growth template enabling epitaxial growth, or an agent providing an active interface. Oxide-support interactions in terms of chemical bonding and charge transfer create (and stabilize) novel non-native oxide phases that have no relation to the respective bulk compounds. In a way, the support-overlayer coupling may create a hybrid system, in which the interface largely determines the support-oxide properties (Netzer *et al.*, 2010; Netzer, 2010). Obviously, the choice of the support is of paramount importance in the latter case. Lattice-matched oxide surfaces have been successfully used for the epitaxial growth of oxide thin films (Chambers, 2000) (see also Chapter 4, Section 4.3.4), but metal single-crystal surfaces are very popular as substrates for the preparation of 2-D systems because of their flexible chemical behaviour, structural variety, and experimental accessibility.

The flexibility of 2-D oxide lattices is manifested by their easy deformability and the advent of surface rumpling. In metal-supported 2-D oxides, charge transfer between the support and the overlayer may occur, the direction and extent of which is determined by the electronegativity of the support (see Chapter 4). As a structural response to the electric field created by the charge transfer at the interface, a separation between the atomic planes of cations and anions occurs in the 2-D oxide layer leading to a non-planar arrangement, i.e. surface rumpling (Goniakowski and Noguera, 2009; Pacchioni, 2016). The local structural relaxation as a response to a charged adatom at the surface may also give rise to a polaronic distortion in the oxide layer which stabilizes the adatom, but this effect is not observed in thicker oxide films or on surfaces of bulk oxides.

The epitaxial growth of a thin film on a crystalline substrate usually requires a match of lattice constant and lattice symmetry between the substrate and the overlayer, forming a coherent interface. It appears that this concept is relaxed in the case of 2-D oxide overlayer growth (Obermüller *et al.*, 2017). The growth of well-ordered 2-D oxide layers has been observed in many interface scenarios, ranging from pseudomorphic to incommensurate lattices—see examples of oxide structures discussed in the following. For commensurate structures, the coupling of the oxide to the substrate has to be strong enough, so that structural parameters of the substrate are imprinted in the growing oxide layer; this coupling provides extra stability to the oxide monolayer and detrimental strain effects in the overlayer may be accommodated by the flexibility of the 2-D lattice. Incommensurate ordered oxide phases are less common; as the coupling to the substrate has to be weak, the latter provides merely mechanical strength. Nevertheless, novel non-native phases may form, in which confinement and surface energy must create extra terms for stability. An example is the observation of a well-ordered 2-D WO_3 layer on Ag(100), which is incommensurate and rotationally disordered with respect to the substrate (Obermüller, 2015). Recent DFT investigations have disclosed an oxide structure that is related to a 2-D sheet of α-MoO_3, non-native for WO_3. Thus, despite weak interactions, the Ag surface appears to stabilize a metastable structure for 2-D WO_3 (Negreiros *et al.*, 2019).

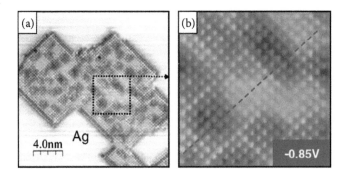

Fig. 5.3 STM images of a NiO(100) (1 × 1) island embedded in Ag(100). (a) (20 ∗ 20 nm^2); (b) (5 ∗ 5 nm^2). Reprinted from Steurer *et al.* (2012) Copyright 2012, with permission from Elsevier.

5.2.3 Classification and examples of metal-supported 2-D layers

A categorization of structure types of metal-supported 2-D oxide systems has been attempted in recent reviews (Surnev *et al.*, 2012; Netzer and Surnev, 2016). In brief, planar oxide monolayers, oxide bilayer and trilayer structures, and complex structures incorporating various M-O structure motifs with different symmetries have been identified. One should nevertheless keep in mind that the belonging of a 2-D oxide layer to a given category is strongly dependent on its interaction with a particular substrate. For example, MO or M_2O_3 2-D layers which are flat when free-standing (monolayer structure) may become rumpled as a result of the strain exerted by the substrate or of an electron exchange with the substrate, leading to their qualification as bilayers or trilayers. In the following, some illustrative examples of different structure types of metal-supported 2-D binary oxides are presented to give an impression of the range of structures that have been observed.

Planar oxide monolayers. The NiO(100) monolayer on Ag(100) is an example, in which a single (100) plane of rock-salt structure is stabilized by a substrate. The NiO(100) plane has a relatively good lattice match with the Ag(100) surface and forms a coherent (1 × 1) structure with little residual strain (Thomas and Fortunelli, 2010). Figure 5.3 shows STM images of a NiO(100) nanoisland on Ag(100) formed by the deposition of 2/3 of a monolayer of Ni atoms in 1 x 10^{-6} mbar O_2 on a clean Ag(100) surface at room temperature and subsequent annealing at 600 K in oxygen. The NiO island in Fig. 5.3(a) is embedded into the top Ag layer and its morphology reflects the square symmetry of the system; the embedding into the Ag surface provides energetic stabilization to the NiO monolayer by additional coordination of the NiO edge states (Steurer *et al.*, 2011; Steurer *et al.*, 2012). Figure 5.3(b) displays a high-resolution STM image of the NiO island, where the square NiO lattice is clearly recognized; according to DFT simulations, the Ni atoms are imaged as bright protrusions in this STM image (Steurer *et al.*, 2011).

Another example is the ZnO monolayer on Pd(111), which represents a 2-D phase with hexagonal symmetry (Weirum *et al.*, 2010). The stable structure of ZnO in the

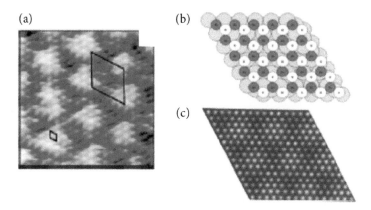

Fig. 5.4 (a) STM image (6.5 ∗ 6.5 nm²) of the (6 × 6) ZnO phase on Pd(111). (b) DFT structural model and (c) STM simulation of the (6×6) ZnO structure (Zn: blue; O: white; Pd: light grey). Reprinted from Weirum *et al.* (2010) with permission. Copyright 2010 American Chemical Society.

bulk is the wurtzite lattice, but as already mentioned previously, in 2-D ZnO, a BN-type hexagonal lattice becomes the most stable phase. Figure 5.4(a) shows a high-resolution STM image of the (6 × 6)ZnO structure on a Pd(111) surface, which has been interpreted in terms of a hexagonal monolayer. The (6 × 6) periodicity on the ZnO/Pd(111) surface is formed by a coincidence lattice, where 5 times the oxide unit-cell constant fit onto 6 times the Pd unit-cell constant. Figure 5.4(b) displays a DFT derived structure model. The DFT simulated STM image in Fig. 5.4(c) is in very good agreement with the experimental one in (a), giving credence to the proposed structure model.

   In contrast to the NiO(100)/Ag(100) system discussed previously, the (100) planes of the rock-salt structure 3d transition metal monoxides MO (M = Mn to Co) have a large lattice mismatch to the Pd(100) surface (∼ 7–14%); epitaxial growth of (100)-type oxide monolayers is therefore not possible because the misfit strain is too large. These systems have been investigated in detail because Pd is an excellent catalyst material for a diversity of chemical reactions and the oxide-metal interface is of relevance for catalytic activity. Under appropriate thermodynamic conditions, i.e. metal concentration/coverage and chemical potential of oxygen, a c(4 × 2) surface structure has been reported for these MO layers on the Pd(100) surface (Franchini *et al.*, 2009*a*; Allegretti *et al.*, 2010; Agnoli *et al.*, 2005). With the help of DFT calculations, these structures have been rationalized as derived from a rock-salt structure single (100) plane, into which one quarter of cation vacancies is introduced. These cation vacancies are ordered and define the c(4×2) periodicity with respect to the Pd(100) plane, yielding an overall M_3O_4 stoichiometry—see Fig. 5.5 for the Mn-oxide. As indicated by the DFT calculations, the introduction of cation vacancies and the associated lattice relaxation are effective means to release the interfacial strain (see the lateral displacement of Mn cations in Fig. 5.5) (Franchini *et al.*, 2009*a*). These MO c(4 × 2) monolayers have interesting magnetic structures as discussed in Section 5.5.

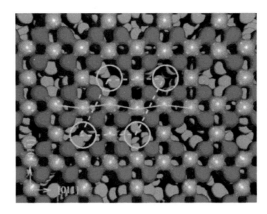

Fig. 5.5 Top view of the geometrical model of the c(4 × 2) Mn_3O_4 phase on Pd(100) (O: red; Mn: light grey; Pd: green). Dashed lines indicate the 2-D unit-cell, full line the Mn lateral displacement; circles highlight the position of vacancies. Reprinted from Franchini *et al.* (2009*a*) with permission. Copyright 2009 by the American Physical Society.

Oxide bilayer structures. A prototypical example of 2-D oxide bilayer structures is FeO on Pt(111) (Galloway *et al.*, 1994; Ritter *et al.*, 1998; Weiss and Ranke, 2002), which consists of two densely packed hexagonal planes of cations and anions, stacked in a rock salt structure [111] direction. The FeO(111) wetting monolayer on Pt(111) forms a high-order coincidence structure due to the lattice mismatch, where the O plane is at the outer surface and the Fe plane at the interface to the Pt(111) surface. In STM images, the structure displays a characteristic Moiré pattern (Ritter *et al.*, 1998). The Fe-O interlayer distance is reduced in the bilayer phase compared to its value in the rock-salt phase of bulk FeO (Kim *et al.*, 1997*b*), but the intralayer distance is enhanced by strain and electron exchange with the Pt support, compared to its flat structure when unsupported. The FeO bilayer cannot grow much beyond two monolayers because of the polar character of the structure, which introduces an electrostatic instability as the slabs become thicker (Tasker, 1979); a transformation to a Fe_3O_4 island phase occurs upon further growth (Ritter *et al.*, 1998).

While the growth of a hexagonal oxide bilayer on a hexagonal metal substrate is intuitively intelligible, the formation of a hexagonal oxide bilayer on a square substrate is more surprising. Indeed, the FeO c(2 × 10) structure on Pt(100) (Shaikhutdinov *et al.*, 2000), FeO c(8 × 2) on Pd(100) (Kuhness *et al.*, 2016), CoO c(10 × 2) on Ir(100) (Ebensperger *et al.*, 2010), or CoO (9×2) on Pd(100) (Gragnaniello *et al.*, 2011) phases have been rationalized in terms of hexagonal rock-salt (111)-type close packed bilayer structures. Here we present the FeO c(8 × 2) structure on Pd(100) for the purpose of illustration (Kuhness *et al.*, 2016). Figure 5.6 presents STM and LEED data together with a schematic structure model of the c(8 × 2) FeO phase on Pd(100). The c(8 × 2) layer was prepared by oxidation of one monolayer of Fe atoms at 400°C in a moderate oxygen pressure of $5 * 10^{-8}$ mbar. Figure 5.6(a) shows that the c(8 × 2) layer wets the entire Pd surface, the sharp LEED pattern in Fig. 5.6(c) confirms the good long-range order (with hexagonal (yellow) and primitive (white) unit-cells indicated). The

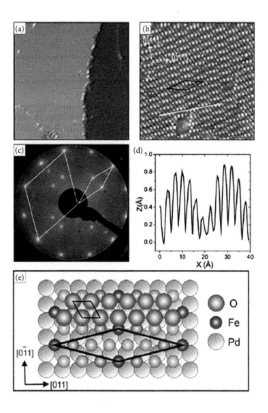

Fig. 5.6 STM images of the FeO c(8×2) phase on Pd(100): (a) $50 * 50$ nm^2 and (b) $7 * 7$ nm^2. The hexagonal and primitive c(8×2) unit-cells are indicated in (b); (c) corresponding LEED pattern with the hexagonal (yellow) and primitive c(8×2) (white) reciprocal unit-cells of the Fe-oxide layer indicated; (d) height profile taken along the [011] direction (line indicated in the STM image (b)); (e) hard sphere model of the c(8×2)-FeO(111) bilayer, consisting of a hexagonal Fe layer (blue) in contact with the Pd substrate (gray), terminated by a hexagonal O layer (red). Dark and bright blue in the colouring of the Fe atoms corresponds to lower and higher positions due to the different registry with the substrate. Reprinted from Kuhness *et al.* (2016). Copyright 2016, with permission from Elsevier.

high-resolution STM image of the c(8×2) structure (Fig. 5.6(b)) displays rows along the Pd [011] direction, consisting of four bright protrusions separated by three weaker maxima. This is also seen in the STM line profile (along the white line in panel (b)), which reveals an atomic separation of ~3.1 Å, a sine wave height modulation of 0.8 Å, and a periodicity of ~ 22 Å, corresponding to 8 Pd(100) surface lattice vectors. The c(8×2) superstructure is the result of the interference of the hexagonal structure of the overlayer with the square lattice of the (100) substrate surface. The schematic structure model in Fig. 5.6(e) gives more details of the structure: a close-packed plane of Fe atoms (blue spheres) is in contact with the Pd surface, while the terminating O atoms (red spheres) occupy half of the three-fold hollow sites of the Fe lattice. In the [011] direction, 7 Fe atom spacings match 8 Pd lattice vectors, resulting in

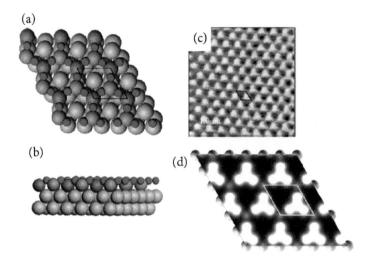

Fig. 5.7 Top (a) and side (b) views of the DFT derived structure model of the (2×2) FeWO$_3$ phase on Pt(111) (Pt: grey; W: blue; Fe: green, O: red). (c) High-resolution STM image of the (2×2) FeWO$_3$ layer (5 * 5 nm^2). (d) Simulated STM image. The (2×2) unit-cell is indicated in (a), (c) and (d). Reprinted from Pomp *et al.* (2016) with permission. Copyright 2016 American Chemical Society.

a (7×8) coincidence lattice and a unit-cell vector of 3.143 Å along [011]. The Fe atoms change periodically between four-fold hollow and near-bridge sites at the Pd surface (the colouring of the blue spheres in Fig. 5.6(e) reflects their relative heights). In order to produce the $\times 2$ periodicity in the [0$\bar{1}$1] direction, the second unit-cell vector must measure $a_{Pd}/\cos 30° = 3.175$ Å. The c(8 $\times$ 2) phase is thus a slightly distorted quasi-hexagonal FeO (111)-type bilayer, with a somewhat expanded lattice with respect to the FeO(111) bulk plane (a = 3.04 Å). DFT calculations have validated this structure model and revealed that the O anions are imaged bright in the STM due to their greater height. The STM brightness in this phase is therefore produced by a topographic contrast rather than an electronic structure effect.

Hexagonal bilayers of 2-D oxides on metal surfaces may also occur in more open network configurations: honeycomb and Kagome lattices have been observed (Netzer and Surnev, 2016). The 2-D V$_2$O$_3$ phase on Pd(111) (shown in Fig. 3.1, Chapter 3) is an example of a honeycomb lattice, which forms a (2×2) structure and consists of V atoms located in the threefold hollow sites of the Pd(111) surface with bridging O atoms in the outer layer. A Kagome lattice has been reported for the Ti$_2$O$_3$ bilayer on Pt(111) (Sedona *et al.*, 2005), for which a similar structure model as for V$_2$O$_3$ has been proposed (Barcaro *et al.*, 2009). Similar honeycomb structures have also been observed in the case of Ti$_2$O$_3$ and Nb$_2$O$_3$ deposited on Au(111) (Wu *et al.*, 2011*a*; Wang *et al.*, 2019*a*).

As an example of the flexible design capabilities of metal-supported 2-D oxide layers, the formation of a 2-D ternary oxide layer with honeycomb geometry is cited. A 2-D iron tungstate FeWO$_3$ phase has recently been fabricated by a solid-state reaction

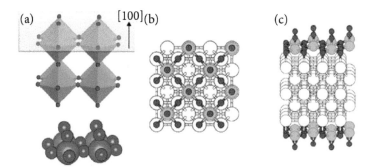

Fig. 5.8 (a) Cubic WO_3 bulk structure: a 2-D sheet consisting of a bilayer of a WO_2 plane plus an O plane is highlighted (top); bottom: rigid-sphere model of the 2-D WO_3 layer. Top view (b) and side view (c) of the DFT validated 2-D WO_3 model on Pd(100). Reprinted from Doudin *et al.* (2016) with permission. Copyright 2016 American Chemical Society.

of the 2-D FeO(111) bilayer on Pt(111) with $(WO_3)_3$ clusters deposited from the gas phase (Pomp *et al.*, 2016). The DFT-derived structure model is shown in Fig. 5.7(a,b), corresponding to a buckled mixed layer of Fe^{2+}/W^{4+} in a hexagonal honeycomb lattice, terminated by oxygen atoms in Fe-W bridging positions. Figure 5.7(c) and (d) show the experimental and DFT simulated STM images, which confirm the proposed structure model. This 2-D $FeWO_3$ layer is a non-native oxide with no analogue to bulk iron tungstate phases, has a novel stoichiometry, and a predicted ferromagnetic ground state (see Section 5.5).

The oxide bilayer geometry is not restricted to hexagonal lattices. A 2-D WO_3 sheet has been reported on a Pd(100) surface, which is characterized by a square bilayer structure (Doudin *et al.*, 2016). This bilayer may be visualized as a 2-D sheet, cut out parallel to the (100) plane of a cubic WO_3 bulk structure, as depicted in Fig. 5.8(a). The sheet consists of a square-symmetry WO_2 plane with an O layer on top of it, yielding a stoichiometric 2-D WO_3. This model has been confirmed by DFT theory to constitute the lowest energy state of 2-D WO_3 on Pd(100)—see Fig. 5.8(b,c) for the DFT model. This $(WO_2 + O)$ bilayer forms a well-ordered (2×2) superstructure on Pd(100), which has been observed experimentally by STM and LEED. The WO_3 layer on Pd(100) is however strained by $\sim 2.8\%$, which gives rise to a characteristic domain pattern with antiphase domain boundaries to release the tensile stress (Doudin *et al.*, 2016).

Oxide trilayer structures. As mentioned, the hypothetical MO_2 trilayer honeycomb structures investigated theoretically (Ataca *et al.*, 2012) have not yet been prepared experimentally, but MO_2 trilayers in hexagonal close-packed configurations (corresponding to a rock-salt (111)-type O-M-O stacking) have been reported for MnO_2, FeO_2, and CoO_2 supported on Pd(100) (Franchini *et al.*, 2009a; Kuhness *et al.*, 2016; Gragnaniello *et al.*, 2010), and FeO_2 on Pt(111) and Pd(111) surfaces (Giordano *et al.*, 2010; Zeuthen *et al.*, 2013). The cations in the trilayers have a higher formal oxidation state than in the bilayer structures, and this is in line with the higher oxidation

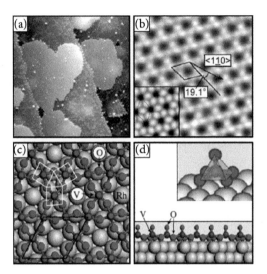

Fig. 5.9 The V_3O_9 ($\sqrt{7} \times \sqrt{7}$)R19.1° phase on Rh(111): (a) Large-scale STM image (100∗100 nm²). (b) High-resolution STM image (5∗5 nm²). The $\sqrt{7}$ unit-cell and the Rh(111) substrate direction are indicated. The insert shows a DFT-simulated image. (c) DFT model of the $\sqrt{7}$ phase. The unit-cell and structural units are indicated (V: green; O: red; Rh: grey). (d) Side view of the V_3O_9 model. The insert shows a detailed view of the pyramidal $O_4V{=}O$ unit. Reprinted from Schoiswohl *et al.* (2006) with permission. Copyright © 2006 IOP Publishing Ltd. All rights reserved.

potentials required in their preparation procedures. Incidentally, a hexagonal trilayer is also the most stable structure for 2-D layers of ceria (Surnev *et al.*, 2012). Ceria in the bulk crystallizes in the fluorite structure, the O-Ce-O stacking is the most stable configuration with the (111) plane as the most stable surface termination. The stability of supported (111)-type CeO_2 trilayers is therefore not unexpected. It is also noted that the intrinsic surface oxide phases of diverse group VIII metal surfaces consist of such trilayer systems (see Chapter 2, Section 2.1.3).

Complex 2-D oxide systems. The variety of complex 2-D oxide structures is large, making it a difficult choice for illustrative presentation. We have taken the ($\sqrt{7} \times \sqrt{7}$)R19.1° (in short $\sqrt{7}$) phase of vanadium oxide on the Rh(111) surface as the demonstration object (Schoiswohl *et al.*, 2004). At submonolayer coverages, the $\sqrt{7}$ layer grows in the form of 2-D islands (Fig. 5.9(a)), but the islands then coalesce at higher coverages and form a smooth wetting layer. The high-resolution STM image of Fig. 5.9(b) shows the structure details of a hexagonal Kagome lattice, with three bright protrusions per unit-cell. Figs. 5.9(c,d) present the established DFT model: a V_3O_9 (= VO_3) oxide phase containing pyramidal $O_4V{=}O$ building blocks (indicated by the squares in Fig. 5.9 (c)), with four bridging O atoms in the basal plane, a V atom in the center, and a vanadyl V=O group forming the apex (Fig. 5.9 (d)). The pyramids are joined together via the O atoms at the interface, which results in three VO_3 units per $\sqrt{7}$ unit-cell. The basal O atoms are shared with the Rh substrate atoms, which

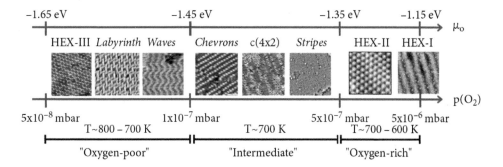

−1.65 eV   −1.45 eV   −1.35 eV   −1.15 eV

μ_o

HEX-III *Labyrinth* *Waves*   *Chevrons* c(4x2) *Stripes*   HEX-II HEX-I

$p(O_2)$

5×10^{-8} mbar 1×10^{-7} mbar 5×10^{-7} mbar 5×10^{-6} mbar

T~800 – 700 K   T~700 K   T~700 – 600 K

"Oxygen-poor"   "Intermediate"   "Oxygen-rich"

Fig. 5.10 Schematic phase diagram of the 2-D Mn oxides on Pd(100), presented as a function of the oxygen pressure $p(O_2)$ and of the oxygen chemical potential μ_O. The nominal coverage of Mn on Pd(100) is 0.75 ML. Reprinted from Li *et al.* (2009*a*) with permission. Copyright © 2009 IOP Publishing Ltd. All rights reserved.

provide charge transfer to the oxide; the formal V_3O_9 stoichiometry is therefore not incompatible with the maximum oxidation state of 5+ of the vanadium atoms. The structure model of the $\sqrt{7}$ V_3O_9 phase is confirmed by the good agreement between experimental images and theoretical STM simulations (insert of Fig. 5.9(b)), and the calculated phonon spectrum which reproduces the characteristic phonon loss of the V=O stretching vibration (at ∼ 130 meV) in HREELS spectra (Schoiswohl *et al.*, 2004). Another example of a complex structure arrangement is the alumina bilayer, which is formed by oxidation of the NiAl(110) alloy surface, as discussed in Chapter 2.

Oxide systems with multiple stable cation oxidation states (e.g. Ti-, V-, Mn-, Co-oxides, etc.) may involve a number of 2-D phases with variable stoichiometries, depending on the chemical potential of oxygen (i.e. oxygen pressure and temperature) applied during the preparation procedure. In these cases, complex phase diagrams with several low-energy structures and near-degenerated structural ground states have been observed. For example, vanadium oxide on Pd(111) and Rh(111) surfaces (Surnev *et al.*, 2001; Schoiswohl *et al.*, 2006), cobalt oxide on Pd(100) and Ir(100) (Allegretti *et al.*, 2010; Gragnaniello *et al.*, 2010; Biedermann *et al.*, 2009), titanium oxide on Pt(111) (Sedona *et al.*, 2005; Barcaro *et al.*, 2009), or manganese oxide on Pd(100) (Li *et al.*, 2009*a*) may be cited as oxide systems in this context. The 2-D Mn oxide on Pd(100) is presented here as a benchmark system in Fig. 5.10, in which the range of stability of different Mn oxide phases, in the form of respective STM images, as a function of the chemical potential of oxygen μ_O is portrayed. Nine major 2-D oxide nanophases have been detected in the accessible range of the chemical parameter space. The phases in the different regions of μ_O (e.g. 'oxygen-poor', etc.) have related structures and often coexist at the Pd(100) surface. In the 'oxygen-rich' region, the hexagonal MnO_2-type trilayer structures are observed, at 'intermediate' μ_O the c(4×2) Mn_3O_4 phase plays the central role, which can be easily transformed into the adjacent phases by a vacancy propagation mechanism (Franchini *et al.*, 2009*a*; Surnev *et al.*, 2012), whereas the atomic details of the structures in the 'oxygen-poor' region are less clear (Netzer and Surnev, 2016).

Kinetic effects, particular mechanisms for polarity compensation (Goniakowski *et al.*, 2007*a*), and geometric flexibility of 2-D oxide lattices (Freysoldt *et al.*, 2007), as well as variable connectivities of metal-oxygen building blocks all conspire in creating these complex phase diagrams. The role of defects as structure stabilizing concept in the formation of novel oxide structures has also been recognized (Sedona *et al.*, 2008; Barcaro *et al.*, 2012).

To summarize, the geometric structures of 2-D oxide systems range from robust bulk-like 2-D sheets derived from layered solids to novel non-native phases. The latter are particularly found if fabricated on metallic substrates, due to the effects of the coupling to the substrate and the introduction of an active oxide-metal interface. The substrate thus adds extra design capabilities: engineering of the elastic strain in epitaxially coupled systems, the stabilization of kinetically formed meta-stable phases, and the promotion of novel 2-D oxide-metal hybrid phases with properties determined largely by interface effects.

5.3 Electronic properties and applications

5.3.1 Fundamental properties

As discussed in Chapter 4, the electronic characteristics of oxide films depend on the geometric structure and the bonding characteristics of the atoms involved, and it is intuitively logical that, in the extreme 2-D limit, they are strongly modified as compared to their respective 3-D bulk counterparts. The electrical behaviour of 2-D oxides ranges from insulating and semiconducting to metallic. In the former cases, the band gaps determine the electronic excitation spectrum and the optical properties. As a general trend the 2-D band gaps tend to be smaller than in the bulk due to the reduction of the Madelung potential acting on the cation and oxygen atomic levels. However, in transition metal oxides, and especially in Mott–Hubbard oxides, the change in the d level splitting by the 2-D crystalline field may lead to a gap increase. Examples of the former behaviour are unsupported rocks-salt oxides like MgO(001) (Goniakowski and Noguera, 1995; Schintke *et al.*, 2001), CaO (001) (Protheroe *et al.*, 1983) or NiO (001) (Dudarev *et al.*, 1997). For an isostructural 2-D material, i.e. a material with a bulk-plane geometry, the band gap converges rather rapidly to the bulk value with layer thickness, as shown for MgO ultra-thin films, which reach the bulk gap value at a thickness of 3-4 monolayers ($\sim$ 1 nm) (Schintke *et al.*, 2001). In contrast, in more covalent oxides, like ZnO, there is a tendency to an increase of the gap upon reduction of dimensionality. In MO_2 (Ataca *et al.*, 2012), M_2O_3 or $MM'O_3$ (Le *et al.*, 2018; Goniakowski and Noguera, 2018) transition metal oxides, the gap in the 2-D oxide may be larger (WO_2, $VFeO_3$) or smaller (Cr_2O_3, Fe_2O_3) than in the bulk. In some cases, a metallic bulk may even transform into a semiconducting 2-D oxide (CrO_2, MoO_2, Ti_2O_3). Ataca *et al.* (2012) have calculated the band structures of a range of 2-D transition metal dioxide MO_2 compounds in hexagonal honeycomb structures (M = Sc to Ni, Nb, Mo, W). They found a rich diversity of electronic and magnetic properties, ranging from semiconducting to ferromagnetic to non-magnetic metallic. In the calculated series, the band gap increases as the M atom varies from Sc to W.

In addition to the band gap size, band gap modifications include changes from indirect to direct band gaps (or vice versa) with decreasing number of atomic layers. This property had been highlighted by optical spectroscopy in MoS_2 ML-thick films (Mak *et al.*, 2010), leading to the interesting outcome that the freestanding monolayer exhibits an increase in luminescence quantum efficiency by more than a factor of 10^4 compared to the bulk material (it is recalled that optical transitions—photon absorption/emission—involve vertical, i.e. direct, excitations within the reduced band scheme).

When 2-D oxides are grown on a metal support, the interaction with the latter has a particularly strong influence on the global electronic structure. As long recognized at metal-semiconductor interfaces and recalled in Chapter 4, the mixing and merging of semiconductor and metal electronic states (Metal Induced Gap States) can produce what can be termed a metallization of the overlayer (Tersoff, 1984; Monch, 1990; Bordier and Noguera, 1991). This may result in a finite oxide conductivity, with conventional charge transport (Surnev *et al.*, 2012). For example, in the NiO on Ag(100) case, it has been shown that the band structure of the NiO overlayer is influenced by acquiring charge from the metal support up to at least three monolayers (Thomas and Fortunelli, 2010; Steurer *et al.*, 2012). The sign of the charge transfer is mainly determined by the oxide and metal band offsets (see Chapter 4, Section 4.3.3). While relatively small when $s - p$ states are involved, it may become substantial and interpreted as a change in the oxide cation oxidation state, in the case of Mott–Hubbard oxides, as shown in the case of Ti_2O_3, V_2O_3 (Goniakowski and Noguera, 2019) or Nb_2O_3 (Wang *et al.*, 2019*a*) on Au(111).

The work function Φ plays a key role in phenomena connected with electron transport, such as conductivity and charge transfer, and also adsorption. For the important class of metal supported 2-D oxides, the oxide overlayer modulates the work function of the support, most of the time by reducing it. More precisely, three major effects impact the metal work function Φ, via formation of interfacial dipoles. One is the electrostatic compression of the metallic surface electronic density, which leads to a reduction of Φ. A second one is the interface charge transfer (which can go in both directions), causing an increase or decrease of Φ. These two contributions to the work function change upon deposition of a thin oxide layer on a metal support are illustrated schematically in Fig. 5.11. Furthermore, a structural response of the oxide lattice to the interfacial charge transfer in the form of surface rumpling can influence the work function of the oxide-metal system (Goniakowski and Noguera, 2009; Pacchioni, 2016). For example, a negative charging of the support induces an electrostatic repulsion on the oxygen atoms, which are pushed outwards, transforming flat unsupported MLs to rumpled ones when supported. Table 5.1 contains a collection of work function values of metals and of oxide-on-metal systems, as obtained from DFT calculations (Giordano and Pacchioni, 2011).

The charging of adsorbates on the surface of a 2-D oxide supported on a metal is an impressive example of spontaneous charge transport via electron tunnelling through the oxide layer, which is determined by the oxide-metal work function and the adsorbate HOMO-LUMO positions; its consequence is a strongly modified electronic and chemical behaviour of the adsorbate. According to Table 5.1, the work function can be

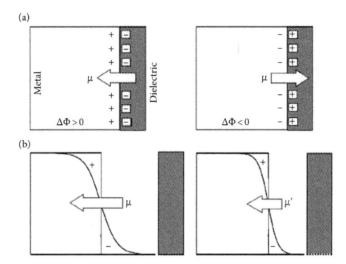

Fig. 5.11 Two contributions to the change of the work function Φ upon deposition of a thin oxide dielectric layer on a metal surface: (a) charge transfer at the interface; (b) compressive electrostatic effect. Reprinted from Prada *et al.* (2008) with permission. Copyright 2008 by the American Physical Society.

Table 5.1 Work functions Φ (eV) of metals and oxide thin films on metals from DFT calculations (Giordano and Pacchioni, 2011). Copyright 2011 American Chemical Society.

	Φ(metal)	Φ(oxide/metal)
FeO/Pt(111)	5.8	5.8–6.4
TiO$_2$/Pt(111)	5.8	5.6
SiO$_{2.5}$/Mo(112)	4.2	4.9
NiO/Ag(100)	4.3	3.9
MgO/Ag(100)	4.3	3.0
BaO/Au(100)	5.1	2.3
MgO/Mo(100)	4.2	2.1
BaO/Ag(100)	4.3	2.0
BaO/Pd(100)	5.2	2.0

selected by choosing the appropriate oxide-metal system, and this allows one to tune, within certain limits, the charge transfer and the state of adsorbates and in further consequence the chemical reactivity patterns of surface species (see Section 4.6).

The doping with foreign atoms is a proved means to introduce additional functionalities in oxide materials, where small changes in carrier concentration and structure can effectuate significant modifications of the band structure and consequently large changes in the physical properties (Wang *et al.*, 2012*a*; Nilius and Freund, 2015) (see Chapter 4, Section 4.5.2). In the 2-D case, the dopant atoms can be adsorbed at the surface or located at the metal-oxide interface (Jerratsch *et al.*, 2009); incorporation into the 2-D layer is also possible in special cases of 2-D architectures (Ulrich *et al.*,

2009; Pacchioni, 2016). The introduction of defects, e.g. O vacancies, may be regarded as a form of electronic doping, in which defect levels are introduced in the oxide band gap. For example, an O vacancy in 3-D MgO bulk (F centre) can be associated with two electrons (F^0 state), or with one (F^+ state) or zero electrons (F^{2+} state). F centres result in impurity levels in the band gap of MgO. On a metal-supported 2-D MgO layer, spontaneous charge transfer between the support and the F centres can occur, depending on their energetic position with respect to the metal Fermi level (Giordano and Pacchioni, 2011). The stability of defect charge states and electronic doping levels thus depends on the work function of the combined oxide-metal system. In transition metal oxides, oxygen vacancies lead to reduced cation states and, depending on their concentration and localization, to isolated states in the band gap or a band gap reduction, with concomitant repercussion on the work function (Greiner *et al.*, 2012).

5.3.2 Applications

The emergent electronic properties of 2-D oxides provide a rich and complex field of scientific activity. Their exploitation in applications of sub-nanometre type device technology is a very exciting and nascent area of applied research, where proof-of-concept results and prototype systems are emerging. Insulating and semiconducting 2-D oxide materials are being considered as dielectric alternatives in next-generation nanoelectronics: as insulating gate dielectric in field effect transistors (FET), as electrochemical (super-)capacitor materials, as ultra-thin dielectric layers in tunnel junctions, or as resistive memory (memristor) components.

The traditional gate dielectric in FETs with metal-oxide-semiconductor (MOS) structure, SiO_2, is reaching its physical limit with a thickness of ~ 1.5 nm, as the critical dimension of FET channel length approaches ~ 60 nm. At a thickness less, or of the order of 10 atoms, the leakage current due to quantum mechanical tunnelling through the SiO_2 layer increases beyond acceptable limits and dielectric breakdown may occur. The desired further downscaling of transistor size thus requires replacing SiO_2 (permittivity $\epsilon_r = 3.9$) with higher κ dielectrics (ϵ_r typically larger than 20–30), so that the same gate capacitance can be obtained with thicker dielectric gate layers. Metal oxides with centrosymmetric (i.e. non-polar) atomic arrangements such as Al_2O_3, TiO_2, HfO_2, Ta_2O_5 and lanthanide oxides ($\epsilon_r < 100$) have been proposed as possible high κ gate insulators compatible with Si MOSFET technologies. Ferroelectric oxides with residual dipoles in the unit-cell, typically perovskite-based compounds, have also been considered as very high κ materials ($\epsilon_r > 1000$). However, the materials have to be thermodynamically stable when brought in contact with the reactive Si at elevated temperatures—which is required for the device processing—and this can be a problem. High κ 2-D oxide nanosheets have been proposed as gate insulators in single-nanosheet devices. Osada *et al.* (2006*b*) have used a solution-based approach to fabricate $Ti_{0.87}O_2$ nanosheets and a corresponding nanofilm on a Si substrate (see Fig. 5.12). The $Ti_{0.87}O_2$ layer is high κ with $\epsilon_r > 100$, features a large band gap (3.8 eV), and good thermal stability. Sekizaki *et al.* (2017) have subsequently employed this 2-D titania layer in a proof-of-concept device and have demonstrated FET operation of a molecularly thin anatase phase, produced through the solid-state transformation from

(a)

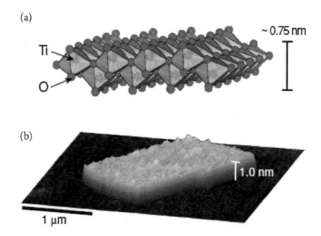

(b)

Fig. 5.12 (a) Structure model and (b) AFM image of a 2-D $Ti_{1-\delta}O_2$ layer supported on a SiO_2 substrate. TiO_6 octahedra are connected via edge-sharing to produce the 2-D lattice with a thickness of ~ 0.75 nm. Reprinted from Osada and Sasaki (2012) with permission. Copyright © 2012 Wiley-VCH Verlag GmbH and Co. KGaA, Weinheim.

the $Ti_{0.874}O_2$ layer. The anatase FET exhibits interesting transport characteristics with a high carrier mobility and high on/off ratio (Sekizaki *et al.*, 2017).

Capacitor components are central elements in electronic devices and their downscaling is an important step in the advancement of nanoelectronics. The use and integration of ultra-thin high κ dielectrics is a way to reach high capacitance density at reduced size and increased charge/discharge speed. High-energy-density capacitors also play a crucial role in a variety of energy storage devices. In real capacitors, the overall dielectric performance depends on the permittivity (ϵ_r) and leakage currents, and it is useful to define a figure of merit (FOM) to compare different materials. FOM may be defined as FOM $= \epsilon_0 \epsilon_r E_{br}$, with ϵ_0 the absolute permittivity of the vacuum and E_{br} the breakdown field (i.e. the breakdown voltage divided by the dielectric thickness d: $E_{br} = V_{br}/d$). FOM corresponds to a maximum charge that can be stored in the capacitor. Figure 5.13 shows FOM values for various oxide nanosheets and high κ thin films. The dots in Fig. 5.13 refer to 2-D oxide nanosheets, which clearly afford high capacitances as a result of high κ values and molecularly thin thickness (Osada and Sasaki, 2012). The high dielectric constant of 2-D materials is also advantageous for screening disorder and defects, which results in an increased electronic mobility (Balendhran *et al.*, 2013a) and useful applications in high mobility electronics (Kalantar-zadeh *et al.*, 2016).

Resistive switching memories (memristor), consisting of an active oxide layer sandwiched between two metal electrodes, exhibit a modified conductivity as a result of a voltage pulse across the electrodes, displaying a strong hysteresis of the I–V characteristics. The high/low resistance states of the structure constitute the on-off two-level characteristics of the system. Several processes have been invoked to explain the resistive switching phenomena, including migration of anion or cation components of the oxide layer or from the electrodes, redox processes, vacancy propagation and formation

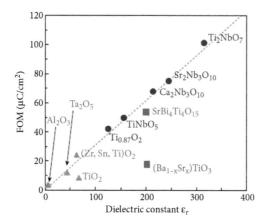

Fig. 5.13 Figure of merit (FOM) values of oxide nanosheets (dots) and high κ thin films. The dotted line is a guide to the eye. Reprinted from Osada and Sasaki (2012) with permission. Copyright © 2012 Wiley-VCH Verlag GmbH and Co. KGaA, Weinheim.

of conductive filaments to provide the low resistance state. The latter should dissolve again upon bias reversal to yield the high resistance state of the device. 2-D transition metal oxides show great promise for memristor applications (Kalantar-zadeh *et al.*, 2016). Indeed, a quasi-2-D $WO_3.H_2O$ nanosheet as active switching layer has shown remarkable performance for a resistive random access memory (RAM) in terms of low operating voltage, large resistance on/off ratio and long retention time (Liang *et al.*, 2015).

The modification of the electronic structure upon adsorption of molecules forms the basis for the application of 2-D oxides in chemical sensors, specifically gas sensors. In a way, the adsorption results in a kind of doping and the corresponding electronic density changes are reflected in a modified electronic or optical response, which can be detected as a change of resistivity or optical absorption. The benefits of using 2-D materials in sensors is their enhanced sensitivity due to the enhanced surface-to-volume ratio (Pacchioni, 2012). Some oxides change colour upon doping, e.g. upon exposure to reducing gases (MoO_3, WO_3) giving rise to a gasochromic response (see Chapter 8). Plasmonic gas sensing has been proposed for substoichiometric 2-D MoO_3 flakes, which show a tunable 2-D plasmon resonance in the visible light absorption region (Alsaif *et al.*, 2014*b*).

At this point, we would like to digress briefly to mention the LEGO concept, which is an intriguing idea for the design of new materials and new generations of device systems (see Geim and Grigorieva (2013) for a perspective article). Its principle is borrowed from the LEGO construction kit of small children, where LEGO building blocks are stacked one on top of another. Stacking different 2-D crystals on top of each other in a superlattice heterostructure, held together by van der Waals forces, each block defined with one-atomic-plane precision, can produce artificial materials with new properties, advanced functionalities, and ultimate interface control. Figure 5.14 illustrates this approach for 2-D dielectric nanosheets in diverse device architectures.

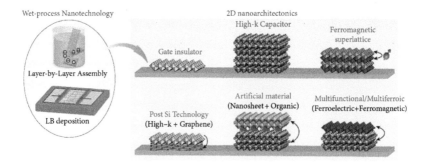

Fig. 5.14 Illustration of the LEGO concept for the design of materials via nanoarchitectonics, i.e. the stacking of 2-D nanosheets for creating new dielectric nanomaterials. Reprinted from Osada and Sasaki (2012) with permission. Copyright © 2012 Wiley-VCH Verlag GmbH and Co. KGaA, Weinheim.

5.4 2-D nanoribbons and nanoislands

Compared to 2-D layers, the specificity of 2-D nano-objects—nanoribbons (NR) and nanoislands—comes from their limited width, which enhances the importance of under-coordinated edge states, and introduces new properties and functionalities, as demonstrated for graphene nanoribbons (Yagmurcukardes *et al.*, 2016; Dutta and Pati, 2010; Acik and Chabal, 2011).

The discovery of edge states and 1-D conductivity in graphene has aroused excitement for 2-D nano-objects in the oxide community. However, while oxide nanoislands readily form during the first stages of growth on a substrate, the reproducible synthesis of well-characterized oxide nanoribbons remains a difficult task. To our knowledge, there exists a single report of ZnO zigzag nanoribbons synthesized via thermal evaporation of zinc powder followed by an oxidation under a H_2O_2 atmosphere (Wang *et al.*, 2010*a*). Nickel oxide nanoribbons have also been obtained by reactive deposition MOCVD method inside carbon nanotubes, but their dimensions—about 800 nm long and 4 nm thick—make them more 3-D than 2-D objects (Matsui *et al.*, 2001). Considering the difficulty in producing 2-D oxide nanoribbons, it is not surprising that most information we have come from atomistic simulations.

5.4.1 Nature of edges in M_xO_y oxide nanoribbons and nanoislands

Because of the amalgam which has been made in the theoretical literature between the nature of graphene edge states and that of 2-D oxides, it is useful to first make a thorough analysis of edge electrostatic characteristics. Indeed, at variance with the edge states in graphene NR, which are built from non-bonding π orbitals on low-coordinated neutral carbon atoms, oxide NR may have polar or non-polar edges, depending on the oxide stoichiometry, their in-plane orientation, and the orientation of their edges.

In rock-salt MO oxides (FeO, NiO, CoO, MgO, etc.), the square structure of (001)-oriented MLs provides two dense edge directions, as represented in Fig. 5.15. The atomic rows parallel to [100] are neutral with as many oxygen ions as cations. [100] edges are thus non-polar. By contrast, rows parallel to [110] contain a single type of

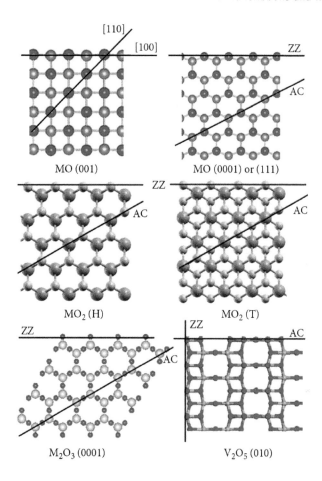

Fig. 5.15 Top panels: ball representations of MO NRs cut from a (001) or a (0001)/(111) ML, with non-polar [100] or armchair (AC) edges and polar [110] or zigzag (ZZ) edges. Middle panels: ball representation of MO_2 trilayers of H (left) and T (right) structure, with indication of the AC and ZZ edge directions. Large gray and small yellow balls represent M cations and oxygen atoms, respectively. The latter are on top of each other in the H structure. Bottom panels: ball representation of (left) a M_2O_3(0001) ML and (right) a slightly tilted V_2O_5 (010) monolayer, with indication of the AC and ZZ edge directions. The unsupported M_2O_3 ML is flat (M cations in blue, oxygen atoms in red), while the V_2O_5 one is corrugated (oxygen atoms in red, vanadium atoms in gray), with some oxygen atoms on top or below vanadium cations.

charged atoms, alternatively cations and oxygen ions. [110] oriented NRs thus display a non-zero in-plane polarization perpendicular to the edges, responsible for their polar character, which is the 2-D counterpart of 3-D polarity at surfaces. Similarly, the hexagonal structure of rock-salt (111) MLs or of wurtzite (0001) MLs allows the formation of armchair (AC) or zigzag (ZZ) edges, as in graphene (Fig. 5.15). While AC edges are neutral and non-polar, ZZ ones are charged and polar. The polar character of [110] edges has been early recognized (Ferrari *et al.*, 2005*a*), but that of ZZ edges has long been overlooked in the theoretical literature (Wang *et al.*, 2012*c*; Goniakowski *et al.*, 2013; Güller *et al.*, 2013; Noguera and Goniakowski, 2012).

Interestingly, in two monolayer-thick MO(0001) or (111) NRs, oxygen atoms and cations belonging to different layers are on top of each other. The polarization in one layer is thus opposite to that in the other one, leading to a cancellation of polarity at ZZ edges. This reasoning may be extended to several layers, evidencing an odd-even alternation of the ZZ polar character as a function of thickness.

As mentioned in Section 5.2, unsupported MO_2 trilayers may adopt an H or a T structure in which the two oxygen atoms per formula unit are on top of each other or shifted with respect to each other, respectively. Both layer types may possess armchair and zigzag edges, but their electrostatic characteristics are different (Fig. 5.15). In H-type MLs, ribbons with AC edges are non-polar due to the $MO_2/MO_2/...$ row stacking, while those with ZZ edges are polar ($O_2/M/O_2/M/...$ row stacking). In the T structure, the $MO_2/MO_2/...$ row stacking for AC edges and $O/M/O/...$ row stacking for ZZ edges is such that none of them is polar.

Unsupported flat $M_2O_3(0001)$ MLs adopt a honeycomb structure, which displays twelve member rings containing an equal number of oxygen and metal atoms. NRs with AC and ZZ edges may be cut out of them, as shown in Fig. 5.15. The row stacking is $M_2O/O_2/...$ in the former case and $O/M/O_2/M/...$ in the latter, showing that both AC and ZZ edges are polar.

$V_2O_5(010)$ MLs are built from distorted VO_5 square pyramids. NRs with AC or ZZ edges may be formed (Tang *et al.*, 2011), with non-polar $V_2O_5/V_2O_5/...$ and polar $VO_2/VO_2/O/...$ row stackings, respectively, as shown in Fig. 5.15.

The polar or non-polar character of the various edge types are summarized in Table 5.2. Since it is determined by the value of the in-plane polarization, it is not modified by the corrugation (vertical atomic displacements) which takes place when the layers are supported.

Table 5.2 Non-polar or polar character of ribbon edges in the various M_xO_y MLs. AC and ZZ refer to armchair and zigzag edges, respectively.

M_xO_y layer orientation	non-polar edges	polar edges
MO (001)	[100]	[110]
MO (111) or (0001)	AC	ZZ
MO_2 (H)	AC	ZZ
MO_2 (T)	AC, ZZ	-
M_2O_3 (0001)	-	AC, ZZ
V_2O_5 (010)	AC	ZZ

5.4.2 Intrinsic characteristics of non-polar edges

As a general statement, nano-objects with non-polar edges are expected to be the most stable and indeed have been frequently observed. This is the case for the square islands with [100] edges which form in the first stages of growth of MgO on Ag(001) (Schintke *et al.*, 2001; Ferrari *et al.*, 2005*b*), on Mo(001) (Gallagher *et al.*, 2003) and

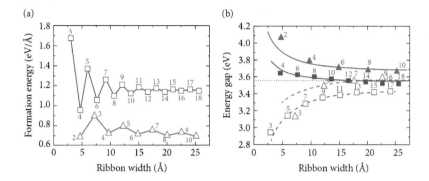

Fig. 5.16 Calculated formation energy (a) and gap width (b) of 1 trilayer TiO_2 nanoribbons (T structure) with AC (square symbols) or ZZ (triangle symbols) edges, as a function of ribbon width. Note the decreasing (resp. increasing) behaviour of the gap width in ribbons with an even (resp. odd) number of rows. Reprinted from He *et al.* (2010) with permission. Copyright 2010 American Chemical Society.

on Au(111) (Pan *et al.*, 2012), and during the growth of CoO on Ag(001) (Sebastian *et al.*, 1999).

MgO, ZnO, TiO_2 and V_2O_5 unsupported nanoribbons with non-polar edges have been the subjects of first principles simulations, which showed that they are semiconducting and non-magnetic, similarly to the infinite layers of these oxides. Compared to the latter, their specific characteristics nevertheless rely on the existence and density of under-coordinated atoms at the borders, the electrostatic interaction between opposite edges and the quantum confinement effects associated to their finite width.

The simulation of free-standing NRs cut out of a TiO_2 T layer, with both AC and ZZ non-polar edges, exemplifies these concepts (He *et al.*, 2010). As shown in Fig. 5.16, ribbons with ZZ edges are more stable than those with AC edges, consistent with their lower density of undercoordinated atoms. All ribbon formation energies display an odd-even alternation as a function of their width, with lower energies for even widths. This non-monotonic variation was assigned to electrostatic interactions between atoms on opposite edges. The electronic gaps also vary in a non-monotonic manner with the ribbon width. For even numbers of rows, they decrease toward that of the infinite trilayer, a property characteristic of quantum confinement effects. A similar variation has been predicted for free-standing V_2O_5 nanoribbons with AC edges and also assigned to quantum confinement effects (Tang *et al.*, 2011).

Applying an electric field, or passivating edges are various means to manipulate the ribbon electronic properties. This has been, for example, demonstrated for free-standing ZnO AC ribbons (Kou *et al.*, 2009). The electronic polarization resulting from the application of an electric field perpendicular to the edges reduces the gap width and, beyond a critical value which decreases as the ribbon width increases, it induces a transition from a semiconducting to a metallic state. Passivation of all edge atoms by hydrogen leads to an increase of the gap width, but the adsorption of H on edge oxygen atoms only is equivalent to n-doping and drives the ribbons to a metallic state.

5.4.3   Intrinsic characteristics of polar edges

As polar thin films, polar nanoribbons can be represented by a capacitor model, provided that the capacitor 'plates' are 1-D atomic rows instead of 2-D atomic layers and that their charge density $\pm\lambda$ per unit length replaces that per surface area. The specificity of the 1-D dimension is that, in the absence of compensating charges, in the large N width limit, the potential difference across a polar ribbon diverges as $\ln N$. The total dipole moment, which increases linearly with N, and the electrostatic potential difference thus follow different N laws.

Polarity may be healed by modification of the outer row charge density. The value of the compensating row charge density $\Delta\lambda$, in the asymptotic limit, obeys the same type of relationship as at polar surfaces (Güller *et al.*, 2013; Noguera and Goniakowski, 2012) (x_i the positions of the ith row perpendicular to the edge, a the length of the repeat unit):

$$\Delta\lambda = \frac{\sum_i^m x_i\lambda_i}{a} \tag{5.1}$$

This expression reduces to $\Delta\lambda = R_1\lambda/(R_1 + R_2)$ for MO ribbons with successive charge densities $\pm\lambda$ and inter-row distances R_1 and R_2 ($R_2 = R_1$ for rock-salt [110] edges and $R_2 = 2R_1$ for honeycomb ZZ edges).

In the absence of other mechanism of compensation, an overlap of the outermost row VB and CB takes place, and the ribbon borders become metallic. A 1-D electron/hole gas is formed with large effective mass at the oxygen edges and lighter one at the cation edges. This metallic character evidenced in simulations of free-standing ribbons with ZZ polar edges has long been interpreted as resulting from dangling bond states analogous to those of graphene. It has even been invoked for the realization of intrinsic quantum anomalous Hall effect in Nb_2O_3 ZZ NRs, which are characterized by the nonzero Chern number and chiral edge states (Zhang *et al.*, 2017). However, in polar ribbons, the metallic character has clearly nothing to do with the low coordination of border atoms, although it may be reinforced by it in small gap systems. Its very origin is electrostatic. Accompanying the band overlap, the oxygen states at Fermi level are strongly polarized, and, due their high density of states, a magnetic instability occurs towards a (generally) ferromagnetic state. This scenario has been proposed in MgO(111) (Güller *et al.*, 2013), BeO(0001) (Wu *et al.*, 2011b), ZnO(0001) (Botello-Méndez *et al.*, 2008; Topsakal *et al.*, 2009), and V_2O_5 (Tang *et al.*, 2011) ZZ NRs.

The electronic and magnetic structures of the edges may be modified in several ways. In close similarity to polar surfaces, the most obvious one is edge reconstruction in which an adequate number of edge ions are removed (e.g. one half at MO [110] edges or one third at MO ZZ edges, or more complex distribution). The reconstructions substantially lower the energies of polar ribbons, especially those which require the largest charge compensation (for example, MO [110] compared to MO ZZ ribbons in which $\Delta\lambda = \lambda/2$ or $\Delta\lambda = \lambda/3$, respectively). Reconstructed edges then recover a semiconducting, non-magnetic ground state similar to non-polar objects (Güller *et al.*, 2013; Goniakowski *et al.*, 2013).

As another way to modify polar edges, a tensile strain exerted along the edge direction has been suggested to induce a metal to semiconducting transition in unsupported polar ZnO ribbons, associated to a hexagonal to square structural transformation (Si and Pan, 2010). Furthermore, the application of an electric field was theoretically shown to tune the ribbon magnetism by either enhancing or reducing the magnetic moments, depending on its direction relative to the intrinsic ribbon polarization (Kou *et al.*, 2010). Edge passivation is also a means to manipulate the electronic and magnetic structure of polar ribbons (Chen *et al.*, 2009).

5.4.4   Relative stability of polar and non-polar ribbon edges

While, as a general trend, in free-standing NRs, unreconstructed polar edges are less stable than non-polar edges, their relative stability may substantially change when the ribbons or islands are grown on a metal substrate. For example, a thorough theoretical study of TiO nanostripes on a large number of transition metals has concluded that, in a majority of cases, ZZ ribbons are more stable than AC ones (Sandberg *et al.*, 2018). However, several factors come into play.

Obviously, one concerns the thermodynamic/kinetic conditions under which the objects are formed. The STM observation of NiO islands on Ag(001) with polar [110] borders, embedded into the Ag substrate and confined by Ag [110] steps is a striking example (Caffio *et al.*, 2006; Steurer *et al.*, 2011; Steurer *et al.*, 2012). While thermodynamics favours non-polar edges (Ferrari *et al.*, 2005b), the stabilization of the NiO polar borders in that case may likely be the result of kinetic hindrance effects associated to the particularly high stability of [110] Ag steps. Under different growth conditions, MgO islands with [100] non-polar edges on the same substrate have been obtained and shown to display a structural transition from non-polar to polar edges under an excess of Mg (Xu *et al.*, 2019).

The intrinsic characteristics of polar edges, i.e. the coordination of their atoms and the importance of the compensating charge, are other decisive factors. While the role of a metal, whether Ag(100) (Ferrari *et al.*, 2005a) or Au(111) (Pan *et al.*, 2012), appears insufficient to stabilize stoichiometric MgO [110] borders with respect to [100] ones, ZZ ribbons have been predicted to become more stable than AC ribbons on Au(111). This difference was assigned to the initially smaller energy difference between ZZ and AC borders in unsupported ribbons, as compared to [110] versus [100] (Goniakowski *et al.*, 2013).

Finally, the interaction strength between the oxide and the metallic substrate is a crucial factor. Qualitatively, larger metal work functions are expected to induce stronger interactions, and thus be more efficient to stabilize polar edges. The observed shapes of CoO islands grown on Au(111) and Pt(111) obey this trend. AC edges are dominant on Au(111) while on Pt(111) the islands have a triangular shape with ZZ edges (Fig. 5.17). The stabilizing role of metallic substrates results from the participation of the metal to the compensating charges. Indeed, the analysis of the Bader charge distribution below a polar MgO ribbon has evidenced that an electron excess takes place in the substrate below the cation edge, and an electron depletion under the oxygen edge, allowing much weaker charge modifications in the oxide than in the free-standing case, as shown in Fig. 5.18. The energy gain comes from the fact that locally increasing or decreasing the number of electrons in a metal requires only a small shift of the Fermi level and thus much less energy than modifying the occupation of the

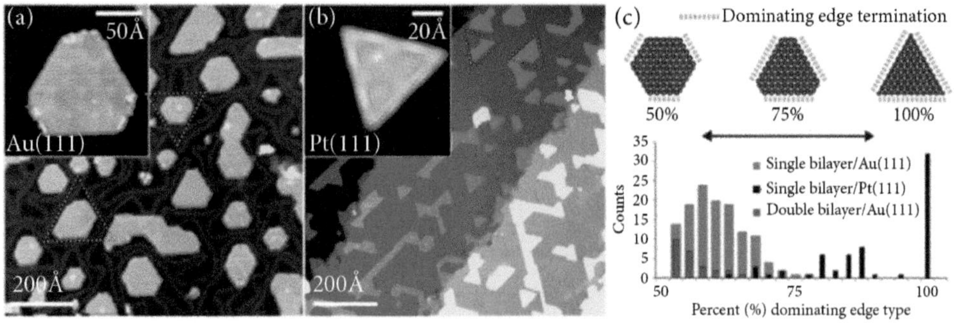
(a)
50Å
Au(111)
200Å
(b)
20Å
Pt(111)
200Å
(c)
Dominating edge termination
50%
75%
100%
Single bilayer/Au(111)
Single bilayer/Pt(111)
Double bilayer/Au(111)
Counts
Percent (%) dominating edge type

Fig. 5.17 CoO single bilayer islands on Au(111) and Pt(111): Overview STM images (a, b) and island shape distribution (c) from 216 islands analysed from STM images in terms of percent dominating edge type, where 50 and 100% corresponds to a regular hexagon and a perfect triangle respectively. Blue dotted triangles indicate two possibilities of island orientations with respect to the substrate. Reprinted from Fester *et al.* (2017) with permission from Springer. Copyright 2017

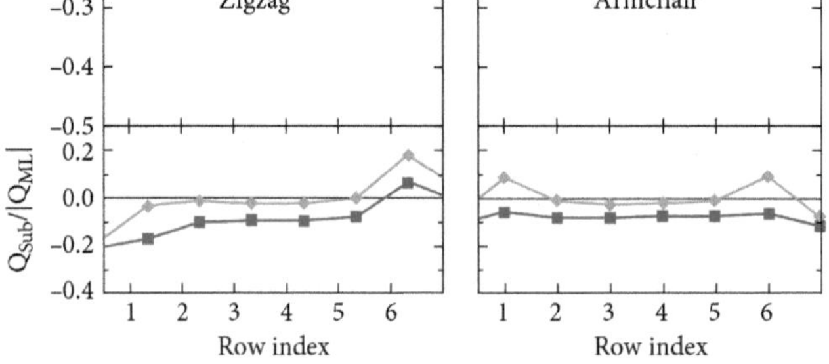
MO MO MO MO MO MO
MO MO MO MO MO MO
Zigzag
Armchair
δ|Q|/|Q_ML|
Q_Sub/|Q_ML|
Row index
Row index

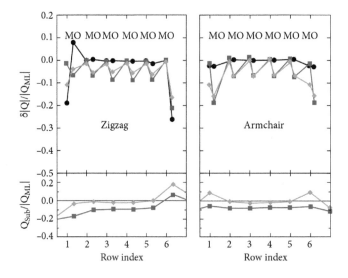

Fig. 5.18 Bader charge distribution in unsupported (black) and Au(111) (red) or Mo(110) (green) supported ZZ and AC MgO(111) nanoribbons: top parts: variations of Bader charges in the ribbons compared to the full ML; lower parts: charge distribution in the metal atoms located under the ribbons. Note the reduction of the compensating charges in the supported polar ribbons, especially at the cation edge. Reprinted from Goniakowski *et al.* (2013) with permission. Copyright 2013 by the American Physical Society.

VBs or CBs in a semiconductor (Goniakowski *et al.*, 2013). The same argument had been put forward for the stabilization of polar films on metal substrates (Goniakowski and Noguera, 1999). However, the polarity screening by the metal occurs at a single film termination, the one in contact with the support, while both ribbon edges are in contact with the support, which makes the stabilizing effect comparatively more efficient.

As at polar surfaces, hydroxylation is an efficient way to compensate NR polarity. It seems to have been decisive in the synthesis of ZnO ZZ ribbons (Wang *et al.*, 2010*a*). On Au(111), the orientation of MgO islands was shown to sensitively depend on the oxygen and water partial pressures in the preparation chamber. At high oxygen partial pressure, square MgO(100) islands displayed non-polar [100] borders, while at lower oxygen partial pressure and in the presence of small amounts of water, MgO(111) islands with ZZ borders and triangular shapes prevailed (Pan *et al.*, 2012). Nevertheless, it should be kept in mind that the extremely low coordination of border sites, by itself, could induce water or hydrogen dissociation, regardless of the edge orientation (polar or non-polar) (Ferrari *et al.*, 2007). Moreover, since the metallic support already provides a large stabilizing effect, more efficient than for thin films, the added value of hydroxylation in 2-D polarity healing remains an open question (Goniakowski *et al.*, 2013).

5.5   Ferroic behaviour

Ferroic systems are characterized by a hysteretic dependence of an order parameter upon an external stimulus below a critical temperature. In response to magnetic, electric, or stress fields, phases with ferromagnetic, ferroelectric, or ferroelastic behaviour develop. Materials exhibiting simultaneously more than one ferroic order are called multiferroic; the most common ones couple ferroelectricity with antiferromagnetism or weak ferromagnetism. In the field of oxide materials, the class with perovskite structures is prevalent in exhibiting ferroic behaviour.

According to the Mermin–Wagner theorem (Mermin and Wagner, 1966), spin ordering in 2-D systems is predicted to collapse at non-zero temperatures as a result of thermal excitations, unless the presence of a magnetic anisotropy opens up a magnon excitation gap, which resists the thermal agitations and results in finite Curie temperatures (Gong and Zhang, 2019). For real-world 2-D materials, magnetic anisotropy is always present due, among other causes, to the ubiquitous presence of interfaces, and ordered magnetic phases should be stable (Shabbir *et al.*, 2018). This has indeed recently been experimentally confirmed for ferromagnetic single layer CrI_3 (Huang *et al.*, 2017) and $CrGeTe_3$ (Gong *et al.*, 2017). In the field of 2-D oxides, spin-ordered phases have been predicted theoretically, but experimental confirmation is mostly missing up to date.

The magnetic properties of various single-layer compounds have been investigated theoretically (Van Gog *et al.*, 2019). In the first transition metal series of M_2O_3 stoichiometry and honeycomb structure, all compounds are predicted to be antiferromagnetically ordered, except V_2O_3 and Mn_2O_3 which are ferromagnetic (Goniakowski and Noguera, 2018). These magnetic orders are preserved when the layers are deposited on an Au(111) support apart from Cr_2O_3 which becomes ferromagnetic (Goniakowski and Noguera, 2019). In single-layer MO_2 compounds, a ferromagnetic metallic behaviour for VO_2, MnO_2, and FeO_2 is predicted in the H structure and semiconducting ferromagnetic properties for MnO_2 and NiO_2 in the T structure (Fig. 5.2) (Ataca et al., 2012). ScO_2 was found ferromagnetic semiconducting in the H structure and half-metallic in the T structure. The ferromagnetic nature of MnO_2 in the T structure has also been confirmed in the calculations of Kan et al. (2013). Incidentally, the preparation of single-layer nanosheets of MnO_2 has been achieved by Omomo et al. (Omomo et al., 2003) following a wet chemical exfoliation procedure of a layered protonic manganese oxide. A hexagonal wurtzite phase of MnO has been prepared by thermal decomposition of Mn-acetylacetonate on a carbon template by Nam et al. (2012); the wurtzite phase has a higher total energy compared to the normal cubic rock-salt MnO phase, by 0.1 eV, but the carbon template apparently induces the formation of the hexagonal wurtzite MnO via a chemical surface interaction. Experimental results have revealed a paramagnetic behaviour of the wurtzite MnO, although the most stable structure was calculated to have AFM ordering (Nam et al., 2012). However, in the 2-D case, wurtzite MnO has been predicted by DFT to transform spontaneously into a planar BN-like structure with ordered spin arrangement (Kan et al., 2013). According to these calculations, the AFM ordering of single layer hexagonal MnO can be switched into half-metallic ferromagnetism by small hole doping, and the Curie temperature for hole-doped MnO (0.25 holes per Mn ion) is estimated to be larger than 300 K. Here, an interesting parallel can be drawn with the Mn_3O_4 monolayer phase supported on a Pd(100) surface (Franchini et al., 2009a), which has already been presented in Section 5.2 (Fig. 5.5). The Mn_3O_4 layer has a c(4 × 2) structure (with respect to Pd(100)), which consists of a MnO(100) plane with 1/4 of the Mn ions removed: the Mn vacancies form the ordered rhombic distribution that specifies the c(4 × 2) unit-cell. Top views of the geometric and magnetic models for the Pd supported c(4 × 2) Mn_3O_4 structure are displayed in Fig. 5.19. Focussing on the magnetic structures, besides trivial ferromagnetic ordering (FM), two antiferromagnetic orientations can be set up in the c(4 × 2) unit-cell, AFM1 and AFM2, which differ in the spin direction alignment along the [100] direction: AFM1 has a FM ordering (Fig. 5.19: left panel), whereas AFM2 has an AFM arrangement (middle panel). The DFT HSE hybrid functional calculations have established that the FM order in the RH1 adsorption geometry forms the lowest energy structure (Franchini et al., 2009a). This FM ground state of the Mn_3O_4 monolayer on Pd(100) may be related to the hole-doped FM hexagonal single layer of Kan et al. (2013), since the introduction of Mn vacancies into the cubic MnO(100) monolayer to form the c(4 × 2) structure may be regarded as a hole- doping process (0.25 holes per Mn ion). In the case of the epitaxial Mn-oxide growth on the Pd(100) surface, this hole doping is prompted by the release of the lattice mismatch strain (Franchini et al., 2009a). A similar geometry and spin structure (RH1-FM) has also been proposed for the isostructural c(4 × 2) Co_3O_4 monolayer on Pd(100) on theoretical grounds (Allegretti et al., 2010).

The magnetic properties of the Mn_3O_4 c(4 × 2) structure have been investigated on a stepped Pd(1,1,21) surface, a vicinal of Pd(100), by Mn $L_{2,3}$ X-ray magnetic circular dichroism (XMCD) (Altieri et al., 2013). The stepped Pd(100)-type (1,1,21) surface has been used as a substrate in these experiments, because the c(4 × 2) Mn_3O_4 phase

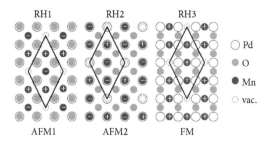

Fig. 5.19 Geometrical and magnetic models of the rhombic (RH) c(4 × 2) Mn_3O_4 phase on Pd(100). Models RH1–RH3 are distinguished by different adsorption site registries of the Mn_3O_4 layer on the Pd surface atoms underneath. Plus and minus signs indicate the orientation of the Mn spins perpendicular to the surface, pointing inward or outward, respectively. FM designates ferromagnetic ordering, AFM1 and AFM2 two distinct antiferromagnetic arrangements. The [100] surface direction is along the short side of the panels. Reprinted from Franchini *et al.* (2009a) with permission. Copyright 2009 by the American Physical Society.

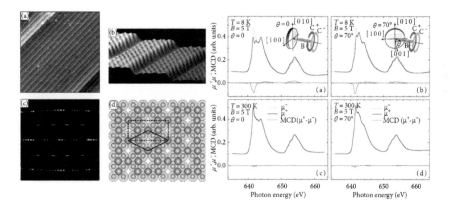

Fig. 5.20 Left: (a) Large-scale (50∗50 nm^2) and (b) high-resolution (7∗8 nm^2) STM images of c(4 × 2) Mn_3O_4 nanostripes on a stepped Pd(1,1,21) surface. The protrusions in the 3-D representation of the STM image (b) correspond to the Mn atoms at the corners and at the center of the c(4 × 2) unit-cell (see model in (d)). (c) Corresponding LEED pattern. (d) Schematic model of a c(4×2) nanostripe on a Pd(100) substrate; primitive (solid) and centred (dashed) unit-cells are indicated (Pd: large gray; Mn: medium blue; O: small red). Right: Mn $L_{2,3}$ XMCD spectra and relative XMCD signal (μ^+ - μ^-) recorded with right (μ^+) and left (μ^-) polarized photons in a field B = 5T at $\theta = 0°$ for temperatures (a) $T = 8$ K and (c) $T = 300$ K, and at $\theta = 70°$ for temperatures (b) $T = 8$ K and (d) $T = 300$ K. Insets: Experimental XMCD geometry. Reprinted from Altieri *et al.* (2013) with permission. Copyright 2013 by the American Physical Society.

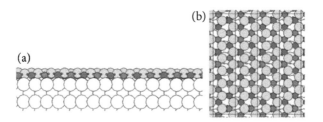

Fig. 5.21 (a) Side view of the (9×2) cobalt oxide unit-cell, showing a Pd-Co-O stacking. (b) Top view of four unit-cells of the (9×2) phase. (Pd: white; O: yellow; Co: blue and red, reflecting the AFM3 ordering of the CoO monolayer). The $[0\bar{1}1]$ direction is along the short side of the unit-cell. Reprinted from Gragnaniello *et al.* (2011) with the permission of AIP Publishing.

can be grown in 28 Å wide single domain nanostripes, which extend virtually defect-free over lengths of approximately 100 nm, as shown by the STM images in Fig. 5.20(a,b) and by the sharp LEED pattern in Fig. 5.20(c). The clear magnetic anisotropy in the temperature and field dependent Mn $L_{2,3}$ MCD spectra (see Fig. 5.20, right panel) has been related by MCD sum rule analysis to the presence of a largely unquenched orbital moment in the paramagnetic manganese oxide nanostripes. The observed magnetic anisotropy has been ascribed to a partial depletion of the 3d shell induced by the hole doping via Mn vacancies in the Mn_3O_4 nanostripes. However, in the absence of hysteresis in the magnetization, the predicted FM of the c(4x2) Mn_3O_4/Pd(100) layer (Franchini *et al.*, 2009*a*) could not be experimentally confirmed. Possibly, the Curie temperature of the c(4×2) Mn oxide layer is below the lowest temperature (8 K), which could be attained in the experiments of Altieri *et al.* (2013), but this question remains unanswered at the present time.

For the pseudomorphic NiO(100) monolayer on Ag(100), the AFM2 spin order (see Fig. 5.19, middle panel) has been calculated as the lowest energy configuration using a DFT+U approach (Steurer *et al.*, 2011). The 2-D Co oxide (9×2) phase on Pd(100) has been interpreted in terms of a coincidence lattice, consisting of a CoO(111)-type bilayer with significant symmetry relaxation and height modulations to reduce the polarity in the overlayer (Gragnaniello *et al.*, 2011). Magnetically, the most stable structure displays an unusual ZZ type AFM ordering (AFM3). Figure 5.21 shows side (a) and top (b) views of the (9×2) Co oxide structure model on Pd(100). In the AFM3 order, Co ions with parallel spin orientation are arranged in a ZZ fashion running parallel to the $[0\bar{1}1]$ direction, which is the short side of the unit-cell.

The 2-D ternary iron tungstate ($FeWO_3$) layer on Pt(111) is of interest, because it forms a mixed (2×2) honeycomb lattice (Pomp *et al.*, 2016). It has been prepared by solid-state chemical reaction of the well-defined FeO(111) monolayer on Pt(111) and cyclic $(WO_3)_3$ clusters from the gas phase. This 2-D ternary oxide consists of a mixed layer of Fe and W atoms, formally Fe^{2+} and W^{4+} species, sitting in the face-centered cubic (fcc) and hexagonal close-packed (hcp) hollow sites, respectively, of the Pt(111) surface; this Fe-W layer is terminated by oxygen atoms in Fe-W bridging sites, forming a buckled honeycomb lattice as depicted in Fig. 5.7. The honeycomb geometry

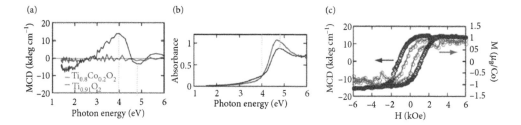

Fig. 5.22 Magneto-optical properties of titania nanosheets: comparison of doped $Ti_{0.8}Co_{0.2}O_2$ and $Ti_{0.91}O_2$. (a) Polar Kerr MCD spectra measured at 10 kOe (1 kOe $\sim$ 0.1 T) and 300 K for $(PDDA/Ti_{0.8}Co_{0.2}O_2)_{10}$ and $(PDDA/Ti_{0.91}O_2)_{10}$. (b) Optical absorption spectra of the films of (a). (c) MCD signal at 4.1 eV and magnetization as a function of applied field, measured at 300 K. Reprinted from Osada *et al.* (2006*b*) with permission. Copyright © 2006 Wiley-VCH Verlag GmbH and Co. KGaA, Weinheim.

is specific for many 2-D materials—graphene is the most prototypical example—and it is particular from the magnetic point of view, because it leads to spin frustration when the system contains AFM interactions. The antiparallel alignment of spins in AFM phases is impossible for a triangular lattice, because one of the spins in each triangle cannot align simultaneously with both of its neighbours. Similarly, in a honeycomb lattice some spins remain frustrated, and this may promote ferromagnetic ordering. Indeed, DFT calculations predict that the 2-D $FeWO_3$ on Pt(111) should exhibit a FM ground state, with an estimated Curie temperature of 95 K (Pomp *et al.*, 2016). It is noted parenthetically that a related $FeWO_4$ bulk phase displays AFM behavior (Garcia-Matres *et al.*, 2003). Other mixed oxide monolayers of $MM'O_3$ composition displaying a honeycomb structure have been investigated theoretically. For M and M' in the first transition metal series (M, M' = Ti, V, Cr, Fe), both cations bear a magnetic moment when the layers are free-standing and the magnetic order is predicted to be ferromagnetic, except in $CrFeO_3$ (Goniakowski and Noguera, 2018). In the latter case, because both cations bear a magnetic moment, the spin lattice is hexagonal (not triangular), there is no frustration, and an AFM order is achievable.

The doping of non-magnetic oxides with magnetic atoms is another proven approach to induce spin functionality in oxide materials (Matsumoto *et al.*, 2001; Fukumura *et al.*, 2004; Ogale, 2010). The dielectric titania $Ti_{0.87}O_2$ nanosheets, prepared by delamination of layered titanates by Sasaki and Watanabe (1998, Osada and Sasaki (2012)—see Fig. 5.12 for a structure model—can be doped by substitution and incorporation of other elements into the lattice of the layered starting compounds during the solid-state synthesis. By delaminating Co substituted $K_{0.8}Ti_{1.6}Co_{0.4}O_4$, $Ti_{0.8}Co_{0.2}O_2$ nanosheets have been obtained (Osada *et al.*, 2006*a*; Osada *et al.*, 2006*b*), which showed a ferromagnetic behaviour even at room temperature. In the Co substituted titania sheets, the Co^{2+} ions substitute for the Ti^{4+} ions. The magnetization of the $Ti_{0.8}Co_{0.2}O_2$ nanosheet is anisotropic due to its 2-D structure, and the magnetic moment per Co ion is greater than the spin moment of Co expected theoretically for low-spin Co^{2+} and found in Co-doped 3-D anatase. Similar ferromagnetic properties have been reported for Fe- and Mn-doped titania sheets (Osada and Sasaki, 2012). The

2-D titania nanosheets can be stacked into multilayers by soft chemical electrostatic self-assembly processes using PDDA (poly-diallyl-dimethyl-ammonium polymer) as counter polycation, which displayed giant magneto-optical effects. Figure 5.22 compares the magneto-optical properties of $Ti_{0.91}O_2$ and $Ti_{0.8}Co_{0.2}O_2$. Whereas the magnetic circular dichroism (MCD) spectrum of $Ti_{0.91}O_2$ is featureless, the $Ti_{0.8}Co_{0.2}O_2$ shows pronounced MCD structures around the absorption edge at 4.1 eV and 4.6 eV (Fig. 5.22(a)), which are related to features in the absorption spectrum (Fig. 5.22(b)). The magnetic field dependence of the MCD signal at 4.1 eV shows clearly ferromagnetic behaviour at room temperature. The doping of oxide nanosheets thus is an attractive method to create nanoscale 2-D magnetic layers with robust magneto-optical effects. Their easy manipulation and assembly capabilities bear potential for applications of design of spintronic and magneto-optical devices.

Multiferroic oxides exhibit simultaneously several ferroic orders, amongst them ferromagnetic, ferroelectric (see Chapter 4), and ferroelastic. The coupling between ferroic orders enables the manipulation of a given order parameter by another external stimulus, for example magnetoelectric coupling allows the manipulation of the electric polarization by a magnetic field or the magnetization by an electric field. Multiferroic oxides are rare, $BiFeO_3$ is a room temperature multiferroic and the rare earth manganites ($REMnO_3$) present rich phase diagrams with potential multiferroic phases (Bibes *et al.*, 2011; Lu *et al.*, 2015). A 2-D monolayer of a group IV monoxide, α-SnO, has recently been proposed on the basis of calculations to feature ferromagnetism and ferroelasticity; it has been argued that the material displays an unusual 'Mexican-hat' band dispersion, which seems to be at the root of the multiferroic behaviour (Seixas *et al.*, 2016). The coupling of two ferroic 2-D oxides in heterostacks following the principles of the LEGO concept (see Fig. 5.14), e.g. by combining ferromagnetic and ferroelectric sheets, is a promising route to create artificial multiferroic structures.

5.6   Catalysis aspects

The application of 2-D oxides (one unit-cell thick) in catalysis has introduced new paradigms in the understanding of catalytic reactivity. New functionalities develop due to the particular geometric and electronic structures of 2-D systems, and effects mediated by the enhanced mechanical flexibility of the 2-D lattices and the possibilities of charge transport through the oxide layer (in the presence of a substrate) contribute to support novel and highly active reactivity patterns. As a result of the prevalence and importance of oxide-metal combinations in practical real-world catalysts, 2-D oxide layers supported on metal surfaces have been mostly investigated in fundamental studies, and we focus on these systems here. The so-called oxide monolayer catalysts, designating high-surface-area polycrystalline oxide powders covered with a monolayer phase of another oxide (e.g. vanadia on Al_2O_3, SiO_2, TiO_2, etc.) (Roozeboom *et al.*, 1979; Bond and Tahir, 1991; Briand *et al.*, 2004; Wachs, 2005; Wachs, 2013), prepared by wet chemical impregnation techniques are beyond the scope of this section and will not be discussed. The catalysis applications of intrinsic thin films of 2-D oxide nanosheets are mainly in photocatalysis and electrocatalysis, where the high surface area and the concomitant high number of reaction sites while minimizing the catalyst loading are taken to advantage (Kalantar-zadeh *et al.*, 2016). The homogeneity of

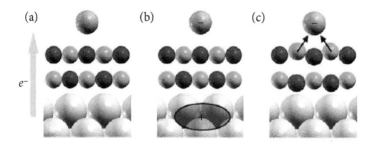

Fig. 5.23 Electron transfer phenomena through a 2-D oxide layer: (a) electron tunnelling; (b) polarization of the substrate (image charge); and (c) polaronic distortion of the oxide. Electron transfer in the opposite direction can also occur. Reprinted from Giordano and Pacchioni (2011) with permission. Copyright 2011 American Chemical Society.

geometric and electronic structure of oxide nanosheets and their uniform catalytic site distribution are other benefits of these materials. The photocatalytic water splitting reaction and the photoreduction of CO_2 into hydrocarbon fuels are among the popular reaction systems that have been addressed (Tan *et al.*, 2017; Chen *et al.*, 2012).

Chemical catalysis is a complex technology that involves a catalyst material of often multicomponent nature, e.g. metal and oxide, surface phases in complex morphologies, the chemical reactant-product mixtures, and the process parameters (p, T). In order to reach an atomistic understanding of catalyst processes and to enable the application of advanced experimental characterization techniques, well-defined reductionist model systems are necessary that can mimic the properties of the real catalysts. 2-D oxide layers supported on metal surfaces are such excellent model systems for catalysis (Freund and Pacchioni, 2008; Freund, 2011). Moreover, the previously mentioned new functionalities of 2-D oxides add extra features to the catalytic reaction behaviour and create novel catalysts of their own right. On a 2-D oxide layer, which is supported on a metallic substrate, the interaction between an adsorbate and the oxide-metal interface has to be considered, which can lead to spontaneous electron transfer through the oxide layer. The crucial parameters determining the direction and extent of charge transfer are the work function of the oxide-covered metal (see Section 5.3) and the electron affinity or the HOMO-LUMO position of the adsorbate—the latter may also be modified by the interaction with the oxide surface (Freund and Pacchioni, 2008; Giordano and Pacchioni, 2011). In response to the charge on the adsorbate, a polaronic distortion of the oxide 2-D lattice occurs in the form of a local structure relaxation (Goniakowski *et al.*, 2009). The polaron formation is enabled by the mechanical flexibility of the 2-D lattice and is beneficial to the total energy balance. Figure 5.23 is a sketch illustrating the electron transfer through and the polaron formation in a 2-D oxide layer. Since neutrals, cations, and anions have different adsorption properties and consequently a different chemical reactivity, specific oxide-support combinations can be designed with specific adsorbate charge states, thus allowing to activate adsorbed molecules or metal particles and to design catalysts for particular chemical reactions. For example, Au adatoms charge negatively on 2-D MgO on Mo(100), and O_2 molecules on 2-D CaO on Mo(100) form O_2^- superoxo anions, which may act as

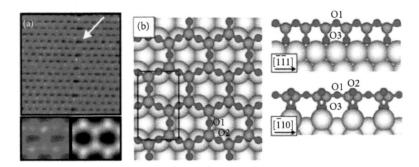

Fig. 5.24 (a) STM images of the silica layer with hexagonal structure on Mo(112). (b) Model of the SiO$_2$/Mo(112) structure (top and side views). Reprinted from Freund and Pacchioni (2008) with permission. Copyright 2008 by the Royal Society of Chemistry. Permission conveyed through Copyright Clearance Center, Inc.

electron donors and reactive intermediates in oxidation reactions. Au nanoparticles on 2-D CaO/Mo(100) have been shown to grow as flat 2-D islands, in contrast to their 3-D particle morphology if grown on bulk oxide surfaces (Nilius and Freund, 2015) (see Chapter 4, Section 4.5.3).

The oxide-metal work function can be modified by doping, and adsorbed species can be activated or deactivated on the respective surfaces by the induced charge transfer. 2-D oxides with porous molecular sieve character, e.g. SiO$_2$ on Mo(112) (Weissenrieder *et al.*, 2005),—see Fig. 5.24 for STM images and a structure model of SiO$_2$/Mo(112)— can incorporate adsorbed (doping) atoms into their nanopore cages, depending on the charge and size of the atoms (Freund and Pacchioni, 2008; Giordano and Pacchioni, 2011). For example, Pd atoms are able to penetrate the openings in the defect-free oxide surface, while Au atoms are too large and bind on SiO$_2$ line defects (Ulrich *et al.*, 2009).

The surface of a precious metal catalyst may form a 2-D surface oxide coating under reaction conditions (cf. Chapter 2, Section 2.1.3), which then becomes the catalytically active phase. This has been demonstrated first for the CO oxidation reaction on a single crystal Ru(0001) surface, which under realistic CO pressure conditions is covered by a 2-D RuO$_2$ layer as the catalytically active phase (Over *et al.*, 2000): the undercoordinated Ru cations at the oxide surface have been postulated to be particularly significant for the catalytic reactivity. In this context, the strong metal-support interaction (SMSI) effect can be mentioned, reported some 40 years ago (Tauster *et al.*, 1978; Tauster *et al.*, 1981; Van Delft and Nieuwenhuys, 1985), in which metal catalyst particles (such as Pd, Pt) supported on a reducible oxide (such as titania) become deactivated upon heating. This has been ascribed to the encapsulation of the metal particles with an oxide monolayer (Dulub *et al.*, 2000), which leads to a reduced adsorption capacity and thus a reduced catalytic activity. The mechanism of oxide migration over a metal surface and the dynamic behaviour of a 2-D V-oxide phase in wetting-dewetting a Pd surface have been revealed by in-situ variable temperature STM experiments in combination with a DFT calculated surface oxide phase diagram (Surnev *et al.*, 2002): the stability of the different V-oxide phases as a function of the

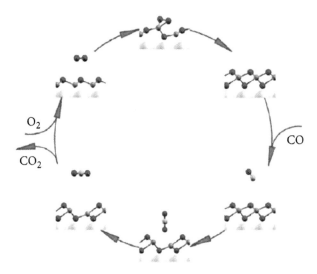

Fig. 5.25 Individual steps in forming the active FeO_2/Pt(111) phase starting from FeO/Pt(111) and its reaction with CO to form CO_2, as revealed by DFT calculations (Fe: blue; O: red; Pt: gray; C: yellow). Reprinted from Giordano and Pacchioni (2011) with permission. Copyright 2011 American Chemical Society.

chemical potential of oxygen and the V concentration in the different oxide phases leads to a reversible wetting-dewetting process when cycling the thermodynamic conditions of oxygen pressure and temperature. Such processes may also occur during encapsulation of metal particles by an oxide overlayer, which is driven by the dynamics of oxide redox reactions. For Pt particles on a Fe_3O_4(111) ultra-thin film, the encapsulating oxide layer after heating to 850 K has been identified as a FeO(111)-type bilayer (Qin *et al.*, 2008) (see Section 5.2 for the FeO bilayer).

The study of the CO oxidation reaction on the FeO bilayer on Pt(111) has unveiled an interesting scenario. The 2-D FeO layer is unreactive for CO oxidation under UHV conditions but becomes very active at 300–450 K at ambient pressure conditions (~ 1 atm) (Sun *et al.*, 2009). The FeO(111) bilayer on Pt(111) can structurally rearrange at higher CO and O_2 pressure conditions and form a new trilayer phase of FeO_2 stoichiometry. The scenario as identified by DFT calculations (Giordano *et al.*, 2013) is illustrated in Fig. 5.25. The oxygen reacts with the FeO layer by pulling out an Fe atom, which lowers the work function locally. The subsequent electron transfer generates a O_2^- superoxo transient species, which dissociates and forms a higher coverage local O-Fe-O trilayer. The FeO_2 trilayer oxidizes CO from the gas phase to CO_2 in a Mars–Van Krevelen mechanism leaving an oxygen vacancy in the oxide, which is reoxidized again at the high oxygen pressure. At lower oxygen pressure, the FeO_2 trilayer is unstable and the reduced catalytically inactive FeO ends the cycle.

A real metal-oxide industrial catalyst typically consists of a highly dispersed (often precious) metal component (preferably in the form of metal nanoparticles) on a high-surface-area oxide support material. The so-called 'inverse catalyst' concept has been introduced as a model system to study the role of the oxide-metal phase boundary in catalytic processes and to identify its catalytically active sites (Williams *et al.*, 1990; Boffa *et al.*, 1994; Leisenberger *et al.*, 2000). An 'inverse catalyst' consists of a well-defined metal single crystal surface that is decorated by a submonolayer coverage of crystalline 2-D oxide nanoislands or oxide nanoparticles: it contains the bifunctional site properties of the metal-oxide interface and its planar geometry allows one to apply surface science microscopy and spectroscopy tools to obtain atomic scale insights into the catalytic processes (Schoiswohl *et al.*, 2005*b*; Schoiswohl *et al.*, 2007; Rodriguez *et al.*, 2009; Fu *et al.*, 2010). For example, on a V-oxide decorated Rh(111) inverse catalyst surface an increased reactivity toward CO oxidation has been detected, which has been attributed to catalytically active low-coordination sites at the oxide-metal boundary on the basis of atomically resolved STM imaging and XPS measurements (Schoiswohl *et al.*, 2005*a*). The active role of the oxide-metal boundary in the water-gas shift (WGS) reaction ($CO + H_2O \Leftrightarrow CO_2 + H_2$) has been confirmed on inverse catalyst systems containing CeO_x and TiO_x nanoparticles on Au(111) surfaces (Rodriguez *et al.*, 2009). Coordinatively unsaturated active sites (CUS) at the oxide-metal interface of an FeO_x/Pt(111) inverse catalyst have been proposed by Fu *et al.* (2010) to explain the high activity for CO oxidation via activation of molecular oxygen at the CUS sites. However, this interpretation is challenged by the work of Sun *et al.* (2009), in which the formation of a transient FeO_2 phase has been postulated to be responsible for the high CO oxidation activity (see previous discussion and Fig. 5.25).

Surface defects generally play a key role in the chemical reactivity of oxide surfaces, because they act as sinks or sources of electrons, which promote oxidation or reduction. In 2-D oxides, oxygen vacancy formation energies are lowered with respect to the bulk because of the possibility of unusual cation oxidation states, which can be stabilized at reduced dimension (Carrasco *et al.*, 2006; Ganduglia-Pirovano *et al.*, 2007). This is yet another element that contributes to a high catalytic activity of 2-D oxide phases.

5.7 Summary

The study of two-dimensional materials is at the forefront of present-day condensed matter physics and chemistry. This also applies to 2-D oxide materials. In contrast to other transition metal chalcogenides, which in the bulk constitute so-called van der Waals solids, with layered structures featuring strong intralayer covalent bonding but weak interlayer van der Waals bonding, only few oxides occur in van der Waals type layered structures in the bulk phase. For van der Waals solids, the individual layers may be easily separated by physical or chemical means generating sheets of 2-D materials, which can then be isolated. The fabrication of 2-D oxides poses extra challenges. Ultra-thin films at the 2-D limit, of one or only a few unit-cells thick, are an established method for preparing 2-D oxides, but they exist mostly on solid substrate templates. However, wet chemical procedures, with sometimes very complex multistep protocols, have made great progress in recent years in generating free-standing (quasi-)2-D oxide nanosheets.

As with ultra-thin films, a richness of structural varieties has been detected in 2-D oxide systems, with electronic properties reflecting quantum confinement and interface effects at the ultimate limit. The detection of magnetic properties of 2-D oxides is in an emerging state, but much theoretical prediction still awaits experimental scrutiny. 2-D oxide layers with lateral confinement, oxide nanoribbons or nanoislands, have also been considered. Theoretical analysis of different edge geometries has suggested interesting emergent phenomena of electronics and magnetism. In general, theoretical prediction in 2-D oxide systems is much ahead of experimental realization. This is due to difficulties in the preparation and experimental handling of 2-D oxide materials. Nevertheless, interesting applications in diverse nanotechnology fields are envisioned and successful proof of concept experiments have been reported. Moreover, the application of the LEGO concept, i.e. the stacking of different 2-D oxide crystals on top of each other forming oxide heterostructures with atomically thin components opens up new preparation avenues for novel artificial nanostructured materials.

6
Oxide nanoparticles

Oxide microparticles and nanoparticles (NPs), such as clays or silica, are ubiquitous in the natural environment. Clays are the most ancient materials used by man for his habitat and as artistic or cultural supports (see Chapter 7). Small silica particles compose the major part of sand but also constitute the building blocks of opal gemstones. Those are natural photonic crystals with beautiful iridescence properties due to the regular arrangement of silica nanospheres. Since the most remote times, oxide micro- and nano-particles have also served as pigments. One famous example is ochre, made of iron oxide particles and utilized, together with black MnO_2 particles, to decorate Neolithic caves, as in Lascaux in France, Altamira in Spain, or in la Cueva de los Manos in Argentina. Oxides produce a whole range of colours, from white (TiO_2, ZnO) to red-orange (FeOOH, Fe_2O_3, Pb_3O_4), blue (CoO, Co_3O_4, CuO), or green (NiO). These pigments are still used for the ornament of ceramics or in the food industry (food additive E171 is TiO_2, E172 refers to iron oxides, E551 is SiO_2, E5269 is $Ca(OH)_2$, etc).

Due to their large surface-to-bulk ratio, oxide particles are also efficient catalysts in acid-base reactions, gas phase partial oxidation reactions, or for combustion and depollution processes. Their structure and the nature of their active sites largely influence their catalytic activity and selectivity, which may be even enhanced via defect creation or doping (Vedrine, 2018) (see Chapter 8). In nature, small aluminium oxide particles produced by rocket flights have a significant impact on the atmospheric chemistry (Turco *et al.*, 1982) and silicate dust has been found in a wide range of astronomical environments (Bromley, 2017).

With the advent of nanotechnology, methods of fabrication of ultra-fine particles, in the nanometre range, have been developed. By far, the three nanomaterials which are industrially produced in the largest amounts are TiO_2, SiO_2, and ZnO. Aside from its applications as pigment, which nevertheless represents a production of approximately 5 million tons per year, TiO_2 is known for its photocatalytic properties when in the anatase form. This property is used for example in self-cleaning glasses, for oxidation of NO_2, and remediation of waste waters, or envisioned for the hydrogen production by photocatalytic water splitting. With ZnO, it is used as opacifier and as a protective agent in cosmetic products such as sunscreen creams. ZnO also has optoelectronic and photocatalytic applications due to its large band gap and large exciton binding energy and the versatility of shapes in which it can be produced. It is used as loading

Oxide Thin Films and Nanostructures. Falko P. Netzer and Claudine Noguera, Oxford University Press (2021). © Falko P. Netzer and Claudine Noguera. DOI: 10.1093/oso/9780198834618.003.0006

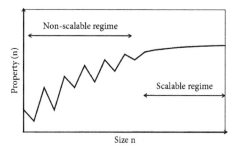

Fig. 6.1 Schematic size evolution of a given NP property, highlighting the scalable and non-scalable regimes.

agent of tires and rubbers. Similarly, SiO_2, which can be produced under various topological phases, finds biomedical applications, can display red, green, blue, or white photoluminescence at room temperature, and SiO_2 nanowires can be used in sensors or actuators for their mechanical properties.

Nevertheless, despite their attractive properties, oxide NPs may have serious drawbacks. Some NPs are suspected of being responsible for immune disorders or precancerous lesions. In particular, care should be taken not to inhale finely divided oxide dust, such as SiO_2 or asbestos (silicate fibres) NPs, which could lead to silicosis or lung cancer.

From a fundamental point of view, NPs bridge the gap between the properties of isolated atoms or molecules and those of bulk condensed phases. They present a large spectrum of atomic and electronic structures, due to their finite size in the three dimensions of space and the fact that their atoms are nearly all 'surface' atoms. Moreover, oxide NPs represent the first stages of nucleation and growth of larger size oxides, and their stability and properties which depend upon the thermodynamic conditions under which they are formed, largely impact the final product (see Chapter 2). It is the aim of the present chapter to describe the specific properties of oxide NPs and analyse their physical origin, both when they are produced in the gas phase (Sections 6.1 and 6.2), or in an aqueous environment (Section 6.3), or when some mixing takes place (Section 6.4).

6.1 Non-scalable regime

6.1.1 Introduction

The size evolution of NP properties presents two regimes. At very small sizes, the NPs, often referred to as clusters, display exotic atomic structures which most of the time bear no resemblance to their bulk counterparts and their properties vary in a non-monotonic way (Fig. 6.1). In this regime, called the *non-scalable regime*, each atom counts. Beyond a critical size (which may depend on the property under consideration), NP characteristics are still size-dependent, but they smoothly converge towards their bulk values. This is the *scalable regime*. In this section, we restrict ourselves to the non-scalable regime for NPs produced in the gas phase.

Most of the experimental information in this regime come from laser ionization time-of-flight mass spectrometry, which records cluster mass abundance spectra. In the eighties, the field had successively focused on rare-gas, metallic, and then semi-conducting clusters, to eventually address oxide clusters at the beginning of the nineties. As was described in Chapter 2, in order to produce oxide NPs in the gas phase, a metal is heated in a crucible and its vapour is mixed with a cold inert gas to which an oxidizing gas (O_2 or N_2O) is added. The clusters are then ionized by a focused laser beam and accelerated in an electric field before entering the detection region for time-of-flight mass analysis. So far, such studies have been restricted to a fairly limited number of oxides. A typical mass spectrum, obtained for MgO clusters, is reproduced in Fig. 6.2. Each detected cluster gives rise to a peak, indicative of its mass. Among them, some have an especially high intensity, which signals a particular cluster stability with respect to fractionation. This is mathematically expressed by the condition that the second energy difference $\Delta_2 E(n) = E(n+1) + E(n-1) - 2E(n)$ involving the energies $E(n-1)$, $E(n)$ and $E(n+1)$ of cluster sizes $(n-1)$, (n) and $(n+1)$ formula units, respectively, is positive. Unfortunately, no structural information can be gained in such experiments and only ionized clusters are detected. Additional information is given by various infra-red (IR) spectroscopy methods. The analysis of vibrational spectra, coupled with numerical simulations, may for example be used to discriminate between competing structures, such as ring, bubbles, or cage structures.

Contrary to the relatively small number of experimental studies, nearly all numerical methods have been applied to the determination of the atomic and electronic structures of clusters from the molecule to the critical size at which bulk-like configurations start appearing. This is true both for the cluster energies, for which classical, semi-empirical or various DFT-based methods have been used, and for the determination of equilibrium structures with methods ranging from simple local optimization, to basin-hopping Monte Carlo methods or evolutionary methods. The main difficulty resides in the fact that, as size increases, the number of competing structures (isomers) exponentially grows and their energy differences get smaller and smaller. This is particularly true close to the cross-over size between the scalable and non-scalable regimes, which explains why some discrepancies still exist in the literature on its value. Nevertheless, over the years, better and better predictions have been made, yielding a rather consistent knowledge of at least the small size regime. All theoretical works on transition metal oxide clusters have been recently reviewed (Fernando *et al.*, 2015).

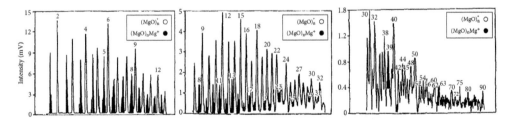

Fig. 6.2 Mass spectra of $(MgO)_n^+$ and $(MgO)_n Mg^+$ clusters produced in the gas phase. Reprinted from Ziemann and Castleman Jr (1991) with the permission of AIP Publishing.

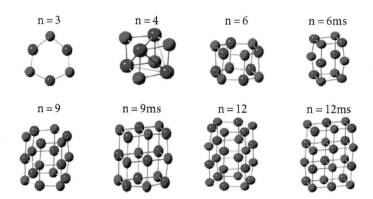

Fig. 6.3 Some low-energy $(MgO)_n$ isomers showing the competition between ring-like structures ($n = 3, 6, 9, 12$) and rock-salt bulk-cut structures ($n = 4, 6ms, 9ms, 12ms$). 'ms' indicates
metastable structures. Reprinted from Chen *et al.* (2014*a*) with permission. Copyright 2014
American Chemical Society.

Several theoretical reviews also exist on non-reducible oxide clusters (Johnston, 2003;
Bromley *et al.*, 2009; Catlow *et al.*, 2010).

6.1.2   Atomic structure

It is not the purpose of this section to make a review of the ground state atomic structures of all oxide NPs. Only a few representative examples will be given to highlight
their sensitivity to parameters such as the ionicity, the anion-to-cation ionic radius
ratio, or the number of under-coordinated atoms.

Surely, MgO has been the most studied oxide in the non-scalable regime, due to the
simplicity of its bulk structure. Early experimental works coupled to simulations based
on a pair potential approach have evidenced the competition between rock-salt bulk-
cut structures and nanotube morphologies made of superimposed $(MgO)_3$ hexagons
(Fig. 6.3). Interestingly, the latter, which, at first sight look quite exotic for MgO, can
also be viewed as cuts from a bulk structure. Indeed, the hexagonal-BN phase, which
is metastable in bulk MgO at ambient or positive pressures but becomes stable at
negative pressures (Fig. 6.4) is made of honeycomb-like layers, and its ground state
energy is less than 0.1 eV per MgO formula unit above the rock-salt one. Nanotubes
with a hexagonal basis are the most stable isomers for n=3, 6, 9, 12, and 15. The
transition towards rock-salt pieces takes place at approximately $n = 24$ (Chen *et al.*,
2014*a*).

Comparison with other rock-salt oxides and NaCl gives hints on how ionicity, ionic
radius ratios, and cation polarizability influence this competition. For example, rigid
ion models of 1:1 binary compound clusters predict hexagonal tubular structures when
ionic charges are assumed equal to $Q = \pm 2$. In contrast, the hypothesis $Q = \pm 1$ always
leads to more compact shapes based on the rock-salt structure (Roberts and Johnston,
2001), which are indeed found in NaCl clusters (Dugourd *et al.*, 1997). The propensity
for tubular shapes results from a stronger electrostatic repulsion between like atoms
when $Q = \pm 2$ which penalizes dense structures. As far as the other alkaline-earth

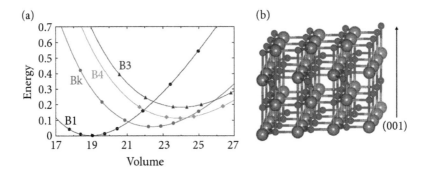

Fig. 6.4 (a) Energy of crystalline phases of bulk MgO (eV per MgO formula unit), as a function of volume (Å^3 per formula unit). B1, B3, B4, and Bk phases are rock-salt, zinc-blend, wurtzite, and hexagonal BN structures, respectively. The energy zero is taken at the B1 ground state. The caesium-chloride (B2) curve is located well beyond the limits of the figure (Goniakowski *et al.*, 2004). Courtesy J. Goniakowski. (b) Structure of the hexagonal-BN phase. All atoms are five-fold coordinated and the (001) plane is made of 6-membered MgO rings. Mg and O atoms are represented by grey and red spheres, respectively.

oxides are concerned, CaO, SrO, and BaO prefer cuboids cut from the rock-salt structure for sizes $n = 3k$ (those represented as metastable states in Fig. 6.3), while the nanotubes are the predicted ground state for MgO. This has been assigned to the smaller polarizability of the Mg cation, compared to other alkaline-earth cations (Escher *et al.*, 2018).

As size increases, the average atom coordination number $< Z >$ also increases, since atoms in the core of the clusters have an environment closer to bulk ($Z = 6$). Total binding energies per atom E_0 and mean anion-cation distances R_0 simultaneously grow (Fig. 6.5), typically as $|E_0| \propto Z * (Z/\alpha)^{-m/(m-1)}$ and $R_0 \propto (Z/\alpha)^{1/(m-1)}$ (α the Madelung constant, m the exponent of the short-range repulsion term, see Chapter 4,

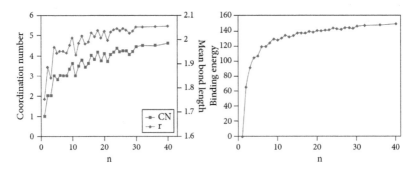

Fig. 6.5 Size evolution of the average coordination number (CN), the mean Mg-O bond length r (Å) and the binding energy per MgO unit (kcal/mole) for the $(MgO)_n$ lowest energy isomers. Reprinted from Chen *et al.* (2014*a*) with permission. Copyright 2014 American Chemical Society.

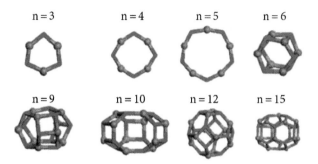

Fig. 6.6 Some low-energy $(ZnO)_n$ isomers showing the competition between ring-like structures and spheroid structures. Reprinted from Catlow *et al.* (2010) with permission. Copyright 2010 by the Royal Society of Chemistry. Permission conveyed through Copyright Clearance Center, Inc.

since α grows less quickly than Z, due to the repulsion between second neighbours. Around under-coordinated atoms, bonds are shorter and stronger which is a typical behaviour when ionic or iono-covalent cohesive interactions are dominant (Noguera, 1996). Applied locally, this reasoning highlights the lower stability of more under-coordinated atoms.

Another oxide which has been the subject of numerous works is ZnO. Despite having the same stoichiometry as alkaline-earth oxides, its larger covalent character, revealed by the atomic four-fold coordination in its wurtzite bulk structure (Phillips, 1970), leads to quite different morphologies at small sizes. Actually, two distinct non-scalable regimes compete, before structures cut from the bulk become stable (Fig. 6.6). Planar rings of increasing sizes are the ground state geometry up to $n \approx 5$–7. Beyond this size, spheroid structures, named bubbles, mainly built from Zn_2O_2 and Zn_3O_3 rings, but sometimes including octagonal or decagonal rings, become the most stable. This second regime, which is not present in alkaline-earth oxides, is assigned to the fact that bubbles involve zinc atoms in three-fold coordination, which is less but not so far from that in bulk wurtzite $Z = 4$, while in alkaline-earth oxides it would amount to breaking half of the octahedral bulk environment. The critical size to reach the scalable regime was estimated of the order of $n = 75$ (Matxain *et al.*, 2000; Catlow *et al.*, 2010; Viñes *et al.*, 2017).

In the first small size regime ($n < 7$), the Zn-O bond lengths display a tendency to decrease as n grows. In the intermediate bubble regime, no monotonic variation is found (Wang *et al.*, 2007), likely because all zinc atoms keep a constant coordination, and only local strains induce some bond-length fluctuations. Actually, this fact is in line with the relaxation processes present at ZnO surfaces, which do not involve bond contraction but rather bond rotation associated to the transformation of sp_3 into sp_2 bonding. Nevertheless, the binding energy per formula unit monotonically grows with n and the second energy difference displays positive values for $n \leq 6$, at $n = 9$, and at $n = 12$.

Despite the differences just mentioned between alkaline-earth oxide and ZnO clusters, these systems share a common feature, which is the absence of terminal oxygen atoms in the lowest-energy (stoichiometric) isomers. This feature is also present in small alumina $(Al_2O_3)_n$ clusters (Li and Cheng, 2012) and thus seems a signature of non-reducible oxides at the nanoscale. However, small $(SiO_2)_n$ clusters do display terminal oxygen atoms (Catlow *et al.*, 2010) named silanones, which are described as $=Si^+$-O^- charge transfer species (Zwijnenburg *et al.*, 2009) and are thus very reactive. Small transition metal oxide clusters like $(TiO_2)_n$ also possess terminal oxygen atoms, usually represented as M=O double bond species (Albaret *et al.*, 1999; Cho *et al.*, 2016).

6.1.3 Electronic structure

A feature shared by small clusters and molecules is the discrete character of their electronic spectra. Understanding how the transition from the molecular to the condensed phase occurs had been the initial scientific stimulus to produce and study small clusters of controlled size.

In the non-scalable regime, many electronic characteristics vary, not only with the number of formula units n, but also with the precise atomic structure, the shape, the presence of under-coordinated atoms, and so on. Among them are the degree of ionicity of the cation-oxygen bond, the HOMO-LUMO gap (HOMO=Highest Occupied Molecular Orbital; LUMO=Lowest Unoccupied Molecular Orbital), or the ionization energy. It has long been assumed that, due to the prevalence of electrostatic interactions, the electronic structure of oxides was little, if not at all, influenced by the local structure. However, the development of concepts such as local electronegativity or site-specific acid/base character, has led to the recognition that, even in highly ionic materials, the electronic structure is not frozen but may vary as a function of the local environment of the atoms (Noguera, 1996). In small clusters, as well as at surfaces, the loss of neighbours with respect to the bulk and the associated structural relaxations perturb the electrostatic potentials, the crystal field splitting, and the orbital hybridization.

By acting on the Madelung potential and the type of hybridization, the local environment of the atoms within the clusters is responsible for the presence of specific states in the electronic spectrum. For example, the optical absorption and luminescence spectra of MgO NPs and surfaces were shown to strongly depend on the oxygen coordination—corners, kinks, steps—with a significant reduction of the exciton excitation-energies and of the luminescence energies of relaxed excitons as the oxygen coordination decreases (Shluger *et al.*, 1999; Chiesa *et al.*, 2005).

The richness of the electronic structure also appears in a particularly striking way in small SiO_2 clusters (Catlow *et al.*, 2010), as shown in Fig. 6.7. Three regimes may be identified in the size dependence of the HOMO-LUMO gap G. For $n \leq 6$, G increases in a weak monotonic way. This is a regime in which the clusters consist of chains of Si_2O_2 rings. Beyond this size, columnar structures are stabilized with terminal Si=O groups. Two symmetric groups are present when n is even, while, for odd n clusters, a single one exists of higher energy. The odd-n minima in the HOMO-LUMO curve ($G \approx 3.5$ eV compared to ≈ 5.7 eV for even n) are produced by localized states around those less-favourable terminations. Finally, above $n = 23$, more two-dimensional structures

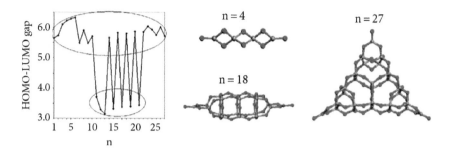

Fig. 6.7 Left: HOMO-LUMO gap (eV) of the lowest energy $(SiO_2)_n$ isomers. Right: examples of ground state atomic configurations. Reprinted from Catlow *et al.* (2010) with permission. Copyright 2010 by the Royal Society of Chemistry. Permission conveyed through Copyright Clearance Center, Inc.

are stabilized, with gaps determined by the electronic states localized on the terminal Si=O groups.

6.1.4   Magicity

The existence of particularly intense peaks in the mass spectra of small clusters has led to the concept of magicity. These high intensities signal particularly stable clusters with respect to fractionation, which means that either they are very stable or that the clusters with one added or one removed unit are particularly unstable. Their sizes are said to be *magic*.

Magic sizes have been observed in nearly all types of clusters, from metals to semi-conductors, insulating compounds, or rare gases. However, the physical mechanism which determines their sequence is specific to each material class. For example, in xenon clusters (Echt *et al.*, 1981), the magic sizes 13, 55, 147 ... correspond to the completion of successive shells around a Mackay icosahedron. In titanium clusters, magic numbers have been related to particularly symmetric structures, as shown in Fig. 6.8. At variance, in alkali or gold clusters, which, due to their quasi spherical shapes could be seen as super-atoms with atomic-like electronic shells $1S$, $1P$, $1D$, $2S$, etc., the magic numbers 2, 8, 20, 34, 58 ... correspond to the completion of closed

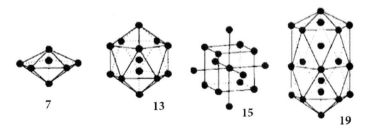

Fig. 6.8 Titanium magic clusters. Reprinted from Neukermans *et al.* (2007). Copyright 2007, with permission from Elsevier.

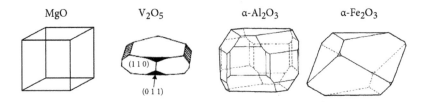

Fig. 6.9 Wulff shapes of some oxide NPs. Reprinted from Sayle *et al.* (1996) (Copyright 1996 by the Royal Society of Chemistry. Permission conveyed through Copyright Clearance Center, Inc.) and Mackrodt *et al.* (1987) (Copyright 1987, with permission from Elsevier) for the vanadium oxide and the sesquioxide NPs, respectively.

electronic shells $1S$, $1S + 1P$, $1S + 1P + 2S + 1D$, $1S + 1P + 2S + 1D + 1F$, and so on. These examples show that magicity may have a structural or an electronic origin, depending upon the type of material.

In oxide NPs, most magic numbers have been associated to particularly stable closed-shell atomic structures, with the smallest possible number of under-coordinated atoms. This is well-illustrated in the mass spectra of MgO clusters shown in Fig. 6.2, in which the magic numbers 2 and 4 are due to the MgO molecule and MgO square configurations, magic numbers 6, 9, 12, 15, 18 to nanotubes with an hexagonal basis, and 16 and 20 to nanotubes cut out of the rock-salt structure, and so on. The HOMO-LUMO gaps in these particularly stable oxide structures are often larger than in the neighbouring $(n+1$ or $n-1)$ clusters which have a larger number of under-coordinated atoms.

6.2   Scalable regime

6.2.1   Equilibrium shapes

Once the scalable regime is reached, the NPs adopt an equilibrium morphology which can be understood as cut from the bulk. Under-coordinated atoms are found on the outermost facets and at the corners and edges which connect them. These sites have higher and higher energies as they correspond to lower and lower coordination, due to the larger number of broken bonds necessary to create them. They play important roles in the NP reactivity.

The equilibrium shape of an NP is determined by the condition that the total energy associated to facets, edges, and corners is minimum at constant volume. Their relative weights scale as $1/N^{1/3}$, $1/N^{2/3}$, and $1/N$, respectively, so that, as size N increases, corners first become irrelevant and then edges. When only facets significantly contribute to the total energy, the Wulff construction allows determining the equilibrium shape from the knowledge of surface energies alone (Zangwill, 1990; Muller and Kern, 2000) (see Appendix 1). Figure 6.9 gives some examples of Wulff shapes of oxide NPs.

For a given crystal structure, surface energies γ strongly depend on the surface orientation. As a rule of thumb, they increase as the surface becomes less dense and involves atoms of lower coordination. For example, in the rock-salt structure, the hierarchy of (non-polar) surface energies $\gamma_{100} < \gamma_{110} < \gamma_{211}$ clearly scales with the number of broken bonds, equal to 1, 2, and 3 per surface atom, respectively. However, the nature (more covalent or more ionic) of the cation-oxygen bonding also plays a role, which explains why oxide NPs with the same bulk structure do not display the same Wulff shapes. This is for example the case of α-Al$_2$O$_3$ and α-Fe$_2$O$_3$ NPs shown in Fig. 6.9, whose Wulff shapes result from the following hierarchy of surface energies: $\{111\} < \{\bar{2}11\} \approx \{011\} < \{01\bar{1}\} \approx \{100\}$ for the former and $\{011\} \approx \{111\} < \{01\bar{1}\} < \{\bar{2}11\} \approx \{100\}$ for the latter (Mackrodt *et al.*, 1987).

The theoretical determination of a cluster structure in the lowest size range of the scalable regime is not an easy task since, on the one hand, the number of atoms is such that a full global optimization may be beyond computer capacity, and, on the other hand, the Wulff theorem does not apply since the energy of corners and edges cannot yet be neglected. Additional structural principles are thus needed, such as considering only dipole-free configurations and/or requiring that all atoms have sufficient coordination. Predictions of anatase cluster structures in this size range have for example been made in that way (Lamiel-Garcia *et al.*, 2017).

Beside surface energies, surface tension τ has significant implications on the internal structure of NPs. It is defined as the reversible work to elastically stretch a unit area of surface. It differs from the surface energy γ by the derivative of γ with respect to strain, and can be positive (tensile stress) or negative (compressive stress). In many materials, the bond breaking at surfaces intrinsically tends to reduce interatomic distances. However, surface atom positions are constrained by the underlying bulk. The resulting positive surface stress induces an excess pressure ΔP inside the NPs—the Laplace pressure—which scales as the inverse of the NP dimension (typically $\Delta P = 2\tau/R$ for spherical NPs of radius R). This supports the general observation of reduced mean atomic distances as size decreases, as measured, for example, by EXAFS.

We should add that, whereas equilibrium shapes are well-defined, the NPs which are experimentally produced may display a bunch of alternative (metastable) structures, which result from a kinetic rather than a thermodynamic bias. ZnO is likely the oxide which displays the largest number of morphologies— nanorods, nanowires, nanotubes, nanocombs, nanorings, nanohelixes/nanosprings, nanobelts, nanocages, hexagonal bipods, tetrapods, spiral structures, or microprisms (Wang, 2004b; Ronning *et al.*, 2005; Spencer, 2012). Efforts have also been devoted to the synthesis of TiO$_2$, WO$_3$ or ZrO$_2$ nanotubes (Roy *et al.*, 2011). Although being only metastable, these oxide structures find a variety of biomedical, photochemical, electrical, gas sensing, and environmental applications.

6.2.2 Polymorphism

Many oxides possess several bulk polymorphs. This is for example the case for ZnO, which, beside the wurtzite (WZ) equilibrium phase at ambient conditions, features several metastable structures, such as the zinc-blend (ZB), the h-BN, the body-centred tetragonal (BCT), the sodalite (SOD) or the cubane structures, shown in Fig. 6.10. Several metastable structures—anatase, brookite, α-PbO$_2$, β-VO$_2$, baddelleyite (ZrO$_2$) among others—also appear in the phase diagram of TiO$_2$ beside the rutile structure which is thermodynamically stable under ambient conditions. Al$_2$O$_3$ possesses a similarly rich phase diagram including α, β, γ, δ, θ, and κ phases in which the Al local

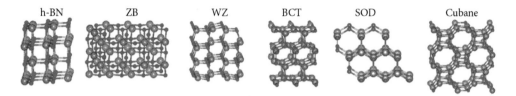

Fig. 6.10 Crystalline structure of several ZnO bulk polymorphs. ZB, WZ, BCT, and SOD refer to the zinc-blend, wurtzite, body-centred tetragonal, and sodalite structures, respectively.

structure includes a variable ratio of octahedral-to-tetrahedral cation sites (Levin and Brandon, 1998).

In the scalable regime, NPs do not necessarily adopt the bulk crystalline structure which is the most stable under the same thermodynamic conditions. The best known example is the TiO_2 NP phase diagram which, according to calorimetric measurements, Fig. 6.11, displays a stability range of anatase below ≈ 40 nm, a stability region for the brookite phase in the range between approximately 40 to 200 nm, the bulk stable rutile phase being only recovered for particles of sizes larger than ≈ 200 nm (Ranade *et al.*, 2002). Similar transitions as a function of size have been demonstrated in other oxide NPs, such as alumina or zirconia (Navrotsky, 2003)

As for thin films (see Chapter 4), the argument for understanding these size-induced structural transitions relies on the competition between the bulk energy of the various polymorphs (always favourable to the most stable bulk structure) and the surface energy contribution which becomes more and more important as size decreases. Consequently, if a polymorph, metastable in the bulk, has an average surface energy which is lower than that of the most stable configuration, it may become the preferred structure at small size. According to calorimetric measurements, this is what happens for TiO_2, the rutile, brookite, and anatase surface enthalpies being found equal to 2.2, 1.0, and 0.4 J/m^2, respectively (Navrotsky, 2003).

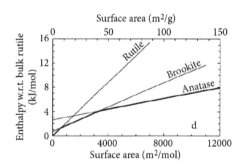

Fig. 6.11 Enthalpy of nanocristalline TiO_2 with respect to bulk rutile (kJ/mole) as a function of specific surface area, from calorimetric measurements. Lower surface areas correspond to larger NPs. Reprinted from Ranade *et al.* (2002) with permission. Copyright (2002) National Academy of Sciences, USA.

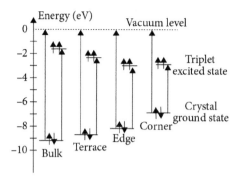

Fig. 6.12 Calculated diagram representing the relative energies of the ground and excited triplet states at different positions in a MgO NP. Reprinted from Shluger *et al.* (1999) with permission. Copyright 1999 by the American Physical Society.

6.2.3 Electronic structure

The electronic characteristics of oxide NPs, concomitant with the changes of morphology and relaxation effects induced by the low atomic coordination, are governed by microscopic mechanisms which are similar to those at work in ultra-thin oxide films (see Chapter 4).

Among them, an important one is the reduction of the Madelung potential on under-coordinated atoms, which brings the oxygen and cation effective levels ϵ_i closer to each other, and in charge transfer insulators induces surface states in the gap region. An illustration of this effect relevant to MgO NPs is shown in Fig. 6.12. The smallest energies necessary to excite electrons from sites of various coordinations Z have been found using both semi-empirical and ab initio Hartree-Fock methods (Shluger *et al.*, 1999) to display a monotonic variation when passing from bulk ($Z = 6$) to terrace ($Z = 5$), edge ($Z = 4$), and corner ($Z = 3$) sites. On this basis, the relative weight of corner and edge ions in size controlled MgO nanocubes could be experimentally deduced from optical properties as measured by UV diffuse reflectance and photoluminescence spectroscopy (Stankic *et al.*, 2005).

Additionally, the small size of NPs is responsible for a quantization of electronic states propagating perpendicularly to the NP facets. This quantum confinement effect produces an overall increase ΔG of the band gap. In a nearly free electron picture close to the band edges, ΔG is of the order of $\hbar^2\pi^2/2\mu R^2$, with R the effective radius of the NP, and μ the reduced electron-hole effective mass.

In NPs, these various effects induce positive or negative variations of the gap width. Their competition is driven by the nature of the oxide (charge transfer or Mott–Hubbard), its degree of ionicity, the nature (bond rotation or contraction), and the strength of relaxation effects, and the NP size.

Although they are not measurable quantities, charges have played a central role in the understanding of insulating oxides. Various evaluation schemes have been proposed in the past, including the conventional Mulliken (1955) and Bader (1991) schemes for partitioning the electronic density. Starting from the purely ionic limit, in which oxygen

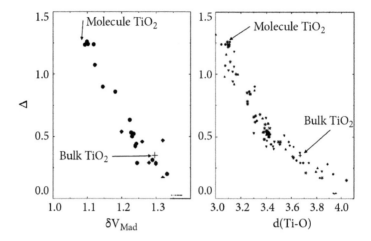

Fig. 6.13 Electron transfer Δ per Ti-O bond as a function of the difference δV_{Mad} of the Madelung potentials acting on the bound Ti and O atoms (in Hartree per electron) (left), and as a function of the Ti-O bond length $d(\mathrm{Ti} - \mathrm{O})$ (in atomic units) (right). Results for neutral stoichiometric clusters, charged clusters, bulk rutile TiO_2 and $TiO_2(110)$ surface are represented by filled circles, stars, plus signs, and diamonds, respectively. Adapted from Albaret *et al.* (1999).

atoms and cations bear formal charges (Q_O^0 and Q_C^0, respectively), the electron sharing due to orbital overlap can be recast as a sum of bond contributions Δ_{OC}, consistently with physical intuition and in line with Pauling's valence bond approach (Pauling, 1929; Albaret *et al.*, 1999; Noguera *et al.*, 2002). The oxygen and cation charges are thus locally modified in a way which depends on their coordination number and the value of the various electron transfers per bond $\Delta_{O_i C_j}$ around them:

$$Q_{O_i} = Q_O^0 + \sum_{C_j} \Delta_{O_i C_j} \qquad Q_{C_j} = Q_C^0 - \sum_{O_i} \Delta_{O_i C_j} \qquad (6.1)$$

The quantities $\Delta_{O_i C_j}$ vanish in the ionic limit (negligible overlap of orbitals between neighbouring atoms, or large anion-cation difference of electronegativity) and increase monotonically with the increasing degree of covalency. Their magnitude thus characterizes the covalent strength of the oxygen-cation bonding and is directly related to the local electronic properties of the two bound atoms. Therefore, their determination provides a simple way to characterize the variable ionic/covalent bond strength in the various environments.

The environment effect lies in the variations of the effective atomic orbital energies ϵ_i with the local Madelung potential V_i. For example, at under-coordinated sites where V_i is reduced, the oxygen and cation effective levels are closer to each other than in the bulk, rendering bonds more covalent. This effect was evidenced in small neutral or charged stoichiometric TiO_2 clusters, bulk rutile, and (110) surface (Albaret *et al.*,

1999). From a partition of the inhomogeneous electron density issued from DFT simulations into Bader's atomic basins (Bader, 1991), oxygen and cation charges Q_{O_i} and Q_{C_j} could be determined and consequently all non-equivalent $\Delta_{O_iC_j}$ by inversion of eqn 6.1. Figure 6.13 displays the values of the $\Delta_{O_iC_j}$ for each bond in these systems, as a function of the difference δV_{Mad} of the Madelung potentials acting on the Ti and O atoms and as a function of the Ti-O bond lengths d. The good correlation with δV_{Mad} confirms the role of the Madelung potential in fixing the covalent character of the bonds, and the correlation with the bond length—a measurable quantity—assesses the relevance of the $\Delta_{O_iC_j}$ to characterize the degree of covalency of the bonds. It is striking to observe that, in a relatively covalent oxide like TiO_2, the covalent character may vary so strongly as a function of the local environment of the atoms (for example, $\Delta = 0.35$ in the bulk around octahedrally coordinated Ti, and $\Delta = 1.24$ in the TiO_2 molecule).

The local Mulliken electronegativity χ—defined as the negative of the first derivative of the total energy of the atom with respect to its electron number—is strongly dependent on charges and local Madelung potentials. It qualitatively varies as $\chi_i = \chi_i^0 + U_i Q_i + V_i$ (i the atom upon consideration, Q_i its charge, U_i the on-site electron-electron repulsion, and V_i the Madelung potential), and is highly dependent on the local environment of the atom. Characterizing the capability to donate or receive electrons, it accounts for the variations of electronegativity as a function of the charge state and of the site environment. On sites of low coordination, the cation electronegativity is expected to be higher and the oxygen electronegativity lower than on atoms in a denser environment. Such an expression may be used as a guideline to predict the *variations* of ionization potentials and electron affinities as a function of the atomic environment, as well as the site basicity or acidity (Noguera, 1996).

6.2.4 Polarity

Polarity refers to the presence of a macroscopic dipole moment in a globally neutral stacking of (alternatively) charged atomic layers $\pm\sigma$, such as is met, for example, at (111) rock-salt surfaces. It results in a monotonic linear increase of the macroscopic electrostatic potential in the direction perpendicular to the stacking, which progressively shifts the local band structure and eventually produces an overlap of the local VB maximum of one surface with the CB minimum of the opposite one. In polar thin films, the overall amplitude of the shift ΔV scales (linearly) with the distance between opposite terminations, leading to the so-called polar catastrophe (Noguera, 2000; Goniakowski *et al.*, 2007*a*) (see Chapter 4).

Compared to semi-infinite surfaces or thin films, the lateral size L of a polar NP facet is finite (Fig. 6.14(a)) and the electrostatic characteristics somewhat change. The variations of the electrostatic potential now depend not only on the thickness H of the particle along the polar direction as in thin films but also on the size L of the polar facet. Simple electrostatic considerations evidence two regimes (Goniakowski and Noguera, 2011). When $L \gg H$, the thin film characteristics are recovered, with a potential difference $|\Delta V|$ which increases linearly as a function of H. However, when $L \ll H$, the electrostatic potential displays interesting new features. First the potential profile close to the surface is much steeper than in thin films (Fig. 6.14(b)). Second,

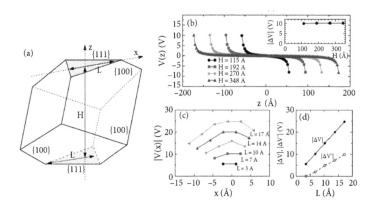

Fig. 6.14 (a) Example of a rock-salt polar NP with two opposite polar (111) facets; the z and x axis are indicated; (b) Variations of the electrostatic potential $V(z)$ at anion sites across the NP when $L \ll H$; the inset shows $|\Delta V|$ as a function of H; (c) $V(x)$ in the NP when $L \ll H$; (d) in the same regime, dependence of $|\Delta V|$ and $|\Delta V'|$ on L. $V(z)$ and $V(x)$ are evaluated on a rigid rock-salt lattice (first neighbour distance 2.8 Å), as in sodium chloride, within a point charge model ($Q = \pm 1$). Reprinted with permission from Goniakowski and Noguera (2011). Copyright 2011 by the American Physical Society.

$|\Delta V|$ no longer depends on H, but rather grows linearly with L. Additionally, in this regime, V is not constant on the polar facet. It goes through an extremum at its center of gravity, and its maximum variation $|\Delta V'|$ also increases linearly with L (Fig. 6.14(c,d)).

Despite these differences with polar thin films or surfaces due to dimensionality effects, a polar catastrophe may occur in NPs requiring a charge compensation $\delta\sigma$ on the outer NP facets. Depending on the oxide, the environment, and the processing conditions, various mechanisms such as nonstoichiometric surface reconstructions, adsorption of charged species, or screening by an interface with a metal deposit may provide the excess surface charge $\delta\sigma$ necessary to heal the polar instability (Noguera, 2000; Goniakowski *et al.*, 2007*a*; Noguera and Goniakowski, 2012) (see Chapter 4).

In the scalable regime, according to Wulff theorem, the equilibrium shape of NPs is determined by the relative values of surface energies along the various orientations. Bare polar surfaces are usually high-energy surfaces even when a non-stoichiometric reconstruction has healed their polarity. However, they are very reactive, and their stability is strongly dependent on the synthesis conditions. They may be totally or partially absent in NPs produced under UHV conditions, while prevailing under other environment conditions. Whatever the type of compensation achieved, such high energy surfaces possess specific structural features and electronic or charged states which are responsible for enhanced physical or chemical properties, rendering them very promising for many different applications.

In the last decade, an important activity has thus been devoted to the control of nanoparticle size and shapes, with the aim of stabilizing high energy surfaces and tuning their properties at will. The issue is to find new synthesis routes to obtain

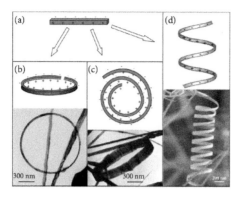

Fig. 6.15 (a) Model of a ZnO nanobelt with polar faces allowing formation of (b) a nanoring, (c) a nanospiral, and (d) a nanohelix. Bottom panels: TEM images of the corresponding ZnO objects. Reprinted with permission from Ref. Wang (2004b). Copyright © 2004 IOP Publishing Ltd. All rights reserved.

artificial materials with enhanced catalytic activity, optical, gas sensing (Han *et al.*, 2012), or photo-degradation properties (Jiang *et al.*, 2010), or to produce new field effect transistors, nanoresonators (Wang, 2004b), or memory devices (Song *et al.*, 2011). In relation to polarity, the oxides which have by far attracted the greatest interest are ZnO, CeO_2, and to a lesser extent MgO.

Two different strategies have allowed stabilizing MgO NPs with polar (111) facets. Synthesis by thermal decomposition of $Mg(NO_3)_2$ in molten $LiNO_3$ produces NPs of octahedral shape with (111) facets, the presence of the ionic liquid allowing compensation by adsorption of charged species (Xu *et al.*, 2008; Jiang *et al.*, 2010). Another route involves precipitation of MgO crystallites via internal oxidation of a Cu-Mg alloy, yielding cubo-octahedral topotaxial MgO precipitates, the shape of which varies from octahedral to cubic as the oxygen activity decreases (Backhaus-Ricoult *et al.*, 2003).

Due to its importance in many fields of applications driven by its reducible character, various synthesis routes of CeO_2 nanoparticles have been followed. In most cases, the particle shape appears as either a truncated octahedron, with (111) side faces and small (100) apical facets, or as a cube limited by (100) faces, depending upon the synthesis conditions. In the CeO_2 bulk fluorite structure, the lowest energy surface is along the non-polar (111) orientation, while the (100) surface is polar. Simulations, based on classical interatomic potentials find that the equilibrium shape of a 10 nm CeO_2 NP in vacuum is a truncated octahedron, with 50% occupation of the oxygen sites on the polar (100) facets, which is what is needed to compensate polarity (Sayle *et al.*, 2004). First principles simulations have found that the oxygen vacancy formation energy on the perfect polar facets is greatly reduced compared to extended surfaces and decreases as the particle size grows. This is rationalized by the existence of oxygen states lying high in the oxide gap as a result of the reduced electrostatic potential at their location (Migani *et al.*, 2010). The electronic properties of these NPs are key factors in the structure-activity dependence of ceria-based Pt catalysts (Vayssilov *et al.*, 2011).

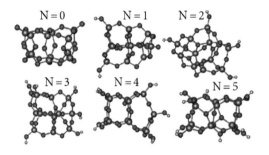

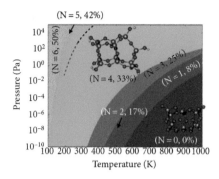

Fig. 6.16 Left: global optimized structures for $(TiO_2)_{12}(H_2O)_N$ NPs for $0 \leq N \leq 5$. Ti, O and H atoms are represented by pink, red and light gray balls, respectively. Right: P-T phase diagrams for hydrated TiO_2 NPs containing 12 formula units. The stability regions for the various degrees of hydration ($N = 1$ to 6) are indicated with colours, from dark blue ($N = 0$) at high temperature or low water pressure, to light blue and green ($N = 4$ to 6) at low temperature or high water pressure. Blue and black dashed lines indicate the equilibrium vapour pressure with ice and liquid water respectively. Reprinted from Cuko *et al.* (2018*b*) with permission. Copyright 2018 by the Royal Society of Chemistry. Permission conveyed through Copyright Clearance Center, Inc.

Among the wide variety of observed ZnO nanostructures, some have been related to the polar character of their exposed surfaces. Wurzite ZnO possesses two low-index polar surfaces $(0001)/(000\bar{1})$, and $(10\bar{1}1)/(10\bar{1}\bar{1})$ each with two non-equivalent terminations. Figure 6.15 gives some examples of ZnO nanostructures and their relationship to the charged $(0001)/(000\bar{1})$ polar surfaces (Wang, 2004*b*).

6.3 Oxide NPs in water or in contact with a humid atmosphere

The presence of water molecules, water films, or liquid water in contact with oxide NPs is the generality rather than the exception. Indeed, to obtain dry NPs, strict UHV conditions or severe calcination treatments have to be implemented. At variance, residual H_2O pressure nearly always subsists in laboratory apparatus, leading to the decoration of NPs with water molecules either in a physisorbed or a dissociated state. The formation of oxide NPs in water or in contact with a humid atmosphere is also relevant both in the lab where co-precipitation is a method of choice for synthesizing oxides or hydroxides (Chapter 2 Section 2.3) and in the natural medium where precipitation is often the result of aqueous fluid supersaturations created by rock weathering. A wide range of NP morphologies, often very different from those found in dry environments, can then be observed. They depend on parameters such as the degree of supersaturation, the acidity of the solution, the presence of specific anions, the temperature, etc., which in the lab can be purposely fine-tuned.

6.3.1 Non-scalable regime

At small sizes, in the non-scalable regime, the presence of strongly under-coordinated atoms favours the dissociation of water molecules. Resulting protons and OH^- groups bind to the NP oxygen atoms and cations, respectively. Per water molecule, two hydroxyl groups OH$^-$ are thus formed. In this size range, most information comes from numerical simulations. For example, as shown in Fig. 6.16, a restructuring of the TiO$_2$ NPs takes place which depends on the degree of hydration (Cuko *et al.*, 2018*b*). The phase diagram as a function of temperature T and water partial pressure P shown in Fig. 6.16 evidences regions of various levels of hydration, and suggests that under ambient conditions, TiO$_2$ NPs are already strongly hydroxylated even in the presence of trace amounts of water vapour. The electronic characteristics are also modified by hydroxylation, especially those determined by under-coordinated atoms. This is the case of the HOMO-LUMO gap. Due to the formation of bonding-anti-bonding states between the under-coordinated oxygen atoms and the protons, the states in the gap which were associated to the former in the dry particle disappear. The HOMO-LUMO gap thus generally increases—although not in a fully monotonic way—with the degree of hydration.

In contact with liquid water, as occurs during the first steps of NP nucleation in an aqueous medium, polyanions, polycations, or neutral entities are formed, among which some are particularly stable and can be considered as 'magic' entities. In silica rich aqueous solutions, for example, as well as in calcium carbonate rich solutions, polymeric entities with various linear, branched, rings, or cage structures have been identified. They have been assigned to 'prenucleation clusters' allowing nucleation to occur with low energy barriers (Gebauer *et al.*, 2008; Gebauer *et al.*, 2014; Van Driessche *et al.*, 2016) (see Chapter 2, Section 2.4). An example is shown in Fig. 6.17 in the case of calcium carbonate.

6.3.2 Scalable regime

On oxide NPs and surfaces, dissociative adsorption of water often occurs. However, a large amount of work has been devoted to precisely assess this point on specific oxide surfaces, and discrepancies still exist in the literature on whether water dissociates totally, partially, or not at all on some surfaces. We send the interested reader to review articles, such as (Thiel and Madey, 1987; Henderson, 2002), or to overviews on particular oxides, such as ZnO (Wöll, 2007), TiO$_2$ rutile (Huang *et al.*, 2014), or TiO$_2$ anatase (Gong and Selloni, 2005).

These studies show that, for a given oxide polymorph, some surfaces are prone to dissociate water while others are nearly non-reactive. This is, for example, the case for the dominant anatase TiO$_2$(101) surfaces on which water adsorbs molecularly, while the minority ones (001) may become hydroxylated (Gong and Selloni, 2005). On TiO$_2$ rutile surfaces, partial water dissociation takes place on the most stable (110) surface while the (001) remains inert with respect to water (Huang *et al.*, 2014). The same is true for MgO. Its most stable (100) surface can only partially dissociate water (Giordano *et al.*, 1998; Kim *et al.*, 2002), while the polar (111) surfaces easily get hydroxylated (Refson *et al.*, 1995).

There is often an anti-correlation between the surface stability and its capacity to dissociate water. This is due to the fact that less stable surfaces are usually less dense and have more under-coordinated atoms. This is particularly true for polar surfaces where the presence of charged hydroxyl groups represents an efficient means to

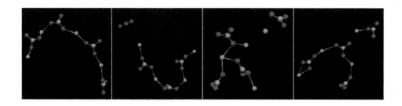

Fig. 6.17 Configurations of CaCO$_3$ entities in water containing four formula units, as observed after 1 ns of molecular dynamics simulations, under conditions, [Ca^{2+}] = 0.4 mM, [HCO$_3^-$] = 10 mM and pH = 10. Atoms are coloured green, blue, and red for calcium, carbon, and oxygen, respectively. Surrounding water and remote sodium, bicarbonate, and carbonate species are omitted for clarity. Reprinted from Demichelis *et al.* (2011) with permission from Nature. Copyright 2011.

compensate the macroscopic dipole moment (Noguera, 2000; Goniakowski *et al.*, 2007 *a*; Noguera and Goniakowski, 2012). For the same reason, the presence of surface defects, whether structural (corners, step edges, etc.) or stoichiometric (oxygen vacancies), increases the probability of water dissociation (Bikondoa *et al.*, 2006).

In all cases, the presence of water on a surface decreases its surface energy. When water is simply physisorbed, the energy lowering is modest, but it is much larger when water dissociates. Experimental results on surface energies are scarce. Some calorimetric measurements have nevertheless been performed on NPs in the scalable regime, leading to surface energy values averaged over the NP facets. For example, experiments performed on monoclinic zirconia NPs show a noticeable decrease of their surface enthalpy upon water adsorption (from 3.4 J/m^2 to 2.9 J/m^2) (Radha *et al.*, 2009). A similar decrease was found for anhydrous and hydrated nano-ceria surfaces, with surface enthalpies equal to 1.16 and 0.86 J/m^2, respectively (Hayun *et al.*, 2011).

Numerical simulations confirm/predict surface energy lowering upon hydration. Interestingly, this lowering is stronger as the face is less stable, indicating that the stability order of the various surface orientations may change upon hydroxylation. This is for example the case for MgO. The relative stability of the dry surfaces from the least to most stable follows the order (111) < (110) < (100). At variance, as shown in Fig. 6.18, under even low water pressure, the hydroxylated (111) surface becomes the most stable (Geysermans *et al.*, 2009). Smoke MgO NPs indeed progressively change shape when immersed in water. The starting cubic shapes, characteristic of the most stable MgO (100) dry surfaces, steadily transform into octahedral habits made of hydroxylated (111) polar facets (Hacquart and Jupille, 2007). Actually, in the natural environment, the MgO mineral, named periclase, mostly exposes (111) faces, due to the humid environment in which it forms. The hydroxylation of the MgO(111) surface represents the first step of its transformation into brucite Mg(OH)$_2$, which would occur under higher water pressure or temperature. Beyond the case of MgO, the change of surface energies upon hydration usually leads to modifications of the Wulff NP shapes.

Another consequence of hydroxylation concerns the rate of nucleation F. According to the classical theory of nucleation (CNT), F is an activated function of the nucleation

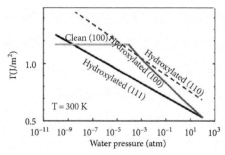 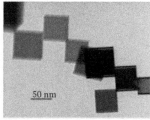 

Fig. 6.18 Left: calculated surface free energies of low-index orientations of MgO as a function of water pressure, estimated at T = 300 K Reprinted from Geysermans *et al.* (2009) with permission. Copyright 2009 by the Royal Society of Chemistry. Permission conveyed through Copyright Clearance Center, Inc. Right panels: TEM image of cubic-shaped crystallites of MgO smoke after 0 days (middle) and their octahedral shapes exposing (111) facets after 7 days (right) in deionized water. Reprinted from Hacquart and Jupille (2007). Copyright 2007, with permission from Elsevier.

barrier ΔG^* which increases as the cube of the mean surface energy γ (Chapter 2 Section 2.4). Nucleation is thus generally facilitated in aqueous media, as compared to dry medium. Mean oxide surface energies extracted from nucleation experiments and interpreted within the CNT are indeed in the 20–200 mJ/m^2 range (Fig. 2.7, Chapter 2) (Söhnel, 1982), rather than in the J/m^2 range expected for dry surfaces.

The modification of surface energies by hydration may also impact the NP structural phase diagram by shifting the cross-over sizes which delineate the various polymorph phases. This has been experimentally demonstrated in the case of ZrO$_2$ NPs which display a tetragonal phase at small size rather than the bulk monoclinic one (Radha *et al.*, 2009). Upon hydration, the surface energies of both polymorphs change. The surface enthalpy for hydrated monoclinic zirconia was shown to be ≈ 0.5 J/m^2 lower than the corresponding value for the anhydrous polymorph while a smaller difference (≈ 0.2 J/m^2) exists for tetragonal ZrO$_2$. These differences induce a stability crossover between the monoclinic and the tetragonal phases at smaller particle size for hydrated than for anhydrous zirconia (≈ 28 nm versus ≈ 34 nm).

6.3.3 Acid-base properties

The acid-base concept has had a long historical development, starting with Arrhenius in the 1880s, and continuing with Brønsted, Lewis, Mulliken, Hudson and Klopman, Parr and Pearson, etc (Noguera, 1996). Most contributions have focused on molecules. However, when a solid particle takes part in an acid-base reaction, new parameters have to be considered, such as the surface orientation, the site coordination number, the structural or stoichiometry defects, some steric factors, and so on. For a surface, as well as for a large molecule, it is no longer possible to talk of *one* acid-base strength. One has to distinguish between non-equivalent surface sites and quantify their reactivity which now not only depends on local factors but also on long-range electrostatic interactions.

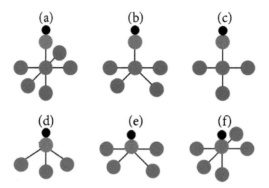

Fig. 6.19 Configurations of OH groups on oxide NPs. (a), (b), and (c): the hydroxyl group is adsorbed on a five-, four-, and three-fold coordinated cation, respectively; (d), (e), and (f): the OH group is attached to three, four, or five cations, respectively. Cation, oxygen, and hydrogen atoms are represented as blue, red, and black balls, respectively. The oxygen atom of the hydroxyl group is coloured in pink.

Following Brønsted's definition, an acid is a hydrogen-containing molecule able to dissociate into a proton H^+ and an anion, while a base is associated to the reverse process. On a hydroxylated oxide, oxygen sites are basic since they can fix a proton, and cations are the acid sites which can fix OH groups. In the former case, the strength of the O-H bond defines the basicity of the oxygen site and as it increases, so does the O-H stretching frequency. Similarly, the strength of the cation-OH bond defines the acidity of the cation site, and as it increases, the OH stretching frequency decreases.

A hydroxylated oxide NP may possess various non-equivalent hydroxyl groups, as represented in Fig. 6.19. The OH properties, and in particular their stretching frequencies, depend on the way they are bonded to the NP cations and the coordination of the latter. Infra-red absorption and HREELS experiments, which give access to the values of the OH stretching frequencies, have highlighted some correlations between their shift and the nature of the adsorption site (Coluccia *et al.*, 1988; Knözinger *et al.*, 1993*b*; Knözinger *et al.*, 1993*a*):

- Higher frequencies are assigned to hydroxyl groups bound to surface cations, and lower frequencies to protons adsorbed on surface oxygen atoms. Both families present many internal splittings on rough surfaces, due to all the possible surface sites, adsorption sites—on-bridge, ternary, etc.,—and lateral interactions between adsorbed groups.

- On a given compound, lower stretching frequencies characterize less basic oxygen sites because they reveal weaker O–H bonds. For hydroxyl groups adsorbed on a surface cation, lower frequencies characterize more acid cations, because a weaker O–H bond is generally associated with a stronger cation–OH bond. This rule is not always obeyed when different compounds are compared.

- The stretching frequency of hydroxyl groups bound to a single surface cation is lower on more under-coordinated cations. The reverse is true for protons adsorbed on top of a surface oxygen.

Such correlations are important for understanding the link between the nature of the active sites and the NP activity and selectivity in catalytic acid-base reactions (see Chapter 8, Section 8.1).

6.3.4   The oxide-water interface

At the surface of oxide NPs immersed in water, a thin film of mixed dissociated and molecular water molecules held together by H-bonds forms. Due to the interaction with the NP surface, the water molecules belonging to this film loose some of their degrees of freedom and thus some entropy, compared to their counterparts in the bulk of the solution. They may be considered as quasi-frozen, with a structure close to ice in near to epitaxy relationship with the oxide surface (Hu *et al.*, 1995; Al-Abadleh and Grassian, 2003).

Under the influence of pH, protonation and deprotonation of -XOH surface sites may take place:

$$-XO^- + H^+ \longleftrightarrow -XOH$$
$$-XOH + H^+ \longleftrightarrow -XOH_2^+ \tag{6.2}$$

The changes of Gibbs energy in these two reactions define the oxide pK_1 and pK_2 values ($pK = -\Delta G/2.3RT$, T the temperature and R the gas constant). At high pH, both reactions are displaced towards the left and the surface becomes negatively charged with a large density of neutral XOH and negatively charged $-XO^-$ species. At variance, at low pH, both reactions are displaced towards the right and positively charged surface sites XOH_2^+ are dominant. This is illustrated in Fig. 6.20 in the case of silica. One defines the Point of Zero Charge (PZC), as the pH value for which the surface possesses as many positively as negatively charged sites. For pH < PZC, the surface is globally positively charged. Above PZC, it is globally negatively charged. The PZC value is related to the pK_1 and pK_2 of the surface by:

$$PZC = \frac{1}{2}(pK_1 + pK_2) \tag{6.3}$$

Parks has compiled PZC values for many oxides, oxyhydroxydes, and hydroxides (Parks, 1965). Typical values are given in Table 6.1, which, for example shows that SiO_2 and MgO display the strongest acidic and basic characters, respectively.

Parks has stressed the importance of the cation characteristics in determining the PZC: their (formal) charge Q_M and their ionic radii r_M. As a function of Q_M, he established the hierarchy of PZC values displayed in Table 6.2.

The oxides with a basic character (high PZC) involve cations with the lowest formal charge. They are alkaline or alkaline-earth oxides, while cations in acidic oxides

Table 6.1 PZC values for a set of oxides, oxyhydroxydes, and hydroxides (Parks, 1965).

SiO_2	2	Fe_3O_4	6.5	ZnO	9
δ-MnO_2	2.8	α-FeOOH	7.8	α-Al_2O_3	9.1
α-$Al(OH)_3$	5	γ-AlOOH	8.2	NiO	10.3
TiO_2 rutile	5.8	α-Fe_2O_3	8.5	MgO	12.4

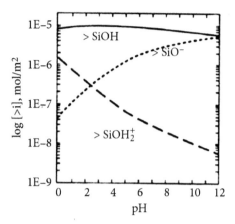

Fig. 6.20 Density of neutral -SiOH, positively charged -SiOH$_2^+$ and negatively charged -SiO$^-$ sites of SiO$_2$ surfaces in contact with an aqueous solution, as a function of pH. The curves for -SiOH$_2^+$ and -SiO$^-$ cross at a pH value of the order of 2, which is the PZC of SiO$_2$. Reprinted from Duval *et al.* (2002) with permission. Copyright 2002 American Chemical Society.

Table 6.2 Range of PZC values for binary oxides of various stoichiometries (Parks, 1965)

oxide	Q_M	PZC
MO$_3$	+6	PZC < 0.5
M$_2$O$_5$	+5	PZC < 0.5
MO$_2$	+4	0.5 < PZC< 7.5
M$_2$O$_3$	+3	6.5 < PZC < 10.4
MO	+2	8.5 < PZC< 12.5
M$_2$O	+1	11.5 < PZC

(low PZC) bear high formal charges. The PZC values saturate at a value close to zero, because of the water levelling effect. Water, which is the solvent in PZC measurements, does not allow one to study species more acidic than H$_3$O$^+$, such as M$_2$O$_5$ and MO$_3$ oxides.

As a function of the cation ionic radii, Parks established that the PZC values are roughly linearly decreasing functions of the ratio Q_M/r_M. This correlation is supported by a simple electrostatic model in which the free energy change ΔG, which determines the PZC value, is mainly due to the work of the electrostatic forces when protons approach or leave the surface. In a picture where the interaction between the oxygen and the closest surface cation prevails, ΔG reads:

$$\Delta G = \frac{Q_O Q_H}{\epsilon r_O} + \frac{2Q_M Q_H}{\epsilon(2r_O + r_M)} + \Delta G' \qquad (6.4)$$

In this expression Q_O and Q_H are the oxygen and proton formal charges, ϵ is the oxide optical dielectric constant, and r_O and r_M are the oxygen and cation ionic radii, respectively. The first term in eqn 6.4 represents the proton attraction by the surface oxygen and the second term its repulsion by the neighbouring cation. $\Delta G'$ is the free energy of non-electrostatic origin, assumed to be independent of the oxide. Since only the proton–cation interaction distinguishes the various oxides in this model, the PZC may be written in the most general form:

$$PZC = A - B\frac{Q_M}{(2r_O + r_M)} \tag{6.5}$$

This equation accounts for the higher acid character of oxides involving cations with a high formal charge and a small ionic radius. It is qualitatively well obeyed in the series MnO_2, TiO_2, and SnO_2 in which the PZC values 4.2, 4.7, and 7.3 scale with the ionic radii 0.42 Å, 0.50 Å, and 0.58 Å, but there are a number of exceptions.

Parks' model was refined and extended to account for the environment of the surface sites which take part in the protonation-deprotonation processes, in the so-called MUSIC model (MUSIC = MUltiSIte Complexation model) (Hiemstra *et al.*, 1989*b*; Hiemstra *et al.*, 1989*a*; Hiemstra and Van Riemsdijk, 1996). The model makes use of valence bond charges $v = Q_M/Z_B$ (Z_B the cation coordination number in the bulk) and the local charge neutralization concept introduced by Pauling (1929), in which an oxygen ion linked to n cations bears a charge equal to $Q_O = nv - 2$, an expression very similar to eqn 6.1. The free energy associated with each elementary protonation reaction on a surface then reads:

$$\Delta G = \frac{Q_O Q_H}{\epsilon r_{OH}} + \frac{Q_H n v}{\epsilon r_{HM}} + \Delta G' \tag{6.6}$$

as a function of the proton–oxygen and proton–cation inter-atomic distances r_{OH} and r_{HM}, respectively. With respect to Parks' approach, the main difference lies in the value of the positive charge, each of the n cations in the neighbourhood of the oxygen atom contributing as v. The pKs associated with protonation or deprotonation reactions thus depend in a linear way on the surface oxygen coordination number n:

$$pK(n) = A - B\frac{nv}{r_{HM}} \tag{6.7}$$

This is the interest of the MUSIC approach which accounts for complexation processes on surfaces possessing various non-equivalent surface sites (hence the term 'multisite' in the MUSIC acronym).

Equation 6.7 applies for both pK_1 and pK_2 values. It shows that the basicity of the surface oxygen sites increases as their coordination number decreases (pK_1 variations), while the OH acidity is stronger as their coordination number increases (pK_2 variations). The B constant is considered to be the same for the two reactions, while the A constant is roughly a factor 2 larger for pK_1 than for pK_2 due to the larger O^{2-} charge compared to the OH^- one. This leads to large differences (typically of the order of 14 pH units) between pK_1 and pK_2 values, showing that, in water, only one reaction among the two written in eqn 6.2 can take place. In the case of gibbsite $Al(OH)_3$, for example, calculated pK values are equal to $pK_2 = -1.5$ and $pK_1 = 12.4$ for two-fold coordinated oxygen atoms ($n = 2$), and $pK_2 = 10$ and $pK_1 \approx 24$ for one-fold coordinated oxygen atoms ($n = 1$). This means that, for all pH values less than 12.4, two-fold coordinated groups only exist under the hydroxo form Al_2-OH:

the gibbsite basal surface which only possesses such groups thus remains neutral. At variance, the lateral faces which possess one-fold and two-fold coordinated groups become positively charged below pH $= 10$ and negatively charged above. The average gibbsite PZC is thus equal to 10 (Hiemstra *et al.*, 1989*b*). The values issued from the MUSIC model are consistent with calculated proton binding energies (Kawakami and Yoshida, 1985) and OH stretching frequencies measured by infra-red spectroscopy (Knözinger and Ratnasamy, 1978).

To summarize, one can note the following points (Jolivet, 2019):

- the surface charge may strongly vary on the various facets of a NP in water, due to the existence of surface groups having different environments, with a direct implication on their reactivity;
- oxygenated surface groups can bind a single proton in the pH range accessible in aqueous solutions, and, consequently, never display an amphoteric character.

When the oxide surface has a non-zero charge, mobile ions in the aqueous solution are attracted or repelled (depending on their charge) by the surface. They may adsorb on the surface, thus contributing to the surface charge, in a way which depends on the pH of the solution and its ionic strength. Within the solution, they may also screen the surface charge and contribute to the electrical potential. A so-called double-layer thus forms, in which the first layer is made of adsorbed ions on the surface and the second is composed of ions from the solution which are attracted by the surface charge. However, due to their loose interaction with the latter, these ions are able to move in the fluid under the influence of the electric potential and temperature. It is thus often called the 'diffuse layer'. Its thickness, its density, and more generally its internal structure have a strong impact on the stability of colloïdal suspensions, since it determines the attraction or repulsion between particles (Grahame, 1947). Several models describing the double layer have been proposed, including Helmoltz, Gouy-Chapman, Stern, Grahame, etc. (Outhwaite *et al.*, 1980; Israelachvili, 2011). The simplest one is schematized in Fig. 6.21.

6.3.5 Control of NP size and shape

At pH values different from the PZC, the existence of a surface charge has repercussions on the surface energies and the NP characteristics: their shape, their solubility, and their size. Changing the pH is thus a lever to control the size and morphology of oxide NPs. In some cases, as magnetite Fe_3O_4 or anatase TiO_2, the particle size decreases as the pH of precipitation moves away from the PZC (Fig. 6.22). For layered materials, like brucite $Mg(OH)_2$ or boehmite $AlOOH$, the reverse is true.

To account for these effects, a semi-quantitative model has been proposed (Jolivet *et al.*, 2004), which shows that the interfacial tension between a given surface and an aqueous solution is maximum at pH $=$ PZC and decreases for both larger and smaller pH values. The lowering of the interfacial tension is expressed by Gibbs' law as a function of the density of adsorbed species, and the latter is determined using the MUSIC model and taking into account the relationship between the surface charge and the electric potential close to the surface. Applied to the case of boehmite and making use of DFT surface energies relevant at pH $=$ PZC (Raybaud *et al.*, 2001), this model predicts variations of surface energies for various surface orientations as a

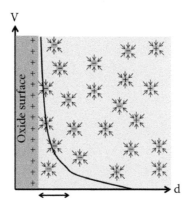

Fig. 6.21 Schematic representation of the double layer which forms close to a charged NP surface. Negatively or positively charged ions in the aqueous solution are represented by green or orange circles, respectively, surrounded by small arrows mimicking the water molecules in their first solvation layer. V is the electrical potential. Its quasi exponential decrease as a function of the distance d from the NP surface defines the effective thickness of the double layer marked by a double arrow.

function of pH as represented in Fig. 6.23. This procedure has allowed rationalizing the modifications of boehmite NP morphologies deduced from the simulation of X-ray diffraction patterns (Chiche *et al.*, 2008), in particular the development of (101) facets as pH increases. The (010) face which remains neutral in the whole pH range and has the lowest energy nevertheless always dominates the NP morphology. The NP solubility also varies in a non-monotonic way as a function of pH. It has been found that the solubility is minimum at the PZC, which can be rationalized in terms of the correlation between surface tensions and solubilities established by Söhnel (Söhnel, 1982) (Fig. 2.7 in Chapter 2).

As far as NP size is concerned, a large decrease of surface energy at pH far from the PZC induces a strong decrease of the nucleation barrier—proportional to γ^3 within

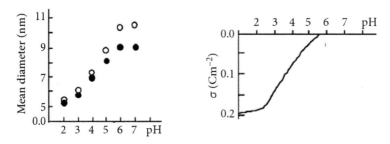

Fig. 6.22 Variation of anatase NP mean diameter (left) and surface charge (right) as a function of pH. Full and empty symbols refer to particles aged for two days and one week, respectively. Reprinted from Pottier *et al.* (2003) with permission. Copyright 2003 by the Royal Society of Chemistry. Permission conveyed through Copyright Clearance Center, Inc.

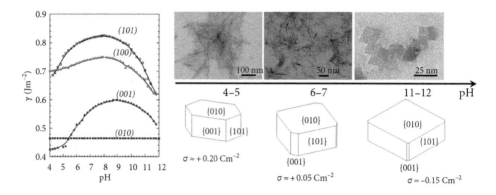

Fig. 6.23 Left: calculated pH variation of boehmite AlOOH surface energies, oriented along (101), (100), (001), and (010). Right: particle shapes in acidic (pH = 4–5), neutral (pH = 6–7) or basic (pH = 11–12) solutions, deduced from the simulation of X-ray diffraction patterns. Reprinted from Chiche *et al.* (2008) with permission. Copyright 2008 American Chemical Society.

the classical nucleation theory (Chapter 2, Section 2.4)—and thus a dramatic increase of the nucleation rate. The content in growth units of the aqueous solution in that case is immediately consumed by nucleation and no ripening takes place. For 3-D particles, the critical nuclei get smaller and more numerous. Those are conditions under which the NP characteristics are determined by thermodynamics rather than by kinetics.

To summarize, at the interface between an oxide NP and an aqueous solution, adsorption and/or dissociation of water molecules or adsorption of charged ions from the solution take place, which produces surface charge. The latter depends on the pH value and the ionic strength of the solution. It also depends on the orientation of the exposed surfaces and the acid-base character of the oxide, through the PZC concept. Strong modifications of surface energies result which are accompanied by morphology changes of the NP. Through the changes in surface energy and solubility, the characteristics of nucleation and growth are modified, which gives a lever to fine tune the size and morphology of the NPs in view of specific applications.

6.4   Mixed oxide nanoparticles

6.4.1   Introduction

Alloying strategies have often been used in chemistry or in metallurgy to purposely improve the materials properties. In metallurgy, for example, alloying elements help increasing the metal strength and hardness or the resistance to corrosion. In another context, the natural environment provides many examples of mixed oxides. Indeed, compared to pure minerals, mixed minerals are the generality rather than the exception. Their formation is often thermodynamically preferred over the separate formation of their parents, as can be judged by the frequent incorporation of trace elements in minerals, the ubiquity of complex non-stoichiometric oxides, silicates, carbonates, or clay minerals, and the scarcity of pure minerals.

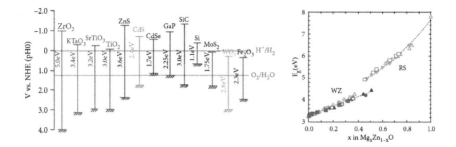

Fig. 6.24 Left panel: relationship between band edge positions of pure oxides and semiconductors, and redox potentials relevant for water splitting. Reprinted from Kudo and Miseki (2009) with permission. Copyright 2009 by the Royal Society of Chemistry. Permission conveyed through Copyright Clearance Center, Inc. Right panel: dependence of the room-temperature band-gap (E_g) in $Mg_xZn_{1-x}O$ films, as a function of the Mg content in the wurzite (WZ) and rock-salt (RS) phases. Reprinted from Ohtomo and Tsukazaki (2005) with permission. Copyright © 2005 IOP Publishing Ltd. All rights reserved.

In the lab, the possibility of doping and/or mixing oxides has long since been used to synthesize artificial compounds with enhanced reactivity or tunable properties (Fig. 6.24). In water splitting processes, for example, pure oxides are not suitable for the production of hydrogen since either their gap width or the positions of their band edges are not correctly located with respect to the relevant redox potentials (Kudo and Miseki, 2009). Similarly, the substitution of Zr atoms by Ce atoms in ZrO_2 leads to a localized level in the electronic gap, transforming ZrO_2 into a photoactive material sensitive to visible light (Gionco *et al.*, 2014). By mixing MgO and ZnO, it is possible to nearly continuously vary the gap width from the low ZnO value to the large MgO one (Ohtomo and Tsukazaki, 2005) (Fig. 6.24). In relation to their antibacterial activity, doping oxide NPs is also a means to stabilize the NPs in aqueous suspensions and enhance their efficiency (Gordon *et al.*, 2011; Stankic *et al.*, 2016).

These few examples emphasize the wide context in which mixed oxides play significant roles. More generally, combining cations of different size, electronegativity and reducibility, gives a lever for modifying the oxide structural, electronic, and reactivity characteristics, especially when cations with several oxidation states are involved (Yuan *et al.*, 2014). In NPs, additional flexibility exists since their characteristics depend on both the mixing characteristics of their bulk parents and on low-dimensional effects. At low-coordinated sites, segregation phenomena may take place, driven by steric effects or by the competition between the surface energies of their parents. In the following, some basic concepts related to mixing will be recalled, restricted to substitutional mixing. Then some specific properties of mixed oxide NPs will be summarized.

6.4.2   Basic concepts of mixing in the bulk

The outcomes of mixing rely on the existence, sign, and strength of the local interactions between the substituent (solute) and host atoms. In weakly interacting systems, fully disordered Solid Solutions (SSs) may form, in which the solute atoms occupy random positions in the host lattice. An interaction increase leads to SSs with short-range order, defined compounds with long-range order, or phase separation. Simple models of mixing account for all these possibilities and allow deriving generic phase diagrams as a function of the thermodynamic conditions—temperature, pressure, composition, contact with an aqueous solution, etc.—under which mixing occurs.

The thermodynamic state (molar volume V, entropy S, enthalpy H, Gibbs free energy G) of a SS of the $A_{1-x}B_xC$ type is conveniently characterized with respect to the thermodynamic functions of its pure AC and BC parents (the end-members of the SS) (McGlashan, 2007). When no chemical interaction takes place at the atomic level, which is the case of a *mechanical mixture*, these quantities are obtained by simple linear interpolation between those of the end members, weighted by their respective mole fractions $X_{AC} = 1 - x$ and $X_{BC} = x$. For example, the molar volume of a mechanical mixture is equal to $V_{MM} = X_{AC}V_{AC} + X_{BC}V_{BC}$, which is the expression of the Vegard's law, and its Gibbs free energy is equal to $G_{MM} = X_{AC}G_{AC} + X_{BC}G_{BC}$.

When actual mixing takes place, the linear relationship between the SS thermodynamic functions and those of its end-members no longer holds. Mixing quantities (ΔV_M, ΔS_M, ΔH_M, ΔG_M) are then defined by difference between V, S, H, and G and those of the mechanical mixture, e.g. $\Delta G_M = G - X_{AC}G_{AC} - X_{BC}G_{BC}$.

A first contribution to ΔG_M, of entropic origin, arises due to the disorder induced by the substitution. In the case of full randomness, it reads (R the gas constant):

$$\Delta S_M^{id} = -R(X_{AC}\ln X_{AC} + X_{BC}\ln X_{BC}) \tag{6.8}$$

ΔS_M^{id} is a positive quantity, which lowers $\Delta G_M = \Delta H_M - T\Delta S_M$ (T the temperature) at finite temperature, and thus stabilizes the SS with respect to the mechanical mixture in the whole composition range. When ΔG_M has no other contribution than ΔS_M^{id}, the SS is said to be *ideal* (hence the superscript *id*).

However, very few SSs are ideal, as shown by numerous results of advanced calorimetric and diffraction measurements (Geiger, 2001; Navrotsky, 2014). This stems from the so-called *excess* mixing terms, which include entropy as well as enthalpy contributions. The excess mixing entropy comprises the vibrational entropy and a contribution due to the decrease, with respect to perfect randomness, in the number of accessible configurations when short-range order exists (Benisek and Dachs, 2012). The excess mixing enthalpy may be of elastic and/or chemical essence. The elastic term results from the lattice distortions which take place upon mixing. Always positive, it destabilizes the mixed compound with respect to a mechanical mixture. The chemical term is associated to effective interactions between the substituent and host atoms. Both contributions to the excess mixing enthalpy for *non-ideal* SSs are accounted for in the following polynomial expansion of ΔH_M proposed by Guggenheim (1937):

$$\Delta H_M = X_{AC}X_{BC}[A_0 + A_1(X_{AC} - X_{BC}) + A_2(X_{AC} - X_{BC})^2 + ...] \tag{6.9}$$

The first term depends on the number and strength W of each type of bond, AA, BB or AB. Its constant A_0 is equal to the energy difference between chemical bonds connecting identical or distinct species $A_0 \propto 2W_{AB} - W_{AA} - W_{BB}$. Usually, the largest

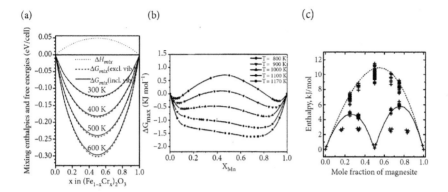

Fig. 6.25 Three simulations of the composition dependence of mixing free energies. (a) $(Fe_{1-x}Cr_x)_2O_3$ nearly ideal solid-solution. Reprinted from Benny *et al.* (2009) with permission. Copyright 2009 by the Royal Society of Chemistry. Permission conveyed through Copyright Clearance Center, Inc.; (b) $Mg_{1-x}Mn_xO$ strongly non-ideal solid solution. Reprinted from Allan *et al.* (2001) with permission. Copyright 2001 by the American Physical Society; (c) $Ca_{1-x}Mg_xCO_3$ non-ideal solid solution with a definite compound—the dolomite mineral—at $x = 0.5$. Reprinted from Vinograd *et al.* (2006) Copyright 2006, with permission from Elsevier.

contribution comes from nearest neighbours but longer-range interactions may sometimes be important. Many body interactions (triplet, quadruplet, etc.) are accounted for in the higher order terms A_1, A_2, and so on. ΔH_M varies with temperature but usually more weakly than ΔS_M^{id}. When only the first term in the Guggenheim expansion is important, i.e. when the strength of the interactions is independent on composition (so-called *regular* SS), ΔG_M is symmetric with respect to $x = 0.5$. Depending on the sign and the value of A_0, ΔG_M displays a single minimum ($A_0/RT < 2$) or two minima ($A_0/RT > 2$) as a function of x, as shown in Fig. 6.25. Higher order terms in the Guggenheim expansion yield an asymmetry of ΔG_M with respect to $x = 0.5$. In the infinite dilution limits ($x \to 0$ or $x \to 1$), $A_0 + A_1$ and $A_0 - A_1$ represent the solubilities of B in pure AC and A in pure BC, respectively. Calorimetric measurements of enthalpies of formation have highlighted a linear correlation between the interaction parameters in the Guggenheim's expansion and the volume mismatch between the two end-members (Davies and Navrotsky, 1983).

Figure 6.25 displays three different behaviours of ΔG_M variations as a function of composition, which illustrate a nearly ideal SS behaviour and two strongly non-ideal ones. In $(Fe_{1-x}Cr_x)_2O_3$, the enthalpy of mixing is positive but small, so that it is overcome at nearly all temperatures of interest by the mixing entropy. A negative free energy of mixing results which is nearly symmetric with respect to composition $x = 0.5$, due to the quasi negligible vibrational contribution compared to the configurational one. At variance, in the mixed $Mg_{1-x}Mn_xO$ oxide, the Gibbs free energy of mixing presents a maximum in a large range of x values, separating two minima at x_1 and x_2 which mark the limits of a *miscibility gap*. Only at very high temperatures, due to the increased weight of the mixing entropy, does the maximum disappear, leading to full miscibility of the two cations. For compositions $x_1 < x < x_2$ in the miscibility gap, phase separation occurs between two SSs of respective composition x_1 and x_2. The non-ideality of the $Mg_{1-x}Mn_xO$ SS was assigned to steric effects arising from the large difference between the ionic radii of Mg^{2+} and Mn^{2+} (0.72 Å and 0.83 Å, respectively (Shannon, 1976)). Relying on the same concept of size mismatch, the CoO-FeO and CoO-MnO SSs were shown to display nearly ideal behaviour for the former and strongly non-ideal behaviour with a miscibility gap for the latter (the ionic radii of Co^{2+}, Fe^{2+} and Mn^{2+} are equal to 0.75, 0.78, and 0.83 Å, respectively) (Pongsai, 2006). In some cases, at intermediate compositions, a definite ordered compound may become more stable than a SS. This happens for example in the mixed $Ca_{1-x}Mg_xCO_3$ carbonate at $x = 0.5$. While the enthalpy of mixing of Ca and Mg ions in the carbonate is large and positive, pointing towards a strongly non-ideal SS, at $x = 0.5$ energy is gained by forming an ordered phase—the dolomite mineral—with alternating layers of Ca and Mg cations.

6.4.3 Specific characteristics of mixed oxide nanoparticles

The ability of finite size NPs to form a homogeneous mixed phase and their thermodynamic phase diagram may strongly differ from the bulk. There are cases where two oxides are insoluble in the bulk while forming stable mixed NPs or vice versa. The first case is well illustrated by mixed titanosilicate NPs (Cuko *et al.*, 2018*a*). A homogeneous mixing of SiO_2 and TiO_2 in the bulk is energetically unfavourable while simulations show that small (less than 1 nm in diameter) mixed $Si_{1-x}Ti_xO_2$ NPs with x between 0.3 and 0.5 are stable. The mixing driving force is the suppression of high energy sites in pure SiO_2 clusters by the replacement of non-bonding oxygen Si centres by Ti, and in pure TiO_2 clusters by the replacement of tetra-coordinated Ti cations by Si. However, as size increases, the energetic contribution of these sites to the total energy becomes less and less important, and a transition toward bulk behaviour takes place at some critical size.

In larger NPs, segregation effects—local variations of composition with respect to the average value—, may be encountered. They have been well-documented at metallic alloy surfaces and in nanoalloys (Ferrando *et al.*, 2008). In NPs, they may manifest themselves in various ways, from full phase separation to partial surface segregation (Fig. 6.26). In the first case, Janus or core-shell NPs form. Janus NPs are composed of two distinct parts of different stoichiometry/structure, potentially displaying different properties (reactivity, hydrophobicity, etc.). In core-shell structures, one component occupies the periphery of the NP while the other one is located in the NP core (Ghosh Chaudhuri and Paria, 2011). Such a configuration in which the most reactive and expensive species would be at the periphery may be searched in heterogeneous catalysis to reduce costs. When partial surface segregation occurs, there is a preferential enrichment, relative to the bulk composition, of the surface region by one constituent.

Segregation effects in oxide NPs are much less documented. It is not always easy to determine which thermodynamic driving force is responsible for one or the other segregation type, because NPs are often synthesized in out-of-equilibrium conditions. Nevertheless, several important factors are identified. The first one is the size mismatch

Fig. 6.26 Schematic representation of several types of segregation effects in NPs. From left to right: Janus NP, core-shell NP, mixed NP with surface segregation.

between the cations: smaller atoms tend to occupy core sites while larger atoms prefer the periphery to escape the internal pressure of the NP. The second factor is the chemical alloying effect potentially leading to phase separation (see preceding section). The third one is surface energy: the end-members with the lowest surface energy tend to segregate to the surface. Beside these three factors, which are also relevant for metallic nanoalloys, there exist preferred environments for cations in a given oxidation state which may favour their location either in the NP core or at the lower-coordinated surface sites. Finally, when NPs are formed in a liquid or gas environment, an additional driving force exists when one component has a preferential bonding to the external ligands. The following case studies exemplify some of these concepts.

Mixed $Ti_{1-x}Ce_xO_2$ NPs have been synthesized by a sol-gel method with nominal compositions $x =10\%$, 50%, and 90% and average particle sizes equal to 2, 6, and 13 nm, respectively (Gionco *et al.*, 2013*b*). This is a case in which steric effects are expected to be important since the Ce and Ti ionic radii largely differ (Ce^{4+}: 1.01 Å; Ce^{3+}: 1.15 Å; Ti^{4+}: 0.56 Å in octahedral coordination). However, the degree of mixing was found to be relatively high, with the presence of both Ti and Ce cations in the surface and subsurface of the NPs. All mixed compounds exhibit a reduced band gap with respect to their two parents, assigned to the presence of a localized Ce^{3+} state in the gap, which reveals the non-stoichiometry of the mixed NPs. Moreover, a definite (oxygen-poor) $Ce_2Ti_2O_7$ compound forms at $x = 0.5$ with a pyrochlore structure, highlighting the existence of chemical interactions between Ti and Ce. The potential applications of such NPs are in heterogeneous catalysis, where ceria is known to be an active species.

Mixed $Zr_{1-x}Ti_xO_2$ NPs of composition up to $x = 0.15$, and sizes in the 30-100 nm range, have been prepared via sol-gel synthesis and then calcined at 1273 K (Gionco *et al.*, 2013*a*). Despite quite different ionic radii of Zr^{4+} and Ti^{4+}, SSs form in this composition range within the ZrO_2 monoclinic phase at low loading, and within its tetragonal phase above $x = 0.15$. A segregation of Ti cations in the subsurface layers is evidenced in the most highly doped material, consistent with a slightly lower surface energy of TiO_2 and the possibility for Ti^{4+} to recover a more favourable octahedral environment. The NPs exhibit a red shift of the valence to conduction band transition proportional to the Ti content, which reaches the value of 1.6 eV for the 10% compound.

Combustion of zinc and magnesium metal vapours at reduced pressures followed by vacuum annealing yields $Mg_{1-x}Zn_xO$ NPs with cubic shape and edge lengths below 25 nm (Stankic *et al.*, 2010). Preferential Zn^{2+} segregation takes place at low-coordinated surface sites of the nanocubes (edges and corners), as revealed by their spectroscopic

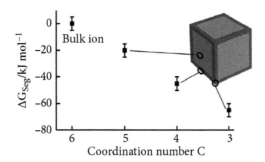

Fig. 6.27 Mean free-enthalpy of zinc segregation on various sites of coordination number C on MgO cubes. The scheme visualizes the expected distribution of Zn-O moieties (darker shade). Reprinted from Stankic *et al.* (2010) with permission. Copyright © 2010 IOP Publishing Ltd. All rights reserved.

signatures. First principles simulations provide the mean free enthalpy of Zn segregation at low coordinated sites of the MgO nanocubes (Fig. 6.27), showing that, at thermodynamic equilibrium, all corner and edge Mg cations in the $Mg_{0.9}Zn_{0.1}O$ NPs should be replaced by Zn^{2+} ions. Since ionic radii of Mg and Zn are close to one another (0.72 and 0.74 Å, respectively), Zn segregation preference was mainly assigned to the difference of cation electronegativity. Due to the less electropositive character of Zn compared to Mg, ZnO bonds are more covalent and thus more stable at low-coordinated sites of the NPs where the electrostatic (Madelung) potential is weaker.

The thermodynamic properties of $Mn_{1-x}Mg_xO$ NPs containing 1,728 to 21,952 ions have been theoretically studied and their phase diagram determined (Purton *et al.*, 2013). In the bulk, the enthalpy of mixing is positive, thus indicating the existence of a miscibility gap in a large composition range at low temperature (see Fig. 6.25). In the NPs, at low-Mg concentrations ($x \rightarrow 0$), due to their smaller ionic radius, the Mg ions dissolve in the core of the MnO particles. However, the Mg-O bonds are stretched with respect to their value in bulk MgO (2.17 Å versus 2.1 Å). This strain is responsible for a positive enthalpy of mixing. In the opposite limit ($x \rightarrow 1$), the preferential location of Mn ions at the MgO particle corners, edges, and facets is driven by the smaller MnO surface energy and the larger Mn ionic radius. The mixing enthalpy is negative in this composition range. The variations of ΔH_M as a function of x at fixed temperature for NPs of various sizes are represented in Figure 6.28. As size increases, the x range in which ΔH_M is positive increases, and its negative minimum becomes shallower. This points to an increase of the miscibility gap and an increase of the consolute temperature (the temperature above which there is complete miscibility).

6.5 Summary

Oxide nanoparticles represent a state of matter which is intermediate between molecules and infinite crystals and constitute a relevant stage in nucleation processes. The presence of a large number of under-coordinated atoms gives them original structural and electronic properties, exploited in many applications.

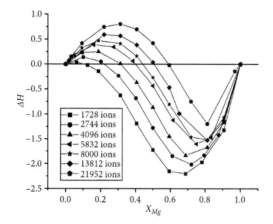

Fig. 6.28 Calculated mixing enthalpy at T = 800 K in mixed $Mn_{1-x}Mg_xO$ NPs, as a function of Mg content, for NPs of various sizes. Reprinted from Purton *et al.* (2013).

At the smallest sizes, in the non-scalable regime, most of the time, their structures, completely differ from the bulk ones and strongly vary with their environment (ultra-high vacuum, gaseous, or humid atmosphere). Some sizes which are especially stable with respect to fractionation are said to be magic.

In the scalable regime, crystalline structures are recovered but the overall NP habit is determined by the relative energies of the various surface orientations, again largely dependent on the environment conditions. For example, polar facets which are unstable in ultra-high vacuum may become the most stable ones in a humid environment, and lead to specific NP shapes. At a given size, the competition between surface and bulk energies may favour crystalline polymorphs which are only metastable in the bulk. In that case, structural phase transitions as a function of size are expected. The NP HOMO-LUMO gaps are sensitive functions of the presence of under-coordinated atoms on the NP facets, edges and corners as well as of confinement effects. The same is true for the NP local electronegativity and acid character. The pH of an aqueous synthesis medium is an efficient parameter to control NP sizes and shapes.

Combining cations of different size, electronegativity and reducibility in mixed oxides allows tuning many of their properties. Mixing may be favourable in the bulk and not in finite size objects or vice-versa. When there is a tendency to phase separation, Janus or core-shell structures are obtained. When actual mixing takes place, segregation in the outermost NP layers usually exists, driven by cation size mismatch, surface energy differences or short-range cation-cation interactions. While all these effects have been well studied in metallic nanoalloys, the research field of mixed oxide NPs is only in its early stages.

7
Clay mineral layers and nanoparticles

7.1   Introduction

Whereas the focus of the previous chapters was on ultra-thin films and NPs made of simple oxides and mainly synthesized in laboratories or industrial companies, the present chapter broadens the perspective and provides an introduction to the properties of naturally occurring oxide ultra-thin films and NPs of more complex composition, namely clay minerals. Clays represent a wide family of nanomaterials, formed from the weathering of primary rocks at the Earth surface, such as feldspar, mica, amphibole or olivine, in the presence of water. For this reason, they are widespread in nature and represent a major component of continental and marine sediments as well as of soils together with sand. In a pedologic context (i.e. the science of soils), all particles with sizes less than 2 micrometers were initially defined as clay particles, whatever their chemical composition. However, such a definition encompasses both simple oxide particles—silica, quartz, iron oxides, etc.—and truly clay minerals. Only the latter are considered here.

Although clay minerals have been known and used since prehistoric times, their structure at the atomic scale was solved in the last century only, thanks to the advent of X-ray diffraction methods. They are alumino-, magneso-, or ferro-silicate layered materials resulting from the association of atomically thin silicon oxide and aluminium/magnesium/iron oxo-hydroxide sheets. They belong to the wide family of phyllosilicates (the Greek prefix 'phyllo = layered'). Their layered structure is responsible for their extreme anisotropy—micrometer size along the layers and nanometre size perpendicularly to the layers, as evidenced e.g. in Fig. 7.1—and their extremely high specific surface area—up to several hundreds of m^2/g—which is at the origin of their unique properties and of their strong interaction with their surroundings, in particular with water.

A whole book would not be enough to present all the fascinating features of clay minerals, their conditions of formation, their properties, their relevance in geophysics and geochemistry and all their applications. We send the interested readers to the numerous general as well as specialized books and review articles which have been written on the subject, among which one can cite *Geology of Clays* by Millot (2013), *Clays* by Meunier (2005), *The Chemistry of Clay Minerals* by Weaver and Pollard (2011), *Handbook of Clay Science* by Bergaya and Lagaly (2013), *Clay Minerals* by Wimpenny (2016) or *Surface and Interface Chemistry of Clay Minerals* by Schoonheydt

Oxide Thin Films and Nanostructures. Falko P. Netzer and Claudine Noguera, Oxford University Press (2021). © Falko P. Netzer and Claudine Noguera. DOI: 10.1093/oso/9780198834618.003.0007

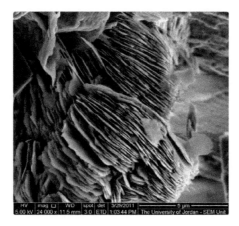

Fig. 7.1 Scanning electron microscope (SEM) image of kaolinite particles. Scale bar is 5 micrometres. Courtesy of Mahmoud Abbas.

et al. (2018). Here, a physico-chemical perspective is chosen, with no claim of covering the whole field. The main focus will be on the most common clays, their structural properties (Sections 7.2 and 7.3), the conditions of their formation (Section 7.4), their interaction with water (Section 7.5), their chemical properties (Section 7.6), and a short account of some of their applications and uses (Section 7.7).

7.2 Structure and composition of the clay layer

The specificity of clays is the existence of alumino/magneso- silicate layers, of well-defined structure but variable composition. Iron may also be present in the layers but with a lower abundance. For the sake of simplicity, we will not consider it in the following.

7.2.1 Structural building blocks

The clay structure consists of atomically-thin layers, made of two types of sheets. The tetrahedral (T) sheet consists in a planar network of SiO_4 tetrahedra linked together by three oxygen corners which delimit the basal ditrigonal cavities (Fig. 7.2(a)). The apical fourth oxygen atoms of the tetrahedra point either upward or downward with respect to the basal plane, depending on the link between the T sheet and the neighbouring sheet.

The octahedral (O) sheets are made of $MO_2(OH)_4$ octahedra linked together by the edges. The smallest structural unit may contain three octahedra. If they are all occupied by a cation, the sheet is called trioctahedral. When one cation is missing, it is referred to as a dioctahedral sheet. The oxygen atoms are bound to three cations in the former case and two cations in the latter. These configurations are reminiscent of those of brucite $Mg(OH)_2$ and gibbsite $Al(OH)_3$ crystals, respectively, and correspond to an occupation of the octahedra centres by divalent (most often Mg^{2+}) or trivalent (most often Al^{3+}) cations (Fig. 7.2(b,c)).

(a)   (b)   (c)

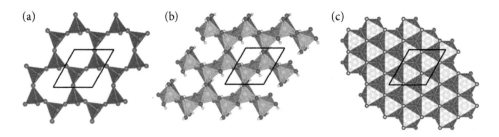

Fig. 7.2 (a) A phyllosilicate SiO_4 T sheet; (b) A $MO_2(OH)_4$ O sheet in a di-octahedral structure, with M a trivalent cation; and (c) A $MO_2(OH)_4$ O sheet in a tri-octahedral structure, with M a divalent cation.

The T and O sheets represent the elementary building blocks of the clay structure. Actually, they never occur separately but are associated in 'layers'. A layer may involve two, three, or more sheets. The most common clays result from the repetition of one T and one O sheet (TO layers in the so-called 1:1 phyllosilicates) or by the repetition of one O sheet sandwiched between two T ones (TOT layers in 2:1 phyllosilicates), as schematized in Fig. 7.3.

In 1:1 phyllosilicates, the two sheets are linked by the apical oxygen atoms of the SiO_4 tetrahedra. One OH group of the $MO_2(OH)_4$ octahedra points toward the SiO_4 sheet, while the three other OH groups stand nearly perpendicular to the outer surface in trioctahedral clays. In dioctahedral clays, only two of them do so, while the third one tilts and lies nearly parallel to the surface. Compared to the ideal structure of isolated T and O sheets, their association induces distortions which lower the symmetry of the structure. The 1:1 layers are about 7 Å thick and have two dissimilar basal terminations: the T sheet exposes siloxane groups while hydroxyl groups protrude from the O sheet, as at gibbsite or brucite (0001) surfaces. For this reason, the basal surface of the O sheet is sometimes named gibbsite/brucite termination.

The 2:1 layers contain two outer T and one inner O sheets and are approximately 10 Å thick. The two T layers are inverted with respect to each other and two-thirds of the OH groups in the O sheet are replaced by oxygen atoms. The layers are symmetric with two identical terminations and no dipole moment.

(a)   (b)

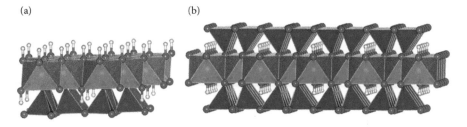

Fig. 7.3 (a) TO and (b) TOT layer of lizardite 1:1 and talc 2:1 phyllosilicates. (Si = blue, Mg = brown, O = red, H = white).

Table 7.1 Classification of the most common phyllosilicates. Q is the layer charge per formula unit, cations in the T sheet are written first and X represents the compensating cation of charge $+n$.

Characteristics	Di-octahedral	Tri-octahedral.
1:1 layers ($Q = 0$)	Kaolinite $Si_2Al_2O_5(OH)_4$	Lizardite $Si_2Mg_3O_5(OH)_4$
2:1 layers ($Q = 0$)	Pyrophyllite $Si_4Al_2O_{10}(OH)_2$	Talc $Si_4Mg_3O_{10}(OH)_2$
2:1 layers $Q \approx$ -0.2 to -0.6	Smectites $(Si_{4-x}Al_x)(Al_{2-y}Mg_y)O_{10}(OH)_2X_{(x+y)/n}$	
2:1 layers $Q \approx$ -0.6 to -0.75	Vermiculites $(Si_{4-x}Al_x)(Mg_{3-y}Al_y)O_{10}(OH)_2X_{(x-y)/n}$	
2:1 layers $Q \approx$ -0.75 to -0.9	Illites $(Si_{4-x}Al_x)(Al_{2-y}Mg_y)O_{10}(OH)_2X_{(x+y)/n}$	
2:1 layers $Q = -1$	Mica Muscovite $(Si_3Al)Al_2O_{10}(OH)_2K$	Phlogopite $(Si_3Al)Mg_3O_{10}(OH)_2K$

7.2.2 Chemical composition

From a chemical point of view, clays may adopt a wide variety of compositions, due to the conditions of their formation by rock weathering in contact with the fluids of the Earth's environment (see Section 7.4). One distinguishes two main families, according to whether the layers are neutral or not. In some minerals, all tetrahedra and all octahedra contain cations of the same valency (+4 in the former, and +3 or +2 in the latter) while in others, isomorphic substitution by cations of lower valencies creates a negative charge Q which is balanced by the presence of compensating cations in the interlayer space. Moreover, in some cases, water enters the inter-layer region and binds to the compensating cations. A simplified classification of the various clay types is given in Table 7.1 (Bailey, 1980). Clays with neutral or charged layers are considered successively, in the following.

Clays with neutral layers. Four clay minerals only contain silicon cations in the T sheets and aluminium or magnesium cations in the O sheets. Kaolinite and lizardite are 1:1 dioctahedral and trioctahedral phyllosilicates, respectively. Pyrophyllite and Talc are their 2:1 equivalents. For all of them, there is a corresponding mineral with Al^{3+} or Mg^{2+} replaced by Fe cations of equal valency. In all these minerals, the layers are neutral and homovalent substitutions are relatively rare.

Kaolinite, of chemical formula $Al_2Si_2O_5(OH)_4$, was named from a Chinese hill from which it has been extracted for centuries. It is a white compound, with a very high melting point, mainly used for the fabrication of porcelain, but also in various modern technologies. It has two nonequivalent (0001) basal surfaces, the siloxane and gibbsite terminations, and its layers bear a non-zero dipole moment.

Lizardite $Mg_3Si_2O_5(OH)_4$ is the trioctahedral counterpart of kaolinite, in which the two Al^{3+} cations are replaced by three Mg^{2+} cations, at constant total charge. It belongs to the large family of serpentine minerals found in metamorphic rocks, in which two other important members are the antigorite and the chrysotile, a form of asbestos. As with kaolinite, lizardite contains two different types of structural OH groups whose

Table 7.2 Ionic radii (Å) relevant for phyllosilicates as a function of coordination Z, according to Shannon (1976). For iron, radii in low spin (LS) and high spin (HS) states are given.

ions	$Z = 4$	$Z = 6$	ions	$Z = 4$	$Z = 6$
O^{2-}	1.24	1.26	Fe^{3+}	0.63	0.79(HS);0.69(LS)
OH^-	1.21	1.23	Al^{3+}	0.53	0.68
Mg^{2+}	0.71	0.86	Si^{4+}	0.40	0.54
Fe^{2+}	0.77	0.92(HS);0.75(LS)			

structure and vibration properties have been analysed using DFT methods (Balan *et al.*, 2002), and, as with kaolinite, its layers bear a dipole moment.

Pyrophyllite $Al_2Si_4O_{10}(OH)_2$ is a 2:1 dioctahedral phyllosilicate with neutral and dipole-free layers. It is a soft mineral, easily carved and machineable which finds various industrial applications. It has a single type of basal surfaces—the siloxane terminations—and only inner OH groups pointing toward the silica rings. Talc, its trioctahedral $Mg_3Si_4O_{10}(OH)_2$ counterpart, is often used as a lubricant and in many cosmetic products. A molecular dynamics simulation of a single layer of pyrophyllite in a wide temperature range, using the CLAYFF classical force field, has evidenced a structural transition around 450 K and its mechanical properties (full elasticity tensor, bending and torsion stiffnesses) were calculated from stress-strain diagrams (Mazo *et al.*, 2008).

Isovalent substitutions may exist in all these minerals when Fe^{2+} replaces Mg^{2+}, or Fe^{3+} replaces Al^{3+}, but the degree of substitution is generally low.

Clays with charged layers. As mentioned previously, clay minerals are rarely stoichiometric. Cations in the T sheet, or in the O one, or in both may be substituted by cations of similar or lower valency. However, not all substitutions are possible, because some would induce distortions of the phyllosilicate skeleton and cost a great deal of elastic energy. The consideration of ionic radii (Table 7.2) and Pauling's rules help us to understand which substitutions are possible in the tetrahedra and octahedra (coordinations Z equal to 4 and 6, respectively).

From geometric considerations and the knowledge of the cation-to-anion ionic radii ratio r_+/r_-, the first Pauling's rule (see Chapter 3) states which polyhedron type is stable around a given cation. Tetrahedra are expected when $0.225 < r_+/r_- < 0.414$ and octahedra when $0.414 < r_+/r_- < 0.732$. Replacement of Si^{4+} in tetrahedra can thus be provided by Al^{3+} and more rarely by Fe^{3+}. More flexibility exists in octahedra where Mg^{2+} and most divalent transition metal cations may replace Al^{3+}.

Heterovalent substitutions mainly occur in 2:1 clays — smectites, vermiculites, illites, and micas — leading to charged layers. Electrical neutrality requires the presence of additional cations which are located in the interlayer space.

Micas are 2:1 di-octahedral phyllosilicates, derived from pyrophyllite by the substitution of one fourth of the Si^{4+} cations by Al^{3+}, and charge compensation by K^+ (K-muscovite) or Na^+ (paragonite) ions. As in zeolite frameworks, two Al tetrahedra generally avoid neighbouring positions, according to the Lowenstein's rule. The compensating cations are located in the hexagonal cavities of the T sheets, with a

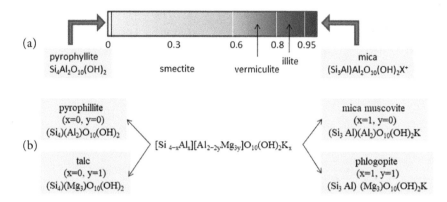

Fig. 7.4 Representation of complex clay compositions as solid solutions with two (a) or four (b) end-members. (a) Binary diagram of $(Si_{4-x}Al_x)Al_2O_{10}(OH)_2$ with substitution of Si^{4+} by Al^{3+} in the T sheets. The two end-members are pyrophyllite ($x = 0$) and mica ($x = 1$) and intermediate compounds are smectites (bedeillites), vermiculites, and illites. (b) Quaternary diagram of $(Si_{4-x}Al_x)(Al_{2-2y}Mg_{3y})O_{10}(OH)_2K_x$, with four end-members pyrophyllite ($x = 0, y = 0$), mica muscovite ($x = 1, y = 0$), talc ($x = 0, y = 1$), and phlogopite ($x = 1, y = 1$). The chemical formula is written assuming K^+ is the compensating cation.

preference for Si_4Al_2 rather than Si_5Al rings (Odelius *et al.*, 1997). The (0001) surface of stacked mica sheets is polar. It becomes charge neutral when half of the potassium ions are removed from the outermost surfaces. Simulations using classical interatomic potentials have shown that the (0001) surface energy is sensitive to the degree of Al/Si ordering, the lowest energy configuration being obtained for complete disorder in the Al/Si substitution (Purton *et al.*, 1997).

Smectites, vermiculites, and illites are 2:1 clays with low ($0.3 \leq Q \leq 0.6$), medium ($\approx 0.70 - 0.75$), or large (≥ 0.8) negative layer charge, respectively. Many different variants of smectites exist, depending on the location and importance of the substitutions. Among them, the montmorillonite and bentonite—both named from the locations where they were first identified, Montmorillon in France and Fort Benton in the United States—, are di-octahedral smectites with substitutions occurring essentially in the O sheets. Beidellite is another smectite with substitutions mainly located in the T sheet. Smectites, vermiculites, and illites exist in di-octahedral and tri-octahedral forms, although this distinction becomes somewhat fuzzy when substitutions take place in the O layer. The layer charge is balanced by monovalent or bivalent cations X^{n+} in the interlayer region. Most of the time, the substituting cations in the sheet skeleton are randomly distributed and the positions of the compensating interlayer cations are correlated to theirs. Compounds of intermediate compositions can thus be considered as solid-solutions formed from two or more end-members. Figure 7.4a shows the example of binary solid solutions $(Si_{4-x}Al_x)Al_2O_{10}(OH)_2X^{n+}_{(x/n)}$ obtained by substitution of Si^{4+} cations in the tetrahedral sheet by Al^{3+} cations, between the two end-members pyrophyllite ($x = 0$) and mica ($x = 1$). Various representations of clay minerals with more than one substitution exist, depending on the number of end-members. The one

(a)   (b)

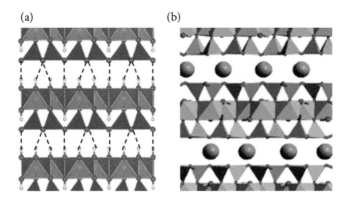

Fig. 7.5 (a) Profile view of the TO layers of lizardite with emphasis on the H-bonds (dashed lines) between them; (b) profile view of the TOT layers of mica muscovite with emphasis on the electrostatic coupling between them via the K ions (in green). Reprinted with permission from Wang *et al.* (2006). Copyright 2006, with permission from Elsevier.

schematized in Fig. 7.4(b) refers to solid solutions with four end-members.

Substitutions may also occur in the interlayer space. They obey mechanisms which are strongly dependent on the possible presence of water and which will be described in Section 7.5.

7.3 Structural properties at larger scales

Clay minerals display various levels of organization, from the nanometre to the micrometre scales, which result from the interactions between the layers and the way they are stacked together.

7.3.1 Interactions between layers

The elementary interactions between layers are of electrostatic, polarization, van der Waals or H-bonding nature, depending on the layer type and composition.

In kaolinite and lizardite, the neutral TO repeat units are held together by weak H-bonds between the OH groups of the gibbsite/brucite termination and the oxygen atoms of the siloxane one (Fig. 7.5(a)), responsible for a low cleavage energy (Geysermans and Noguera, 2009) and a thickness of the interlayer spacing of the order of 3 Å. Interestingly, their layers bear a non-zero dipole moment, so that their stacking is polar. It has been theoretically shown (Hu and Michaelides, 2010) that a polar catastrophe is expected in kaolinite above approximately three layers, while at lower thickness, the formation energy, gap, and density of states present all characteristics of polarity (see Chapter 4). Adsorption of charged species such as those present in the surrounding aqueous solutions may provide polarity compensation.

In pyrophyllite and talc which only have siloxane terminations and in which the layers are neutral and contain no dipole moment, polarization and van der Waals interactions are dominant. In the modelling of these minerals, the use of dispersion

(a)

(b)

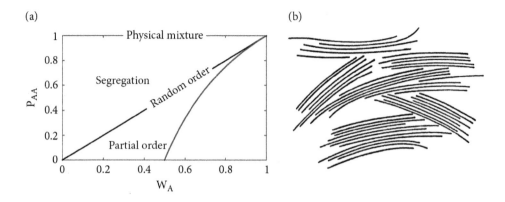

Fig. 7.6 (a) Schematic interstratified illite-smectite phase diagram as a function of the illite content (W_A) and the probability of having two successive illite layers (P_{AA}). The regions of mixing, segregation, and partial order are indicated. The curved line indicates the frontier with maximal order. Courtesy A. Meunier. (b) Schematic view of montmorillonite parti-cles evidencing face/face and face/edge contact points and a complex network of porosities. Reprinted from Lagaly and Malberg (1990). Copyright 1990, with permission from Elsevier.

corrected methods appears necessary to correctly account for the interlayer thickness (Tunega *et al.*, 2012).

In clay minerals with charged layers, the layer interactions are stronger due to electrostatic forces. The inter-layer bonding increases with the layer charge Q, from smectites to vermiculites, but remains weak enough to allow the penetration of sev-eral layers of water molecules in the inter-layer space. This gives rise to the swelling phenomenon (see Section 7.5). In montmorillonites (di-octahedral smectites), simula-tions have predicted that the interlayer spacing varies approximately linearly with the cation radius and that smaller cations (Li^+, Mg^{2+}, and Ca^{2+}) preferentially bind to one surface, whereas larger cations (Na^+ and K^+) remain located close to the middle of the interlayer region (Voora *et al.*, 2011). These electrostatic interactions become even stronger as the negative charge Q approaches -1, as in micas or illites. In these minerals, the compensating cations strongly bind the layers together, preventing water from entering the interlayer space (Fig. 7.5(b)).

7.3.2 Organization at larger scales

At larger scales, ordered or disordered structures may result from these interactions. Those formed by the stacking of several layers are named 'crystals'. As in many layered compounds, the stacking may occur with or without rotations compatible with the layer structure. Rotations of $0°$ or $120°$ rotations are the most frequent when the layers have the same composition, leading to repeat units containing one or two layers (so-called $1M$ or $2M_1$ polytypes). This is not the case when interlayer interactions are weak, as in smectites, in which random rotations induce a turbostatic stacking.

Since different layers have similar or identical two-dimensional periodicity and simi-lar structures, chemically different layers may coexist within one and the same crystal, leading to the formation of mixed-layer minerals, as in illite-smectite interstratified clays (Drits, 2003; Brigatti *et al.*, 2006). This requires that the structural mismatch between the two types of layers is small. Interstratified clays may be regarded as one-dimensional alloys, and, as in an AB alloy, it is possible to rationalize their degree of organization as a function of their composition and the values of the short-range interactions (Brown, 1982; Meunier, 2005; Noguera and Fritz, 2017). Figure 7.6(a) evidences regions of physical mixture ($P_{AA} = 1$), segregation ($P_{AA} < W_A$), partial ($P_{AA} < W_A$), or maximum order ($P_{AA} = 2 - 1/W_A$) of an AB interstratified crystal, as a function of the concentration W_A of its A component, and of the probability P_{AA} of having two successive A layers.

Actually, clay crystals, and in particular smectites, remain small with rarely more than a few layers. The crystal thickness increases with the layer charge as a result of the stronger interlayer interactions. It is thus larger in vermiculites and illites than in smectites, and eventually it leads to more macroscopic crystals in micas. The association of crystals produces 'particles', which may aggregate randomly, as schematized in Fig. 7.6(b). Particles generally display irregular shapes, curved regions and weak contact points and are separated by a complex network of pores of various sizes which may allow the penetration of water (see Section 7.5). The contacts between particles may easily break under stress or in the presence of water, which influences the clay rheological properties.

7.3.3 Malleability–Elasticity

The small size of the clay crystals and their loose organization at larger scales allow them to slide over one another, to agglomerate or to split, depending upon the environmental physico-chemical conditions. This is the basis of their malleability, illustrated in the name 'modelling clay'.

Clay malleability has been widely exploited by men. They have shaped mudbricks for dwelling constructions or large religious architectures. Baked clay has been used to fabricate fired bricks or tiles, or to produce pottery and statuary. When cooking is necessary, illites are preferred to smectites, for they do not incorporate water and thus present no risk of cracking when passing in the furnace. A particularly striking example of clay statuary is the Terracotta Army of Qin Shi Huang, the first Emperor of China who was buried in the late third century BC with more than 8,000 soldiers, 130 chariots with 520 horses, and 150 cavalry horses, all made of clay (Fig. 7.7).

Due to their widespread availability in the river deltas, and thanks to their malleability, clays have served as early writing supports in what formerly was Mesopotamia. Tens of thousands of clay tablets written in cuneiform have been discovered and there are records of the existence of large libraries of clay tablets, as in Ninive under the reign of Assurbanipal.

The mechanical properties of clays are also at the root of many of their modern applications including underground storage of waste materials, drilling activities, and development of novel nanocomposite materials. Due to their layered structures, sheet silicates are among the minerals with the highest elastic anisotropy. The elastic constants of kaolinite, mica muscovite, and illite-smectite crystals have been estimated by DFT simulations (Sato *et al.*, 2005; Militzer *et al.*, 2011). They are extremely

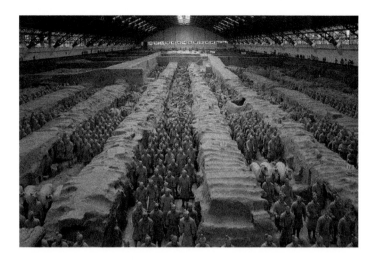

Fig. 7.7 The Terracotta Army of Qin Shi Huang, located near Lishan, central China. Photo produced by David Castor (public domain).

anisotropic, with lower values along the c direction. However, the organization of clay particles at larger scales with porosities of various sizes, as met for example in shales—a mix of clay mud and tiny fragments of other minerals—, makes it difficult to predict their elastic behaviour since an averaging of single crystal properties is insufficient (Militzer *et al.*, 2011). Concerns with the interpretation of seismic wave travel times for example result from this complexity.

The smectite/vermiculite hydration state also impacts their mechanical properties. It was numerically shown that the stiffness perpendicular to the layers largely varies with the hydration level with two maxima associated to the formation of one or two water monolayers in the interlayer space (Ebrahimi *et al.*, 2012).

7.4 Formation and clay size

Clays are ubiquitous in our environment and thus are cheap materials. However, for specific applications which require high purity, synthesis protocols have been developed. Whatever the formation procedure, clay particle sizes rarely exceed the micrometre, which remains, to some extent, a puzzle.

7.4.1 Formation in the natural medium

Three main mechanisms lead to clay formation on Earth (Wimpenny, 2016; Ehlmann *et al.*, 2013).

Clay minerals may be produced by weathering processes of primary silicate minerals at the surface of or close to the Earth's surface. Wind, rain and frost crack rocks and shape our landscapes. Chemical weathering induces a slow dissolution of the rocks in contact with rain or seepage waters. The latter progressively become enriched in elements like silicon or aluminium which are prevalent in the rocks, or in other cations like magnesium, iron, etc., and quickly, oversaturation occurs (see Chapter 2) leading to the precipitation of clay minerals. This mechanism is called neoformation. The clay structure and composition then depend on the type of primary mineral, the fluid composition, its pH and temperature, the water-rock contact time, and the influence of biological agents (Galán, 2006; Appelo and Postma, 2004; Wimpenny, 2016).

Clays may also form by transformation of a previous clay structure when physico-chemical conditions change. This takes place via ion exchange or by alteration of the structural layers. For example, in soils as one moves from the base to the top of a soil profile, one finds that micas have transformed into vermiculite, smectitic mixed layers, and finally amorphous compounds.

Detrital inheritance is a third mechanism of clay formation. Transported by winds or rivers, clay particles accumulate in river beds, bottoms of lakes, in river deltas, and on the sea floor, thus participating to their imperviousness (Eberl, 1984).

Interestingly, clay minerals have also been discovered on Mars. Their presence requires an interaction between water and rock, which the present Martian atmospheric conditions do not allow. Their localization in the most ancient Noachian terrains—of the order of 3.7-billion-years old—raises the question of which climate was prevailing at that time and which mechanism led to their formation. These are still open questions (Ehlmann *et al.*, 2013).

7.4.2   Synthetic clays

Although natural clay minerals have found a large variety of applications since pre-historical times, today's industrial applications require high purity and crystallinity which are not met in natural clays due to the many impurities they contain and their frequent heterogeneous composition. Besides, several interesting clay minerals are not available in sufficient quantities for industrial needs. Synthesis has thus become an alternative means to obtain pure and homogeneous phases (Zhang *et al.*, 2010).

Most early laboratory syntheses were performed under conditions close to those present at the Earth's surface, i.e. at low temperatures, or by transformation of natural minerals. However, they had low yields, they required long reaction times and the desired phases were often mixed with other products. The hydrothermal technique, with the flexibility of varying the synthesis parameters (pH, temperature, pressure and composition) is nowadays widely used. The chosen starting materials are generally aluminosilicates or gels with a composition close to that of the desired product.

An illustrative example is provided by the synthesis of talc $Si_4Mg_3O_{10}(OH)_2$ (Dumas *et al.*, 2013). Talc is used in numerous industrial applications, including paper, paints, ceramics, cosmetics, etc., for its mechanical and lubricating properties. However, natural talc loses its crystallinity when ground below 1 micrometre. Efforts have thus been made to produce an inexpensive, stable, highly pure product of submicronic size. A successful protocol in two steps was developed, with a preliminary preparation of a talc precursor at room temperature with the proper Mg/Si talc ratio, obtained by co-precipitation of sodium metasilicate and magnesium acetate, followed by a hydrothermal treatment of several hours (precursor pH $\approx$ 8.8, $T = 300°C$, $P = 85$ bars). It was found that an increase in the reaction time, temperature, acidity, and salt concentration was beneficial to the talc crystallinity in both the parallel and perpendicular directions to the layers. Such a procedure fulfils the industrial requirements of an inexpensive, convenient, and rapid method of preparation.

7.4.3 Clay size

The structural specificity of clay particles is their nanometric thickness and (sub) micrometric lateral size. This size limitation, which is not met in other minerals which form under the same conditions, to a certain extent remains unexplained.

Indeed, the formation of clays by precipitation from an oversaturated solution obeys the concepts of crystal growth. In a closed system, when the saturation state decreases and thermodynamic equilibrium is reached, the solid phase should consist of a single large particle, gathering all available growth units, which is obviously not the case for clays.

Several hypotheses have been put forward. The first one assigns the limiting factor to the available space for growth, since clay particles are often formed in the pores of rocks, where the fluid flow induces the dissolution of the parent rock and creates oversaturation. However, very often, the clay particles are smaller than the pores, and even under less confined conditions, their dimensions remain micrometric. Additionally, consideration of the available number of ions in the confined space is not sufficient to understand the size limitation.

A second hypothesis (Meunier, 2006) is that growth is limited by the disordered cation distribution in the T and O sheets and the crystal defects in the layer stacking. For simpler elemental crystals, the atom diffusion in the aqueous solution and at the particle surfaces allows an easy integration of growth units into the crystalline network, since there exists an important number of attachment sites compatible with the crystal structure. By contrast, the crystalline skeleton of clays is complex. The building of particles from individual ions faces severe constrains and multiple reasons may explain why ions get stuck outside their expected locations. When defects become too numerous, the barrier for further ion incorporation becomes so high that growth is inhibited, and nucleation of new particles becomes more favourable. This hypothesis is enticing but remains to be validated. Understanding the growth limitation of clays today remains challenging.

7.5 Interaction with water

Due to the conditions under which clays form and occur, the presence of water, its interaction with the silicate layers, and the protonation and deprotonation processes it induces are fundamental to understand the clay macroscopic properties and reactivity. As already mentioned, water can penetrate in the interlayer regions of some clay families (smectites and vermiculites). Its dissociation modifies the surface electric charges, with consequences on sorption properties and stability of colloidal suspensions. The interaction of clay minerals with water depends on their chemical composition, the layer termination (siloxane or gibbsite/brucite), the permanent charge Q they bear and the nature of their compensating cations. In this respect, basal surfaces and edge surfaces behave differently.

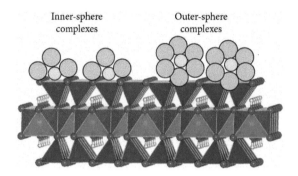

Fig. 7.8 Schematic representation of inner-shell and outer-shell complexes at the siloxane surface of a 2:1 clay layer. Cations and water molecules are represented as yellow and blue spheres, respectively.

7.5.1 Basal surfaces–swelling–confined water

At the gibbsite/brucite terminations of 1:1 clays, the OH groups which point outward from the layer are the most reactive sites. They may form strong hydrogen-bonds with water molecules, thus conferring a hydrophilic character to these terminations. Due to these strong interactions, the structure of the first water layers adsorbed at the kaolinite gibbsite surface is close to that of ice. This may explain why kaolinite particles act as ice nucleating agents in the upper atmosphere.

At the neutral siloxane terminations of kaolinite, lizardite, pyrophyllite, and talc, the silicon atoms are fully coordinated, and the oxygen atoms are engaged in Si-O bonds of a strong covalent character. They have no unsaturated orbitals and present a weak Lewis base (electron donor) character. Their proton affinity is low. The only reactive site is the ditrigonal cavity towards which the hydroxyl groups point, which belong to the O sheet. These terminations are hydrophobic, as evidenced by contact angle and flotation measurements.

In contrast, the siloxane terminations of charged layers are hydrophilic. The compensating cations polarize the surface oxygen atoms. When substitutions take place in the O sheet, the excess negative charge makes it possible for the surface to form reasonably strong adsorption complexes with cations and water molecules. These interactions become even stronger when substitutions occur in the T sheet, since the excess negative charge is localized closer to the siloxane surface (Sposito and Prost, 1982). Depending upon the relative strength of their interaction with the siloxane termination and the water molecules, cations may form inner-sphere or outer-sphere complexes with the water molecules. In the former case, a direct bond to the siloxane surface exists without intervening water molecules. In the latter case, the cations remain fully solvated (Fig. 7.8).

The degree of solvation, i.e. the number of water molecules cations can bind, is a function of their size, charge, and polarizability. Cations with the smallest radii (Li^+, Be^{2+}) can bind four water molecules. This number increases to six for Mg^{2+}, Ca^{2+}, Zn^{2+}, Cd^{2+}, Hg^{2+}, Pb^{2+}, and Al^{3+}, and to eight for cations such as Sr^{2+}. The hydration enthalpy, which is the energy released when one mole of ions undergoes

Table 7.3 Hydration enthalpies (Burgess, 1999) and ionic radii (Shannon, 1976) of some monovalent and divalent cations.

Ion	Hydration enthalpy (kJ/mole) (kJ/mole)	Ionic radius (Å) (Å)
Li^+	−515	0.76
Na^+	−405	1.02
K^+	−321	1.38
Rb^+	−296	1.52
Cs^+	−263	1.67
Be^{2+}	−2,487	0.45
Mg^{2+}	−1,922	0.72
Ca^{2+}	−1,592	1.00
Sr^{2+}	−1,445	1.18
Ba^{2+}	−1,304	1.35

hydration, decreases along a column of the periodic table, as the ionic radii increase, and increases with the cation charge (Table 7.3) (Burgess, 1999).

While these considerations are relevant to the outer surfaces of all clay minerals, they also apply to processes occurring in the interlayer space of smectites and vermiculites (layer charge $|Q| < 0.8$ per Si_4O_{10} units) in which water may penetrate. Their hydration state varies with the relative humidity, in a stepwise manner, typical of phase transitions, as shown schematically in Fig. 7.9. The resulting periods along the c direction in smectites are approximately equal to 10, 12.5, 15, and 17 Å for zero, one, two or three water layers, respectively. Such a volume expansion is named crystalline swelling. The maximum hydration state depends on the layer charge and on the nature of the compensating cations. It increases with the cation hydration enthalpy at given layer charge for monovalent cations. In smectites, it has no limit for Na^+ and Li^+. It corresponds to three water layers for Ca^{2+}, Mg^{2+}, and Ba^{2+}, and to two or one layers for K^+. Associated to hydration/dehydration processes, large hysteretic structural relaxations take place. The swelling properties evolve with temperature and pressure, as attested by the modifications of hydration/dehydration processes upon burial in sedimentary basins. Simulations have found that hydrated smectites are thermodynamically stable down to about 1.5 km depth, while, at greater depths, water is more stable in its (bulk) liquid state than in the inter-layer space (De Siqueira et al., 1997). As a consequence of swelling, soils which contain high concentrations of smectites may undergo as much as 30% volume changes due to seasonal humidity variations. The resulting ground movements—swelling or conversely cracks—are highly detrimental to the stability of structures.

In the interlayer space, the water structure in the two or three layers in contact with the clay termination noticeably differs from that of the bulk liquid phase. This confined water has a higher density and presents some ordering in the plane parallel to the silicate layers, associated to a reduction of orientational and translational entropy (Chang et al., 1995; Boek and Sprik, 2003). However, water does not freeze, in a way comparable to what happens at the kaolinite surface. Its diffusion coefficient remains

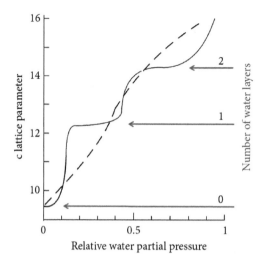

Fig. 7.9 Schematic relation between the *c* lattice parameter (Å) of a smectite crystal and the relative water partial pressure. Dashed line gives the behaviour of mixed-layer minerals. Courtesy A. Meunier.

noticeable but is several times smaller than in bulk water (Arab *et al.*, 2003), due to interactions with the clay surfaces and the compensating cations. The reduction depends upon the degree of substitution, the nature of the inter-layer cations, and the level of hydration.

7.5.2 Edge surface sites and their acid-base character

Due to the thinness of the clay layers, a determination of the edge surface structure, whether by AFM or diffraction techniques, is nearly impossible. Most structural information thus come from numerical simulations, while potentiometric acid-base titration and FTIR methods give insight into the edge surface charge and the protonation/deprotonation characteristics of the terminal OH groups.

Relying on crystal growth theory which states that stable terminations are parallel to continuous chains of strong bonds—the periodic bond chains (PBC)—, stable phyllosilicate edge termination structures have been predicted (Hartman and Perdok, 1955; Hartman, 1973; White and Zelazny, 1988). Their complexity is illustrated in Fig. 7.10(a,b), in the case of pyrophyllite (Churakov, 2006). The dissociative adsorption of water molecules saturates the extremely reactive dangling bonds created by the breaking of Al-O or Si-O bonds. At half coverage (Fig. 7.10(a)), all facets have nearly the same surface energy, while upon further hydration (Fig. 7.10(b)), the (110) and (100) surfaces become by far the most stable (25 and 30 mJ/m², respectively), compared to the (010) and (130) ones (86 and 91 mJ/m², respectively). The predicted Wulff shape of pyrophyllite platelets thus only features the (110) and (100) lateral surfaces, consistent with SEM images of well-shaped crystals of natural pyrophyllite (Fig. 7.10(c,d)). In contrast, on dry pyrophyllite platelets, only (010) and {110} lateral surfaces appear (Bickmore *et al.*, 2003) (Fig. 7.10(e)).

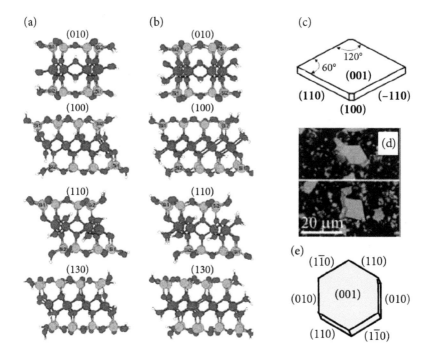

Fig. 7.10 Characteristics of pyrophyllite edges. Left panels: edge structure of the (010), (100), (110), and (130) surfaces with two (a) and four (b) dissociated water molecules. On the (010), (100), (110), and (130) facets, the Si atoms are 3-fold coordinated, the Al atoms are 4-fold coordinated, and there exists a fifth Al-O-Si bond on the (100), (110), and (130) surfaces. Two water molecules per unit cell (a) saturate all Si and a part of the Al dangling bonds. Two additional water molecules complete the saturation of all the Al bonds. Silicon, aluminium, oxygen, and hydrogen atoms are represented as yellow, green, red and white spheres. (c) Wulff shape of pyrophyllite predicted from surface energy calculations at the highest coverage and (d) SEM images of well-shaped crystals of natural pyrophyllite. Reprinted from Churakov (2006) (a, b, c, d). Copyright 2006 American Chemical Society. (e) Wulff shape of dry pyrophyllite platelet. Reprinted from Bickmore *et al.* (2003) with permission from GeoScience World. Copyright Springer Nature 2003.

The terminal OH groups at edge surfaces may experience protonation or deprotonation processes. This represents a second charging mechanism, strongly dependent on the aqueous solution pH, at variance with the pH-independent permanent charge borne by the basal surfaces. The various edge orientations possess different types of functional groups associated to different protonation/deprotonation constants and different site densities. Various approaches have been used to quantify their acid-base character.

The MUSIC model (Hiemstra *et al.*, 1989*b*) which relies on a valence bond approach and the local charge neutralization concept introduced by Pauling (see Chapter 6) has been used to evaluate the elementary protonation/deprotonation free energies of pyrophyllite edge sites, for example (Bickmore *et al.*, 2003). Accent was put on the importance of structural relaxation effects which modify the bond lengths and have implication on the valence bond charges, according to Brown and Altermatt (1985). The calculated pK_a values were found in qualitative agreement with those assigned to analogous functional groups at simple oxide surfaces. However, due to the long-range character of electrostatic interactions, they are influenced by the permanent charge borne by the layers (Avena *et al.*, 2003).

The acid-base character of functional groups may also be quantified by the values of the Fukui functions $f(\vec{r})$, which represent the change of the electron density at each point $\vec{r}$ of space when the total number of electrons varies (Geerlings *et al.*, 2003). Two functions $f^+(\vec{r}) = \rho_{N+1}(\vec{r}) - \rho_N(\vec{r})$ and $f^-(\vec{r}) = \rho_N(\vec{r}) - \rho_{N-1}(\vec{r})$ are defined, associated to the values of the local electronic density $\rho(\vec{r})$ for systems with $N+1$, N, and $N-1$ electrons. They characterize nucleophilic and electrophilic attacks, respectively. The oxygen sites with high partial charge and $f^-(r)$ index favour protonation, whereas the sites with small partial charge and high $f^+(\vec{r})$ index promote deprotonation. Relying on calculated Fukui functions, it was found that, at pyrophyllic edges, the Al-O-Si sites have the highest proton affinity, together with Al-OH groups on the (010) and (130) lateral surfaces, and that Al-OH$_2$ sites have the largest deprotonation tendency, followed by Al-OH and Si-OH groups (Churakov, 2006).

The point of zero charge (PZC) of various clays—or more precisely the point of zero net proton charge (PZNPC)—which is the pH value at which the global proton surface charge is zero (see Chapter 6), has been determined by potentiometric acid-base titration. It is usually obtained by the common intersection point of titration curves corresponding to aqueous solutions of different ionic strengths. In the case of clays, however, its determination presents several difficulties. First, what is measured is not the actual proton charge but an apparent one because the protonation condition at the start of the titration is unknown (Sposito, 1998; Hou and Song, 2004). Moreover, some debate remains as to whether such a common intersection point exists for minerals bearing a structural charge. For example, experiments have shown that in illite and montmorillonite the pH at which the charge vanishes decreases when the ionic strength increases and that the shift is an increasing function of the clay permanent charge. Only approximate PZC values can thus be derived. Typical quoted values are 5.6–6.6 for kaolinite and 9–10 for illite and montmorillonite (Hao *et al.*, 2018), although a value of 5 is also mentioned for the latter (Lagaly and Ziesmer, 2003). A compilation of PZC values for aluminosilicates and other minerals can be found in Kosmulski (2018).

PZC values result from a charge average on all orientations, and their interpretation is challenging due to the structural heterogeneity of edge surfaces which generally display a large number of site coordinations and configurations. They depend on the proton affinity constants of the surface groups but also on the number of each type of surface functional group, i.e. on the edge specific-surface area and the surface density of edge sites (Tournassat *et al.*, 2004).

Experimental studies of growth kinetics have proven that the clay dissolution or precipitation processes exclusively take place on the edge sites of the TOT layers. As observed in situ by atomic force microscopy (Bosbach *et al.*, 2000; Bickmore *et al.*, 2001), in an acidic environment, the dissolution of clay crystallites starts from the

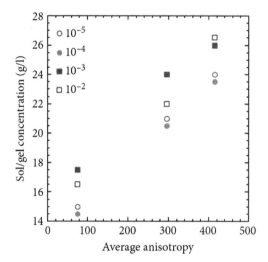

Fig. 7.11 Evolution of the concentration at which the sol/gel transition occurs with the average anisotropy of natural Na-montmorillonite particles in NaCl solutions of various concentrations. Reprinted from Michot *et al.* (2004), with permission. Copyright 2004 American Chemical Society.

edges, with a rate which is limited by the breaking of Al-O-Si bonds, while the basal faces remain essentially inert. By contrast, in the alkaline pH range, dissolution correlates to the deprotonation of the OH groups bonded to at least one Si atom in the T sheet (Tournassat *et al.*, 2004).

7.5.3 Colloidal swelling

While crystalline swelling processes specifically apply to water incorporation between the clay layers, further swelling takes place when the clay particles are immersed in aqueous solutions. This process is known as macroscopic or colloidal swelling and leads to the formation of colloidal suspensions.

According to well-established concepts (Russel *et al.*, 1991; Cosgrove, 2010), colloid stability relies on the competition between short-range attractive—mainly van der Waals—and longer-range repulsive electrostatic interactions between particles, at the basis of the DLVO (Derjaguin, Landau, Verwey, and Overbeek) theory. The shape of the interaction potential is a function of the particle habits (spheres, platelets, etc.), their charge, and their concentration. The ionic strength of the aqueous solution is also an important parameter, since it controls the formation of the double layer close to the particle surface. The double layer comprises the charge surface of the particles and a layer of aqueous counter ions attracted by the former which screen it (see Fig. 6.21 in Chapter 6). The more concentrated the aqueous solution, the weaker the interaction potential, and thus the lower the colloidal stability.

Nevertheless, compared to many colloids, clay particles possess important distinctive features. They are highly anisotropic, often of an irregular and flexible shape which leads to different modes of aggregation (face-face or edge-face), and often have large size distributions. Moreover, their charge may be of different types (permanent charge in the T or O layers, and pH-dependent charge on their edges), and sometimes heterogeneous, even in the layers (Lagaly and Ziesmer, 2003). The concentration at which gelation, i.e. a brutal increase of the fluid viscosity, occurs strongly increases for more anisotropic particles, as shown in Fig. 7.11. It has been found that, when the volume fraction ϕ of the solid phase exceeds some critical value, the mean interparticle distance varies as ϕ^{-1}. In this regime of quasi 1-D local lamellar order, the average lamella thicknesses range from 0.7 to 1 nm, close to those of the individual layers. This evidences a nearly complete exfoliation of the clay layers. Below the critical value, the mean interparticle distances vary as $\phi^{-1/3}$, a characteristic of 3-D swelling of freely rotating objects (Paineau *et al.*, 2011). At a much larger critical volume fraction, gelation or coagulation takes place (Lagaly and Ziesmer, 2003).

As a result of these swelling properties, bentonite—a type of montmorillonite—and other clays form muddy slurries with water. This happens in nature with sometimes devastating consequences—mud flows, mud volcano formation, and so on. But it may also be used by men in the drilling of oil and water wells to seal the walls of the boreholes or lubricate the drill head. At home, the most common use of bentonite absorbent properties is as pet litters.

7.5.4 Permeability

The low permeability of clays—one of the lowest among geologic materials—plays an important role in nature and human activities. The least permeable clays are smectites, in particular those which are not well-crystallized and retain much water in their particles.

The permeability coefficient k (expressed in area units) measures the ability of a material to transmit fluids. In Darcy's law, it relates the fluid flux $\vec{q}$ to the fluid viscosity μ and the pressure gradient $\vec{\nabla}p$:

$$\vec{q} = -\frac{k}{\mu}\vec{\nabla}p \qquad (7.1)$$

In geologic materials, k ranges from 10^{-24} to 10^{-7} m^2, with clays and shales in the 10^{-23} to 10^{-15} m^2 range. The permeability increases logarithmically as the material porosity (i.e. the ratio of the pore volumes over the total volume of the porous material) increases (Neuzil, 2019).

The clay permeability depends on many parameters, such as the particle sizes, their mean distance, and their arrangement with respect to each other, which determine the distribution of pores and the tortuousness of the channels. It is also dependent on the physico-chemical interaction between the fluids and the particle surfaces (Mesri and Olson, 1971). On general grounds, the clay permeability decreases as the particles become thinner and as an applied vertical pressure increases, due to the collapse of the largest pores (Dewhurst *et al.*, 1999).

This property has strong implications on human activities. Due to their low permeability, clays participate to the confinement of ground water in aquifers and protect them from contaminations. They are also responsible for sealing lake, river, and ocean bottoms and thus for the existence of the humid zones of the planet with their ecosystems and their biological diversity. They are used by man, in particular bentonite, to create low permeability liners in landfills, sewage lagoons, or water retention ponds. In several national programmes, they are considered to isolate nuclear waste repositories in the underground. They may block upward migration of injected chemical waste or CO_2, and trap oil and gas in reservoir rocks. They represent water and fertilizer reservoirs for plants, which is a crucial factor for soil fertility.

7.6   Reactivity

An important property of clay minerals is their capacity to fix the cations or anions of the environment, either by exchange with ions belonging to their layers or interlayer spaces, or by complexation on their edges, associated to their high specific surface area. Their intrinsic Lewis, Brønsted, or redox characteristics, their lamellar structure which allows intercalation of active species, or even their inertness are responsible for their multiple catalytic applications.

7.6.1   Sorption at clay surfaces

Clay particles are able to absorb into their lattice, retain or release a great variety of ions or molecules, thanks to their small size, their large surface-to-volume ratios and their porosity. This property is important in soils for the nutrition of plants, and for the migration and retention of metal contaminants in the environment. Modern industrial activities, such as pigment production, mining, metal industry, or electric manufacturing, generate an increasing amount of toxic pollutants which accumulate in soil, water, or air with severe consequences for human health. The following metals or metalloids are considered as toxic beyond a critical dose: Ag, As, Be, Cd, Cr, Cu, Hg, Ni, Pb, Sb, Se, Tl, and Zn. Additionally, the dispersion of radionucleides in the vicinity of geological radioactive waste repositories is a major concern. We will mainly focus here on cation sorption.

Cations may be sorbed in clay mineral via two main uptake mechanisms. The first one is the exchange with the structural or compensating cations at the clay basal surfaces. It is mainly determined by the layer permanent charge and is pH-independent. The second sorption mechanism takes place via surface complexation on the clay edges. It strongly depends on the edge charge, and thus on the pH of the aqueous solution.

Cation exchange capacity.   One defines the cation exchange capacity (CEC) as the number of negative charges able to fix cations at a given pH. It is expressed in moles per kilograms. It depends on the particle size, the temperature, and the ionic strength of the aqueous solution. A monovalent-monovalent cation exchange reads:

$$X^- A^+ + B^+ \longleftrightarrow X^- B^+ + A^+ \tag{7.2}$$

where X^- represents the clay site to which the A^+ or B^+ cation binds. The Gibbs' free energy of this reaction has two main contributions: one from the hydration/dehydration processes of the ions in solution and the other from the cation binding to the clay site. Only ions which are not too tightly bound, essentially the compensating cations in the

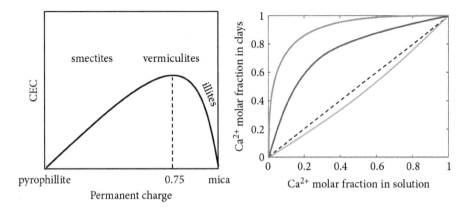

Fig. 7.12 Left: schematic relationship between CEC (in centimole/kg) and the permanent charge (per Si_4O_{10} unit). Right: relationship between Ca^{2+} contents in a Na-montmorillonite and in solutions of various salt concentrations. Blue, red, and green curves refer to 2 M, 0.01 M and 0.001 M Na^+ concentration, respectively. Courtesy A. Meunier.

interlayer, can rapidly and reversibly be exchanged. In clay minerals the most common exchangeable cations are Ca^{2+}, Mg^{2+}, H^+, K^+, and Na^+.

The CEC thus depends on the permanent layer charge, which results from the T or O substitutions, and it is expected to grow with the latter. This is true up to a certain point, but beyond $|Q| \approx 0.8$, in illites and micas, the layers interact so strongly with each other that cations and water can no longer penetrate in the interlayer space and the CEC decreases. The typical variation of the CEC with $|Q|$ is represented in Fig. 7.12. Smectites and vermiculites with their permanent charge Q in the range $[-0.2$ to $-0.75]$ thus are the most prone to cation exchange processes.

As reviewed in Bergaya *et al.* (2013), the propensity of B^+ to replace A^+ is quantified by the selectivity coefficient $_A^B K$:

$$_A^B K = \frac{N_B * [A]}{N_A * [B]} \tag{7.3}$$

in which the symbol $[\,]$ indicates aqueous activities and N_A and N_B are the equivalent fractional occupations of A and B in the mineral. Smectites show a preference of large over small cations with selectivities following the order Cs > Rb > K > Na > Li for monovalent cations or Ba > Sr > Ca > Mg for divalent cations, but for the latter, the trend is reversed in vermiculites. The selectivity of the preferentially adsorbed cations decreases with increasing coverage by this cation. It is also strongly dependent on the ionic strength of the aqueous solution. As shown in Fig. 7.12 (right panel), the more diluted the aqueous solution, the greater the selectivity for Ca^{2+} over Na^+ in a montmorillonite.

Clays are not the only compounds which have a cation exchange capacity. Other fine-grained minerals, such as zeolites, share this property. Despite their more 3-D structure, zeolites present many similarities with clays. They are hydrated alumino-

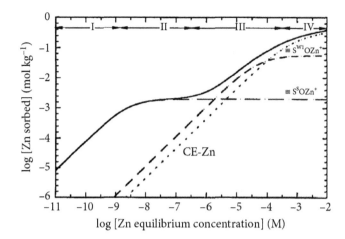

Fig. 7.13 Calculated sorption isotherms for Zn on Na-montmorillonite at pH = 7.5 illustrating the contribution of the different surface species to the overall sorption (concentration of NaClO$_4$ in the aqueous solution equal to $5 * l0^{-2}$ M). CE, S^w and S^s account for cation exchange, weak and strong sites on edges, respectively. At growing aqueous Zn concentrations, complexation of the strong edge sites first takes place, followed by cation exchange in the interlayer space when all strong sites are saturated. Increasing Zn concentration further leads to complexation on weak edge sites. Reprinted from Bradbury and Baeyens (1997). Copyright 1997, with permission from Elsevier.

silicates, with open structures, exchangeable cations, cation selectivity for larger cations, and high CEC (Bish, 2013).

Complexation mechanism at clay edges. The second main mechanism of cation uptake by clays is cation complexation at layer edges. It is induced by the presence of broken bonds (see Section 7.5.2), which undergo protonation/deprotonation reactions. Cations may form complexes (mainly inner-sphere complexes) by exchange with a proton at the terminal S-O, S-OH, or S-OH$_2$ amphoteric groups (S representing a surface edge cation, most often Si, Al, or Mg), whose relative numbers depend on the aqueous solution pH.

To interpret titration and sorption isotherms, both cation exchange and cation complexation processes have to be considered, since they may be simultaneously at work, but with a relative importance which depends on the cation concentration and pH of the aqueous solution. Considering the complexity of clay edges, the first models assumed the presence of 'strong' and 'weak' sites (Bradbury and Baeyens, 1997). Strong sites bind cations more strongly, but their number is smaller than weak site. By recording the pH variations of sorption isotherms, such as the one relevant for Zn sorption on Na-montmorillonite represented in Fig. 7.13, the relative contributions of sorption at strong or weak sites and cation exchange in the interlayer space can be deduced. Later on, relying on the MUSIC model, multiple sites of more atomistic relevance have been accounted for and their contribution to the sorption on clay surfaces has been deduced from titration measurements (Tournassat *et al.*, 2004).

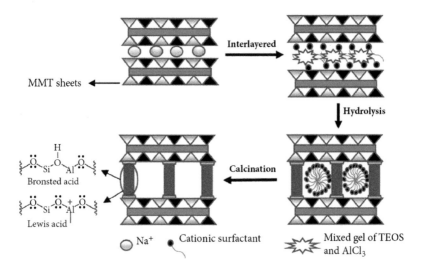

Fig. 7.14 Schematic diagram showing the preparation steps of an Al-containing mesoporous silica-pillared clay. Reprinted from Wu *et al.* (2016*a*) with permission. Copyright 2016 by the Royal Society of Chemistry. Permission conveyed through Copyright Clearance Center, Inc.

The capacity to bind cations at clay edges is a function of several parameters. One is the number of available sites of each type, which depends on the edge area and orientation. Another is the pH of the aqueous solution. At low pH (pH < PZC), edges are positively charged. Their contribution to sorption is weak while that due to cation exchange dominates. At variance, in more alkaline conditions, complexation at edge surfaces has a larger weight.

7.6.2   Clays as catalysts

Historically, clays have been widely used for catalysing organic reactions in the petro-chemical industry, but their potentialities are in fact much wider and are now considered for green and sustainable catalysis (Zhou, 2011; McCabe and Adams, 2013). Their catalytic efficiency relies on several features.

One is the existence of active species in the clay itself. Hydrated edge atoms or cations exchanged in the interlayer space may have Lewis, Brønsted, or redox activity, and thus may be used for certain organic transformations, like the formation of large molecules from smaller ones. This property is invoked by those who suggest that clays may have played a role in the origin of life, an hypothesis still much debated (Brack, 2013). The intrinsic acidity strength may be even enhanced by a treatment with either sulfuric or hydrochloric acid to remove ions from the O layers and thus increase the particle charge.

Smectites are particularly interesting catalysts, due to their variable composition and their flexible interlayer space. By ion exchange or intercalation, active species may be introduced in their structure to enhance or modify their pristine properties. It is for example possible to intercalate large molecules, such as Keggin $Al_{13}O_4(OH)_{24}(H_2O)_{12}^{7+}$

polycations, in the interlayer space, to engineer tunable pore sizes. A calcination treatment strongly binds the molecules to the layers, leading to so-called pillared interlayer clays (Fig. 7.14). The pillars may be active by themselves, produce pores which act as confined nano-reactors in which all sorts of chemical reactions may take place, or allow the intercalation or growth of clusters of homogeneous size distribution, leading to new functional materials.

Finally, thanks to their high surface area (several hundreds of m^2/g), clays may be used as catalyst supports.

7.7   Uses and applications

From prehistoric to modern times, clay minerals have found a wide range of uses and applications, due to their peculiar structural, chemical, adsorption, and insulating properties. Aside from those already mentioned in the preceding sections, some are presented in the following, in relation with the specific properties on which they rely.

First, as raw materials and because of their ubiquity and low cost, clay minerals have long served for building shelters, houses, or even monumental architectures. One reckons that nearly half of the Earth's population lives in a clay habitat.

Due to the strong Si-O, Al-O, or Mg-O chemical bonds which constitute their skeleton, clay minerals have high formation enthalpies, which allow them to withstand high temperatures. Their refractory property is exploited in the metal industry to make crucibles and in the ceramic industry to manufacture refractory ceramics, necessary in cars (exhaust pipes), in spatial rockets (heat shields), in the blast furnaces of the steel industry, amongst other things. Art ceramics also harness this property: a glaze is deposited on baked clays, notably kaolinite, as a basis for subsequent decorative work.

Due to their small size and high dispersion capacity, kaolinite particles are used as fillers and coatings in the paper industry. The orientation of their platelets provides a range of grades and glossy shades to the paper. Similarly, micas are used to produce pearlescent paints, and, in the cosmetic industry, they are incorporated in lipsticks and nail polishes, to give them a micaceous aspect and a high reflectance.

In the pharmaceutical industry, clay minerals are also used as excipients—i.e. as inert components determining consistency, form and volume of pharmaceutical formulations—because of their inertness and the unique rheological properties of their dispersions. For a comprehensive review, see Carretero and Pozo (2009) and Carretero et al. (2013). They may retain drug molecules and control their release (Viseras et al., 2010). Talc, smectites, palygoskite, sepiolite, or kaolinite act as disintegrants, diluents, binders, emulsifying, thickening, and anti-caking agents, thanks to their swelling ability, their quick dispersion in water, their capacity of plastic deformation when moist, and more generally their colloidal properties.

Because they are non-toxic, clays find many applications as active therapeutic agents (Carretero and Pozo, 2010). Smectites, for example, are well-known as anti-acid, gastro-intestinal protector and anti-diarrhoea agents. This use relies on their high sorption capacity and large specific surface area, which allows them to adsorb protons on their (edge) surfaces. For similar reasons as well as for their heat retention capacity, clay minerals are good dermatological protectors and anti-inflammatories.

Several of these properties have long been known and were mentioned in antiquity. Other medical applications comprise the production of dental cements, manufacture of plaster of Paris for immobilization of limbs and fractures, bone grafts, and use in medical diagnosis techniques. Clays have even sometimes been ingested by man, whether for religious purposes or to relieve famine.

Nevertheless, clay minerals could also be potentially dangerous to humans through ingestion, inhalation, or dermal absorption because of their small particle size, limited solubility in the lung, and fibrous habit (Carretero *et al.*, 2013). Inside the lung, clay minerals can cause diverse pathologies, but the toxicity appears to be generally restricted to the presence of quartz, asbestos, and some other fibrous clay minerals.

7.8 Summary

Clay minerals are complex oxides, both from a structural and a chemical viewpoint. While the atomic structure of their layers which associates T and O atomically-thin sheets is rather well-defined, the layer stacking into small crystals and the loose interactions between the latter form disordered structures at larger scale with complex porosity networks. The edges of the layers, of usually small extension, are notably reactive due to the breaking of the strong Si-O, Al-O, or Mg-O bonds.

Few clay minerals have a fixed composition. They are better viewed as solid solutions, due to many possible cation substitutions within their layers, from which charging results which has to be compensated by cations in the interlayer space. When clays are in contact with the fluids of the environment, cation exchange in the interlayer space may occur, as well as cation complexation on the layer edge surfaces.

Due to the conditions of their formation and occurrence, water plays an especially prominent role in the physics and chemistry of clay minerals. It may enter the interlayer space of smectites and vermiculites, giving rise to swelling. It may dissociate on the reactive edge sites, providing additional pH dependent charges to the particles.

All these characteristics explain the important role of clays in nature and in human activities.

8

Surface chemistry, energy conversion, and related applications

In this chapter, we address a variety of seemingly rather diverse topics, which are however linked by the involvement of chemical processes at oxide surfaces and interfaces. Redox reactions and electron transfer processes play an important role in most of the topics presented here. Surface chemistry is obviously at the basis of catalysis, photocatalysis, and corrosion science. It is also involved in biological surface science, where biointerfaces are formed by chemical reactions on oxide surfaces in liquid biological environments (Kasemo, 2002). In solid oxide fuel cells, redox processes of fuels occur at complex solid-gas or solid-liquid interfaces. The doping of oxide materials with foreign atoms is a proved chemical means to introduce new functionalities in a given material. Doping can also be carried out by electrons, and this is what happens in the application of solar energy materials, where oxide thin films play a key role. The discussion in this chapter is with a view towards applications of oxide thin films and nanostructures in a technological device context and in related interdisciplinary fields. For a comprehensive treatment of the application of metal oxides in diverse energy technologies we point the readership to a recent monograph of Korotcenkov (2018). The literature cited in this chapter refers mostly to topical reviews, but some relevant key papers are explicitly mentioned.

8.1 Heterogeneous catalysis

Catalysis is a process by which the rate of a chemical reaction is enhanced by small amounts of a foreign material called a catalyst. It is said to be heterogeneous when the reactants and the catalyst belong to different phases (mainly gas-solid or liquid-solid). The role of the catalyst is to lower the activation barriers along the reaction pathway, to facilitate the adsorption, diffusion, desorption, and rearrangement of the reactants, or, in a multiproduct reaction, to favour the most important one (selectivity property), without itself undergoing chemical change. Actually, this last condition is never completely fulfilled, since real catalyst properties—structure, activity, and selectivity—evolve with time and all catalysts eventually deactivate, hopefully after an extended period of time.

Three main catalytic mechanisms have been identified, as represented in Fig. 8.1. Panel (a) shows the Langmuir–Hinshelwood mechanism, relevant for the vast majority

Oxide Thin Films and Nanostructures. Falko P. Netzer and Claudine Noguera, Oxford University Press (2021). © Falko P. Netzer and Claudine Noguera. DOI: 10.1093/oso/9780198834618.003.0008

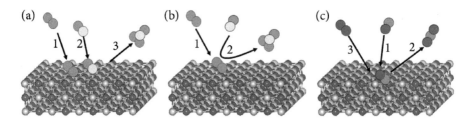

Fig. 8.1 Illustration of the three main catalytic mechanisms: (a) Langmuir–Hinshelwood , (b) Eley–Rideal , (c) Mars–Van Krevelen

of catalytic reactions, in which both reactants first adsorb onto the surface (steps 1 and 2), before a reaction takes place. Surface diffusion allows the adsorbed molecules to meet and interact, before the reaction products desorb from the surface (step 3). In this type of mechanism, the highest activity is obtained when the adsorbed reactants are in the right stoichiometric ratio for the targeted reaction, when they are well dispersed over the surface, and when the desorption of the products is effective. In the Eley–Rideal mechanism, shown in Fig. 8.1(b), only one of the reactants adsorbs on the surface (step 1), while the other one interacts with the adsorbed species directly from the gas phase before desorption of the product (step 2). Finally, in the Mars–Van Krevelen mechanism (Mars and Van Krevelen, 1954) (Fig. 8.1(c)), the surface itself is an active part in the reaction: one reactant forms a chemical bond with the catalyst surface and reacts with a component atom of the surface (step 1). After desorption of the product (step 2), the surface atom is replaced by corresponding atoms from the gas phase (step 3). This mechanism is particularly relevant for oxidation catalysis involving reducible oxide surfaces: the reactants get oxidized by oxygen atoms from the oxide lattice, which are then resupplied from the O_2 gas phase after dissociation at active oxide surface sites (step 3). The oxidation of CO on an ultra-thin FeO_2 layer on Pt(111) described in Chapter 5 (Section 5.6) exemplifies the Mars–Van Krevelen mechanism.

Catalysis is of key importance in environmental chemistry and in many industrial processes. It is encountered in petrochemistry, oil refining, in the pharmaceutical industry, the fine chemical production, and depollution treatments. Although oxide catalysts have never been overlooked, the most important catalysts have long been precious transition metals. In order to reduce their (ever increasing) cost, several strategies have been implemented. One is to highly disperse the precious metal particles on a high-surface area—mostly oxide—support. Another one is to replace the pure metal particles by core-shell particles, in which the active species are at the periphery of an otherwise inert and cheap material. A third one is to use oxides as active catalysts. The recognition of oxide catalyst efficiency has been growing over time, so that oxide materials are now omnipresent in heterogeneous catalysis systems: they are major components as substrates of supported metal catalysts, they may be present in small quantities in conjunction with other catalytic materials as modifiers or promoters, or they may act as active catalysts in their own right (Idriss and Barteau, 2000; Védrine, 2017).

The oxides most often used as catalysts include silica, alumina, TiO_2, VO_x, ZnO, ZrO_2, zeolites, clays, phosphates, and multicomponent or mixed oxides such as perovskites, molybdates, tungstates, MoVTe(Sb)Nb oxide, amongst others. Their performance relies on their grain size, shape, composition, and preparation. They often possess charged defects resulting from the preparation procedures, which, despite low concentrations, may play an important role as active centres in the catalytic process when they segregate at the surface (Ganduglia-Pirovano *et al.*, 2007). Moreover, oxide catalysts are often doped to enhance their activity or selectivity (McFarland and Metiu, 2013). Their shape, size, and surface composition may evolve in the course of the chemical reactions, especially under working conditions. For all these reasons, industrial catalysts are complex materials whose morphology, structure, and composition are often unknown, which makes it difficult to pinpoint the source of their activity and selectivity.

Oxides are efficient catalysts for a wide variety of chemical reactions, including selective or total oxidation, dehydrogenation, or acid-base reactions (Kung, 1989; Vedrine, 2018). Among the oxidation reactions, one can cite oxidation of organic molecules to form aldehydes, ketones, peroxides, alcohols or acids; insertion of oxygen in inorganic molecules and CH_4 to synthesize oxide molecules such as SO_2, NO, or CO; or oxidation of the latter in air depollution processes to reduce the emission of hazardous gases from vehicles or power plants. Dehydrogenation reactions of hydrocarbons induce the formation of unsaturated species, such as olefins. A common feature among these reactions is that the desired products are often not the thermodynamically most favourable ones. For example, oxidation of hydrocarbons straightforwardly leads to water and carbon dioxide, while the desired products are alcohols, aldehydes, ketones, acids, anhydrides, or alkenes and dienes; hence the crucial role of the catalyst.

Key issues of heterogeneous catalysis on oxides involve the redox properties and the acid-base properties of the oxide surfaces/particles, the question of catalytically active sites and their environment, and the structure dependence of the catalytic reaction (Idriss and Barteau, 2000).

There have been many attempts to rationalize the reactivities observed on different oxide catalysts with the acidity or basicity of their surface sites (Metiu *et al.*, 2012). Indeed, these properties favour the adsorption of molecules of opposite character and desorption of acid or basic products, like carboxylic acid or olefins, respectively. Acid sites (cations or protons) have the capacity to *activate* reactant molecules, by favouring for example bond-scission processes of C-H and O-H groups. When proton exchange takes place, the Brønsted acid-base concept applies, which relies on the proton acceptance or release of basic and acidic centres. More generally, one refers to the Lewis acid-base concept, in which the species which receives electrons is a Lewis acid and the one that donates them is a Lewis base. In oxides, cations have a Lewis acid character, which depends on their positive charge and size (although actual charges might not necessarily correspond to formal oxidation states), while lattice oxygen anions are basic, with a strength which depends on the ionic character of the metal-oxygen bonds. Neutral oxygen vacancies with the two electrons left behind by the missing oxygen atom are strong Lewis bases. Hydroxyl groups could be either acidic (Brønsted site) or basic. These properties depend on the local environment of the surface species, since, contrary to simple molecules for which the acid-base strength is unambiguous, oxide powders possess many inequivalent sites, which have each their own acid-base character. Moreover, the intrinsic acid-base character of an oxide particle may be strongly modified by doping or by the segregation of impurities at its surface. For example, in reducible oxides, in the presence of high valence dopants—atoms with a valence higher than that of lattice cations—the dopant may give electrons to the host cation, thus reducing its charge and its acid character. In irreducible oxides, the excess electron(s) remain localized on the dopant, which is thus a Lewis base. For a brief overview of the Brønsted and Lewis acid-base properties of oxide particles and surfaces the reader is referred to Chapter 6 and to Noguera (1996).

In oxidation or dehydrogenation reactions which follow a Mars–van Krevelen mechanism, the oxide efficiency relies on its capacity to lose oxygen atoms. A useful descriptor is thus the typical value of oxygen vacancy formation energies. The latter are high in irreducible oxides, like alkaline or alkaline-earth oxides, but much smaller in reducible ones, like TiO_2, CeO_2, or more generally in oxides involving cations with variable oxidation states. Moreover, doping modifies their intrinsic values. In the presence of low valence dopants—atoms with a valence lower than that of lattice cations—, oxygen atoms in their close vicinity are activated and become more reactive, which means that such doped oxides are better oxidants. In contrast, high valence dopants increase the energy necessary to remove a surface oxygen located close to the dopant, which may hinder oxidation by the Mars-van Krevelen mechanism (McFarland and Metiu, 2013).

Many of these key issues of oxide catalysis can be and have been addressed by surface science techniques, if suitable model systems are prepared that mimic the behaviour of the real complex catalysts in a reductionist manner—this is the so-called surface science approach of catalysis (Henrich and Cox, 1996; Somorjai and Li, 2010). From these investigations of well-defined oxide surfaces, the principles of oxide surface reactivity may be assessed, which form the basis for a more complete understanding of oxide catalytic reactivity.

However, the price to pay for this approach is the existence of a pressure and a material gap, and more generally of a complexity gap (featured on the cartoon of Fig. 8.2), by which catalytic properties (elementary reaction steps, reaction intermediates, etc.) determined under well-defined conditions may not be transferable to realistic reaction conditions. Indeed, the more than thirteen orders of magnitude in pressure between UHV experiments and typical industrial working conditions may strongly influence the structure and composition of the catalyst as well as the activation barriers of the reaction. An illustrative example of the importance of the pressure gap is given by the CO oxidation reaction over the ruthenium surface (Peden and Goodman, 1986). While under UHV conditions the Ru surface remains mainly inactive, it becomes more active than other late transition metals under high pressure and oxidizing conditions. In that case, it has been proved that the catalytic activity is due not to the metal itself but to a thin oxide layer formed at its surface. As far as the materials gap is concerned, industrial catalysts often consist of powders or dispersed nanoparticles on oxides to maximize the interaction area, with consequences on the role of steps, edges, or corners as active sites. Moreover, as discussed previously, oxide catalysts are usually complex,

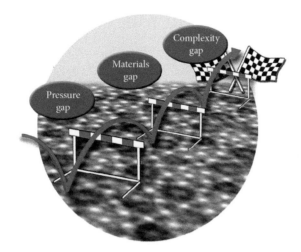

Fig. 8.2 The three gaps between the surface science approach and real catalysis conditions. Reprinted from Schlögl (2015) with permission. Copyright © 2015 Wiley-VCH Verlag GmbH and Co. KGaA, Weinheim.

multi-cation materials, with impurities, structural defects, stoichiometry defects, and dopants. An example is given by the MoVTeNb oxide catalyst (so called M1 phase), active for propene and propane selective oxidation and ammoxidation to acrylic acid and acrylonitrile, respectively. It appeared that the active sites form under propane oxidation conditions and belong to a thin layer enriched in V, Te, and Nb, segregated at the surface of the M1 phase (Hävecker *et al.*, 2012).

Nevertheless, following the successful approach of catalysis by studies on metal single crystal surfaces, the surfaces of oxide bulk single crystals have been employed to disentangle the respective roles of the various parameters in the investigation of catalytic reactions. However, there are several experimental constraints which restrict the use of oxide single crystals in surface science studies: the preparation of well-defined defect-free surfaces is non-trivial and often difficult, and the insulating nature of the samples impairs the application of charged particle probes, as employed in many surface spectroscopy techniques.

An alternative approach has therefore been developed over the last 25 years, which is the use of oxide thin films as model catalyst surfaces, be it as supports for metal particles or as active catalyst surfaces (see (Freund, 2007; Freund and Pacchioni, 2008; Giordano and Pacchioni, 2011) for some more recent reviews). The epitaxial growth of oxide thin films on metal single crystal surfaces yields the catalytically interesting combination of materials of industrially supported metal catalysts, but with the oxide surface and metal-oxide interface in well-defined forms and accessible to characterization with atomic precision. Progress in the epitaxial growth of oxides on metals in the last two decades has produced a rich variety of oxide-on-metal systems that have been characterized at the atomic scale; these have provided detailed insights into the surface chemistry of oxides (see Campbell and Sauer (2013), editors of a volume dedicated to the surface chemistry of oxides).

The thickness of the oxide thin film is an important parameter, which determines its behaviour (see also Chapter 4). For films beyond a critical thickness, the oxide film surface behaves very similarly to a bulk single crystal surface. For thinner films, new functionalities may appear due to the thickness confinement and the influence of the (buried) interface with the metal support. This critical oxide thickness depends on the (combination of) materials, but is typically in the range of 3–10 oxide monolayers (one monolayer is defined here as containing one oxide unit cell). For ultrathin films, thinner than the critical thickness, new phenomena come into play, such as charge transport from the metal support through the oxide to the film surface, structural relaxations, or the formation of completely new oxide phases. Such effects have been discussed in Chapter 5, in conjunction with 2-D oxide materials.

Oxide nanoparticles have been used in traditional oxide catalysis for a long time as components of high surface area oxide powders, but without giving reference to their nanoparticulate form. The progress in the wet-chemical preparation of oxide nanoparticles and nanosheets (see Chapters 5 and 6) has opened up the explicit way to their use in chemical catalysis, whereby complex materials in low-dimensional form have become accessible (see e.g. (Choi *et al.*, 2009) and references therein for zeolite nanosheet catalyst design). As an example, a molecular-beam approach for the fabrication of oxide catalysts in the form of $(WO_3)_3$ and $(MoO_3)_3$ clusters supported on oxide surfaces (TiO_2, FeO) has been reported recently by Rousseau *et al.* (2014) and their catalytic activity has been tested in the dehydration/dehydrogenation of aliphatic alcohols. The oxide clusters have been generated by thermal sublimation from WO_3 (MoO_3) powder samples and directed in a controlled molecular beam onto the substrate surfaces. The electronic structure of $(WO_3)_3$ clusters on TiO_2 and the substrate-cluster interaction have been investigated theoretically (Di Valentin *et al.*, 2013). It has been shown experimentally that the reactivity of the cluster species toward the organic molecules is governed by the terminal M=O oxo groups, with dioxo species O=M=O necessary for the catalytic alcohol dehydration and dehydrogenation (Asthagiri *et al.*, 2016). It has been argued that the stronger Lewis acidity of the W^{6+} sites relative to the Mo^{6+} sites enhances the overall reactivity, but that the more easily reducible Mo^{6+} sites lead to a higher selectivity towards oxidation reaction (e.g. ethanol to acetaldehyde). Quantum size effects in oxide nanoparticles may also have a dramatic influence in redox reactions involving electron excitation across the band gap, due to a variation of the band gap with size. This is particularly important in photocatalysis (see Section 8.2).

To summarize, surface science research that can be realized on oxide nanosystems in the form of thin films, nanoparticles, or supported clusters has contributed important insights into the fundamental issues of oxide catalysis such as defining key descriptors of structure-property relationships, active site recognition, and the description of realistic mechanisms of reaction pathways. Oxide nanostructures as model systems enabling an atomistic description of structural and chemical properties by detailed imaging and spectroscopic studies coupled with theoretical simulations thus represent a key step in the intelligent design of novel catalyst systems with desired activity and selectivity.

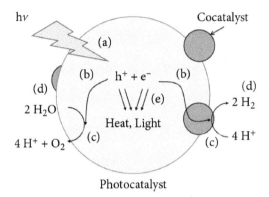

Fig. 8.3 Reaction processes of H_2O splitting on a heterogeneous photocatalyst. (a) Light absorption; (b) charge separation and transfer to the catalyst surface; (c) redox reactions; (d) adsorption, desorption, and mass diffusion of chemical species; (e) charge recombination. Reprinted from Hisatomi *et al.* (2012) with permission. Copyright 2012, Chemical Society of Japan.

8.2 Photocatalysis and photoelectrocatalyis

Photocatalysis denotes an accelerated chemical reaction taking place on a semiconductor surface in the presence of light. The fundamental process on a semiconductor photocatalyst involves the excitation of an electron from the valence band to the conduction band via photoabsorption (UV/Vis light). The excited charge carriers (electrons and holes) separate and are driven to the semiconductor surface, where a redox reaction via charge injection into reactant species can be initiated. The activity of the photocatalytic process relies on the redox power of the photogenerated charge carriers, i.e. the reduction potential of the electrons in the conduction band and the oxidation power of the holes in the valence band. The popular photocatalytic solar water splitting reaction as a promising method for hydrogen generation in a solar energy-based hydrogen economy (Turner, 1999) has been mostly studied and is discussed here as a prototypical example.

Photocatalytic water splitting has been demonstrated first by Fujishima and Honda (1972) on an n-type TiO_2 cathode. The reaction processes of H_2O splitting on a photocatalyst are illustrated schematically in Fig. 8.3. Water splitting is an uphill process, which requires 237 kJmol^{-1}, corresponding to 1.23 eV; this is the minimum energy the absorbed photons have to provide (Wang *et al.*, 2019*b*). In addition, the energies of the VB and CB edges of the photocatalyst have to straddle the reduction and oxidation potentials of water to enable the redox processes. That is, the CB edge has to be more negative with respect to the H^+/H_2 reduction potential (as related to the normal hydrogen electrode (NHE)) to allow the reductive electron transfer, and the VB edge is required to be more positive than the H_2O/O_2 oxidation potential to enable the oxidative hole transfer (with the transferred electrons filling the VB holes). This potential situation is depicted in Fig. 8.4(a).

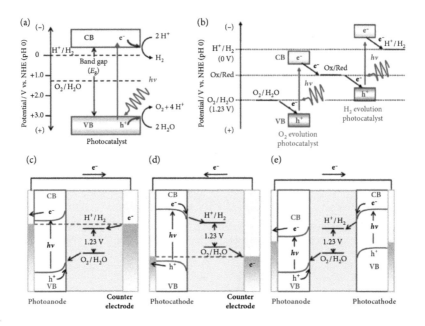

Fig. 8.4 Energy diagrams of photocatalytic water splitting based on (a) one-step excitation and (b) two-step excitation (Z-scheme). Photoelectrocatalytic water splitting diagrams using (c) a photoanode, (d) a photocathode, and (e) a photoanode and photocathode in tandem configuration. The band gaps are depicted smaller in (b) and (e) to emphasize that semiconductors with a narrow band gap can be employed. Reprinted from Hisatomi *et al.* (2014) with permission. Copyright 2014 by the Royal Society of Chemistry. Permission conveyed through Copyright Clearance Center, Inc.

Photocatalysts with suitable band gap energies E_g (1.23 eV $< E_g <$ 3.1 eV— the latter is necessary for UV/Vis use) and appropriate band edge positions for H_2O splitting in a single step process, as shown in Fig. 8.4(a), are rare and difficult to find. Alternatively, two semiconductors can be combined in series with a reversible charge transfer mediator (redox shuttle): this is the two-step process, as illustrated in Fig. 8.4(b), called Z-scheme, which in a way mimics the natural photosynthesis because of the similarity in the excitation and transfer of the photoexcited charge carriers (Bard, 1979; Maeda, 2013). Due to the more relaxed thermodynamic conditions on band gap requirements, many combinations of narrow band gap oxides as O_2 and H_2 evolution photocatalysts are possible in the Z-scheme. An important difference between single-step and two-step process is the photon to H_2/O_2 conversion ratio: it is one photon per H_2/O_2 conversion in the single-step case, and two photons (and two photochemical reactions) in the Z-scheme case.

The photoelectrochemical water splitting process is explained in Fig. 8.4(c,d). The semiconductor electrodes are immersed in an electrolyte solution, where the electron transfer takes place to equilibrate the Fermi levels of the semiconductors with the redox potential of the electrolyte. This charge transfer leads to band bending in the semiconductor, which assists in the separation of light-induced charge carriers. In the

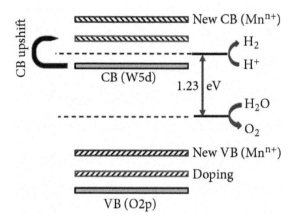

Fig. 8.5 Schematic illustration of the different strategies to modify the band gap and the band positions of WO$_3$. Reprinted from Janáky *et al.* (2013). Copyright 2013, with permission from Elsevier.

n-type semiconductor acting as photoanode, the photoexcited holes are consumed in oxidation reactions at the surface, and the electrons are transferred via the external circuit to the counter electrode, where they induce the reduction reaction (Fig. 8.4(c)). The top of the VB of the photoanode must be more positive than the O$_2$/H$_2$O oxidation potential to enable the oxygen evolution. Conversely, on the p-type semiconductor acting as the photocathode, the CB edge must be more negative that the H$^+$/H$_2$ reduction potential to allow the hydrogen evolution (Fig. 8.4(d)). If the Fermi levels of the photoelectrodes are positioned at undesirable potentials for the redox processes, an external potential between the photolectrodes and the counterelectrodes can be applied to drive the redox reactions. In analogy to the Z-scheme water splitting, the photoanode and photocathode can be connected in a tandem configuration, as shown in Fig. 8.4(e).

Apart from the thermodynamic band gap requirements of the photocatalyst, other factors affect the efficiency of the water splitting process: the transport of the excited charge carriers, the properties of the catalyst surface which influence the kinetics of the redox reactions taking place at the surface, and the chemical stability and photostability of the catalyst. Successful oxide photocatalysts include the d^0, d^{10}, and f^0 oxides, with TiO$_2$ being by far the most studied system (Wang and Sasaki, 2014). WO$_3$ is also an excellent photocatalyst for water oxidation (O$_2$ evolution), with a band gap energy E$_g$ = 2.7 eV, but the CB edge location is too positive with respect to the H$^+$/H$_2$ redox level to enable hydrogen evolution. Band gap engineering, i.e. tuning the band gap arrangement by chemical modification, is an attractive approach to improve the photocatalytic performance. The band gap can be modified by doping with metal and non-metal atoms (introduction of localized states in the band gap) or by compound formation (creation of new VB and CB edges) (Janáky *et al.*, 2013). Figure 8.5 gives a schematic illustration of the modifications of the band gap of WO$_3$ by various strategies.

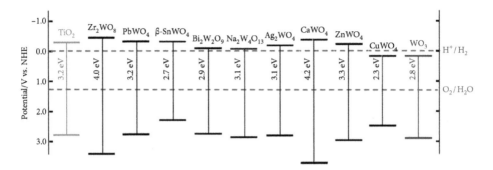

Fig. 8.6 Comparison of the band positions of different W-based ternary oxide semiconductors. Reprinted from Janáky *et al.* (2013). Copyright 2013, with permission from Elsevier.

The doping of WO_3 with hafnium atoms has been proposed by electronic structure theory to improve its photoactivity (Wang *et al.*, 2012*b*). The Hf dopants, replacing W in monoclinic WO_3, induce a shift of both VB and CB edges towards higher energies without losing the vis-light absorption properties, and the new band positions should be favourable for both H_2 and O_2 evolution for efficient water splitting. As mentioned, compound formation is another effective means of modifying the electronic band positions. This is shown in Fig. 8.6 for various ternary W oxides: the band positions of the tungstates are compared to those of TiO_2 and WO_3 and the redox potentials of H_2O are also indicated.

The band gap arrangement can also be modified by physical means. Size quantization in WO_3 nanosheets 4–5 nm thick has been reported to increase the band gap from 2.63 eV to 2.79 eV, with a concomitant up-shift of bands to enable the nanosheets to exhibit activity in the photocatalytic reduction of CO_2 into CH_4 (Chen *et al.*, 2012). Strain engineering can be used to achieve band structure engineering. The optical band gap of thin films of WO_3 has been modified by a uniaxial strain induced by low-energy He implantation, with a reduction of about 0.18 eV per percent expansion of the out-of-plane unit cell length (Herklotz *et al.*, 2017). The morphology of catalysts can also play a role. The morphology of nanoscale semiconductor photocatalysts for solar water splitting has been reported to influence the overall power conversion efficiencies, with increasingly complex structures (1-D wires, 2-D nets, 3-D porous structures) exhibiting superior performance, as a result of structural stability and improvements in the charge transport (Lin *et al.*, 2011). The photocatalytic performance of different exposed facets of nanocrystalline photocatalysts has been addressed recently (Wang *et al.*, 2018). The most basic effect of different exposed facets on photocatalytic reactivity is the varying concentration of active unsaturated sites. For example, the (001) facets of anatase TiO_2 exhibit a higher photocatalytic activity than the thermodynamically more stable (101) facets, which has been ascribed to a higher density of unsaturated Ti atoms (Wang *et al.*, 2018; Yang *et al.*, 2019). A facet-induced spatial separation of the photogenerated charge carriers has also been proposed, i.e. the electrons and holes transfer to different crystal facets due to a slight difference in energy levels between the different facets (Li *et al.*, 2014*b*).

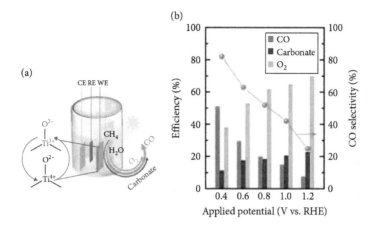

Fig. 8.7 (a) Schematic illustration of selective CH_4 oxidation to CO on a TiO_2 photoelectrode, starting with charge separation between O^{2-} and Ti^{4+} to produce $^-O\text{-}Ti^{3+}$ upon illumination. (b) Dependence of the CO efficiency and selectivity on the applied potentials (left axis: efficiency; right axis: selectivity of CO over all carbonaceous products). The PEC bulk electrolysis was conducted on ALD TiO_2 in a CH_4-saturated 1.0 M NaOH electrolyte at the corresponding applied potentials. Reprinted from Li *et al.* (2018).

Apart from water splitting, the reduction of CO_2 into hydrocarbon fuels and the selective oxidation of CH_4 to CO, both reactions with great environmental and energy relevant potential, have been addressed by photocatalytic means. The strong C-H bond in methane is an obstacle for the selective oxidation of CH_4, which mostly proceeds to the undesired CO_2 once it has become activated. Li *et al.* (2018) have reported a photoelectrochemical (PEC) approach of CH_4 oxidation using a TiO_2 thin film photocathode, whereon the selective oxidation proceeds under light with an applied bias. It was found that the selectivity towards CO is highly sensitive to the applied bias. This is demonstrated in Fig. 8.7(b), where the dependence of the CO efficiency and selectivity on the applied bias is plotted. The first step of the selective CH_4 oxidation to CO over TiO_2 is depicted in Fig. 8.7(a), starting with charge separation between O^{2-} and Ti^{4+} to produce the active $^-O\text{-}Ti^{3+}$ radical upon illumination. The role of the applied potential has been ascribed to the electric field, which facilitates the separation of the photogenerated electrons and holes (Li *et al.*, 2018). As already mentioned, the size quantization induced band gap modification in WO_3 nanosheets has enabled the photocatalytic reduction of CO_2 in the presence of water to CH_4, a reaction that is not possible in bulk WO_3 since the CB edge is more positive than the CO_2 redox potential (Chen *et al.*, 2012).

The light-induced formation of free radicals on oxide photocatalysts, in particular TiO_2, has been used as a pollutant degradation mechanism and photochemical reactions on TiO_2 thin films have led to the discovery of the superhydrophilicity effect, in which the contact angle of water is close to zero (Wang and Sasaki, 2014). The superhydrophilicity effect is applied to prepare anti-fog glass mirrors and to generate self-cleaning surfaces. The mechanism of superhydrophilicity is complex: the decrease

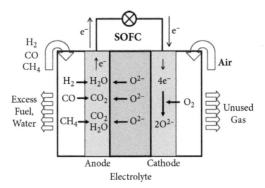

Fig. 8.8 Schematic illustration of a solid oxide fuel cell. Reprinted from Goodenough and Huang (2007). Copyright 2007, with permission from Elsevier.

of the surface tension of H_2O under UV light and the photooxidation of organic contaminants by the strong oxidation power of the VB holes of the photocatalyst have been invoked as reasons for the self-cleaning behaviour (Takeuchi *et al.*, 2005).

8.3 Solid oxide fuel cells

A Solid Oxide Fuel Cell (SOFC) is a very efficient energy conversion device that produces electricity by electrochemically combining a fuel and an oxidant across an ion-conducting oxide electrolyte. A SOFC is not a heat engine and therefore its efficiency is not limited by the Carnot cycle. SOFCs can operate on a variety of fuels, such as H_2, CO or natural gas (CH_4). Hydrocarbons can also be directly converted to a hydrogen-rich gas by internal reforming inside a fuel cell stack. A SOFC contains essentially two porous electrodes separated by a dense oxygen ion-conducting electrolyte. A schematic drawing of a SOFC is shown in Fig. 8.8. The electrodes provide the interface between the chemical energy associated with fuel oxidation and the electrical power and they catalyse the chemical reactions. Oxygen supplied at the cathode (the air electrode) reacts with incoming electrons from the external circuit to form O^{2-} ions according to the reaction: $1/2\ O_2\ (g) + 2\ e^- \rightarrow O^{2-}\ (s)$. The O^{2-} ions migrate to the anode (the fuel electrode) through the oxygen ion-conducting electrolyte. At the anode, oxygen ions combine with H_2 (and/or CO) in the fuel to form H_2O (and/or CO_2), liberating electrons, according to the reactions $H_2\ (g) + O^{2-} \rightarrow H_2O + 2\ e^-$ or $CO\ (g) + O^{2-} \rightarrow CO_2 + 2\ e^-$. The electrons flow from the anode to the cathode through the external circuit (see Fig. 8.8, also Fig. 8.9(a)). The difference of the chemical potentials between air and fuel provides the driving force for the redox reaction and determines the output voltage, which also depends on the overpotentials η at the two electrodes (Fig. 8.9(b)). An individual single cell delivers ~ 1.0 V open circuit voltage, and 0.6–0.7 V under load. Multiple cells can be connected with bipolar plates to form fuel cell stacks, which can provide the desired output voltage.

Classically, that is in the 1990s, SOFCs have used yttrium stabilized zirconia (YSZ) as an electrolyte, perovskite La-Sr-manganite composites as a cathode and Ni-YSZ cermet (ceramic-metal composite) as an anode. The use of YSZ as an electrolyte requires

(a)

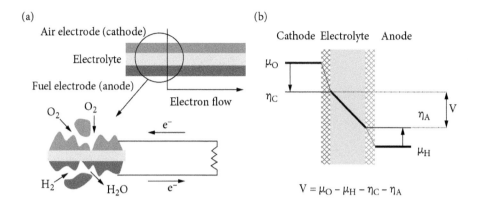

(b)

Cathode Electrolyte Anode

$$V = \mu_O - \mu_H - \eta_C - \eta_A$$

Fig. 8.9 (a) Schematics of a single solid oxide fuel cell; (b) diagram illustrating the drop in oxygen chemical potential across the cell. Reprinted from Jacobson (2009) with permission. Copyright 2009 American Chemical Society.

operating temperatures of around 1,000°C to generate sufficient conductivity in the electrolyte. In recent years, efforts have been directed towards lowering the operating temperature of SOFCs to 500–800°C (Wachsman and Lee, 2011). Lowering the temperature results in a significant reduction of the costs of materials, as well as in enhanced durability, and in less degradation of materials. But a lower operating temperature also means an increase in the resistance of the electrolyte and a decrease in the rates of the electrocatalytic reactions; both factors result in the reduction of cell voltage and output power. The minimum temperature at which a SOFC can be operated depends on the intrinsic ion conductivity of the electrolyte and the minimum film thickness that can be reliably attained in fabrication. Figure 8.10 shows conductivity data versus temperature of three popular electrolyte materials, yttrium stabilized zirconia (YSZ), strontium (magnesium)-doped lanthanum gallate (LSGM), and gadolinium-doped ceria (CGO). Assuming a film thickness of 10 μm and a conductivity of $1*10^{-2}$ Scm^{-1}, the minimum operating temperatures are $\sim$ 700°C (YSZ), $\sim$ 550°C (LSGM), and $\sim$ 550°C (CGO), on the basis of the data of Fig. 8.10.

The transport reaction network in the porous 3-phase SOFC structure is very complex, and how the diverse electrode properties and the processes on many different interfaces influence the overall cell performance is not well understood. The requirements for the electrode materials are stringent and are high electrocatalytic activity and high conductivity. The cathode must catalyse the oxygen reduction reaction, the anode catalyses the fuel oxidation by oxygen ions transported through the electrolyte, and both must transport electrons from and to the external circuit. Moreover, the thermal expansion of all components must be matched, the materials must be thermally and chemically stable at the operating temperature, and porosity is required to allow gas transport to the electrode interfaces. Reviews of SOFC component materials and fabrication are given in (Goodenough and Huang, 2007; Jacobson, 2009; Beckel *et al.*, 2007).

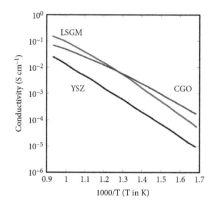

Fig. 8.10 Conductivity data for YSZ, CGO, and LSGM. Reprinted from Jacobson (2009) with permission. Copyright 2009 American Chemical Society.

An actual active direction of research is the utilization of ultrathin film components in SOFCs (Johnson *et al.*, 2009). The nanostructuring of electrodes and electrolytes and the incorporation of nanoparticles appears to lead to improved reactivities and stability (Park *et al.*, 2016). The term *nanoionics* has been coined to emphasize the impact of nano-size effects in electrochemistry (Maier, 2005), taking advantage of short diffusion lengths in nanoporous electrodes, their large surface areas for catalytic reaction, and very low polarization resistance. Micro-SOFCs are miniaturized fuel cell systems for portable applications which rely on the incorporation of nanofilm components as electrodes and electrolyte (Ishihara, 2012; Beckel *et al.*, 2007). With operating temperatures reduced to 300°C and less, nm-thick electrolyte films are mandatory, and the micro-SOFCs bear promise as long-lasting and reliable portable back-up power sources.

Nanosize effects on oxide ion conductivity have been observed in several oxide systems, with large conductivity increases of several orders of magnitude with respect to the bulk values (Ishihara, 2012). These have been ascribed to the interfaces with enhanced conductivity, but the opposite effects—a drastic decrease of oxide ion conductivity in nanocrystalline samples—has also been reported (Ishihara, 2012); the latter has been associated with the accumulation of negative charges at grain boundaries. Presumably, the microstructure and nanoscale morphology of the samples are important parameters in oxide ion conductors, which require rigorous control of fabrication procedures for obtaining reliably reproducible results. Nevertheless, the promise of ion conductivity increase by nano-size effects may be useful for further attempting to decrease the operating temperature of micro-SOFCs to a practical value of lower than 300°C.

8.4 Solar energy materials

Solar energy materials, according to definition by Granqvist (2003), have properties to meet the requirements set by the spectral distribution and intensity of solar irradiation in our natural environment to enable their technological utilization. Examples of

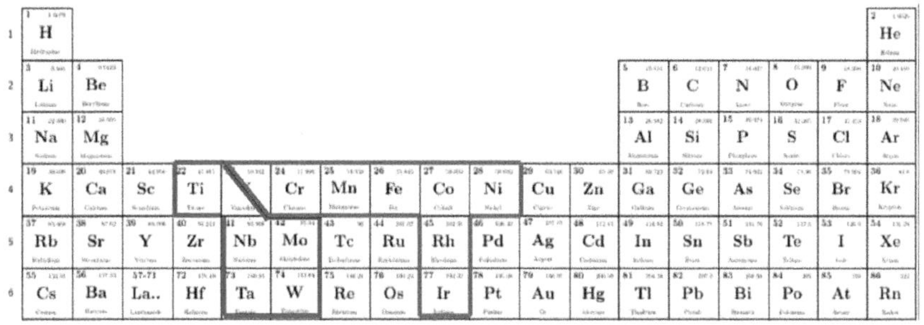

1																	2
H																	He
Li	Be											B	C	N	O	F	Ne
Na	Mg											Al	Si	P	S	Cl	Ar
K	Ca	Sc	Ti		Cr	Mn	Fe	Co	Ni	Cu	Zn	Ga	Ge	As	Se	Br	Kr
Rb	Sr	Y	Zr	Nb	Mo	Tc	Ru	Rh	Pd	Ag	Cd	In	Sn	Sb	Te	I	Xe
Cs	Ba	La..	Hf	Ta	W	Re	Os	Ir	Pt	Au	Hg	Tl	Pb	Bi	Po	At	Rn

Fig. 8.11 Periodic table of the elements. The red and blue boxes indicate transition metals whose oxides display clear cathodic or anodic electrochromism, respectively (Granqvist, 1995).

solar energy materials comprise substances for the thermal and electrical conversion of solar energy, for electrochromism, and thermochromism in passive solar energy applications in energy-efficient architecture design, and for solar photocatalysis. In this section, we concentrate on chromogenic functionalities, that is the reversible coloration and bleaching of a material under the influence of external stimuli. Accordingly, the optical properties of a material can be changed by an external bias in electrochromics, by temperature in thermochromics, by light in photochromics, or by exposure to different gases in chemochromics. Naturally, the latter chemichromic functionality may be employed in chemical sensor technology. Oxide thin films are key materials for chromogenic applications, in particular in electrochromic (EC) and thermochromic devices. In the following, emphasis is put on electrochromic behaviour, which may play an important role in the present day energy saving discussion in our society.

EC materials are characterized by the ability to change their optical properties, specifically light absorption and transmission, reversibly and persistently, when a voltage is applied across them. Electrochromism can be employed in practical windows to regulate the throughput of solar irradiation. EC regulated *smart windows* with tunable properties can be used in advanced fenestration to create energy-efficient buildings with benign indoor climate, in which the potential for energy savings is enormous, given the large amount of the World's primary energy use that is consumed in the built environment (Granqvist, 2012; Granqvist, 2014). Oxide ECs in advanced fenestration via *smart windows* may be regarded as one of the *green* nanotechnologies, which are a hype issue in the current climate change discussion at the time of this writing.

The optical functionality of an EC film involves the change of the optical absorption when ions are inserted or extracted. The colouring or bleaching effect in the EC material is thus effected by a charge transfer process. There are two types of EC metal oxides: i) *cathodic*, which colour under ion insertion; and ii) *anodic*, which colour under ion extraction. Figure 8.11 shows a periodic system of the elements with different coloured boxes indicating transition metal oxides, which display cathodic or anodic electrochromism. A standard EC device incorporates two EC films, shuttling ions between these two films makes both films colour, while shuttling the ions in the other direction makes them both bleach. A suitable combination of cathodic and anodic EC films creates a neutral visual appearance, which is beneficial for fenestration in

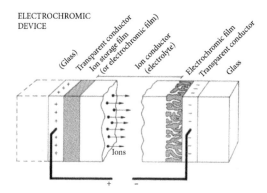

Fig. 8.12 General design of a 'battery'-type EC device. Arrows signify ionic movement in an applied field; electrons move in the opposite direction via the external circuit. Reprinted from Granqvist (2012) with permission. Copyright 2012, with permission from Elsevier.

buildings (such a combination is e.g. WO_3 (cathodic) and hydrous NiO (anionic)).

Most advanced EC technology uses multilayer structures with a fundamental resemblance to a thin film electrical battery with transparent electrodes. A generic design of a five-layer EC device is illustrated in Fig. 8.12. The optical transmittance is modulated when ions from the central electrolyte are shuttled between the conducting thin films acting as anodes and cathodes and inserted (cathodic EC layer) or extracted (anodic EC layer). The effect in the EC active layers is due to the electron flow via the external circuit, which is required to balance the charge generated by the ion flow. In analogy to the battery picture, the charging level corresponds to a degree of optical absorption, which is maintained under open circuit conditions. The change of optical properties is not instantaneous, but takes some time (of the order of minutes). The required dc voltage is ~ 1 V—which makes it convenient to combine an EC device with photovoltaics.

Taking WO_3 and hydrous NiO as cathodic and anodic materials, which are prevalent in EC devices (Deb, 2008), and H^+ as ion shuttle, the functioning of an EC device can be described as follows:

$$[WO_3 + H^+ + e^-]_{bleached} \longleftrightarrow [HWO_3]_{coloured}$$
$$[Ni(OH)_2]_{bleached} \longleftrightarrow [NiOOH + H^+ + e^-]_{coloured} \tag{8.1}$$

The electrons are exchanged via the external circuit. The EC films are mixed conductors, ionic and electronic, while the central electrolyte is a pure ion conductor.

The physical basis of the EC effect is complex and involves the crystalline and electronic properties of the EC materials, their ionic conduction, and the nature of the optical absorption process. The crystalline structure, which is of course intimately tied to the electronic structure, has to sustain ionic transport, and nanostructure effects may play a role here, as mentioned in the following. The structure of most EC oxides—(defected) perovskites, rutiles—are based on TMO_6 octahedra (TM = transition metal), which are connected by sharing corners and/or edges. Their band

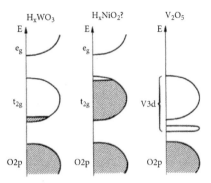

Fig. 8.13 Schematic band structures for different types of EC oxides. Shaded regions denote filled states. Reprinted from Granqvist (1995) with permission. Copyright 1995, with permission from Elsevier.

structures consist of an O $2p$ derived valence band, which is fully occupied, and metal d states, which are split into t_{2g} and e_g states by the ligand field of the octahedral coordination (Goodenough, 1971). Figure 8.13 shows schematic band structures for different types of EC oxides. In the cathodic EC oxides, e.g. in WO_3, the $5d$ states are empty in the pristine state and are separated by a ~ 2.8 eV gap from the O $2p$ band. In the anodic oxides, e.g. $Ni(OH)_2$, the t_{2g} band is fully occupied while the e_g band is empty. If the t_{2g}-e_g gap is large enough, the material is transparent. V_2O_5 is a special case, in which a large distortion of the V-O units from octahedral symmetry—with one much larger V-O bond, so that the structure consists of layers of square pyramidal VO_5 units—leads to a narrow split-off $3d$ band in the gap. This narrow band may be filled or emptied by ion insertion or extraction (i.e. by the respective charge balancing electrons—see Fig. 8.13, right panel). The EC behaviour of V_2O_5 is therefore neither purely cathodic nor anodic, but of a hybrid type. In the cathodic case (H_xWO_3, Fig. 8.13, left panel), the t_{2g} band becomes partially filled by ion insertion (by the charge balancing electrons) and low-energy photon absorption leads to colouring. In the anodic EC oxide (Fig. 8.13, middle panel) ion extraction creates empty t_{2g} states with concomitant photon absorption possibilities and colouring.

The optical absorption mechanism in EC oxides is not well understood, but it has been generally associated with charge transfer. For WO_3, a polaron hopping mechanism has been proposed (Niklasson and Granqvist, 2007). It is believed that the inserted electrons are localized at cation sites ($W^{6+} + e^- \rightarrow W^{5+}$), electron-phonon coupling leads to a local displacement of neighbouring ions—a local lattice distortion—and to the creation of a small polaron—energetically with a state 0.1–0.2 eV below the conduction band. Upon photon absorption, the polaron can be transferred to an adjacent site (polaron hopping). Polaron hopping requires adjacent W^{6+} sites, and site saturation effects are expected at high intercalation levels, which have been experimentally detected (Berggren *et al.*, 2007). Electrochromism is thus essentially an electronic effect associated with electron transfer between neighbouring sites (in the polaron model), but optical modulation also requires ionic effects, namely ion transfer, which is based on the mixed conduction character of the EC materials. Most structures

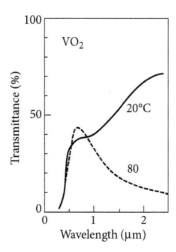

Fig. 8.14 Spectral transmittance through a thermochromic VO$_2$ film at two temperatures. Reprinted from Sobhan *et al.* (1996) with permission. Copyright © 1996 Wiley-VCH Verlag GmbH and Co. KGaA, Weinheim.

of EC oxides involve octahedral TM-O units with spaces in between the octahedral units that are large enough for the transport of small ions (H$^+$, Li$^+$). Aggregates of octahedra may form porous nanoparticles, which introduce nanostructural effects into EC films. Moreover, EC oxides are notable for their substoichiometry. For example, WO$_{3-x}$ or MoO$_{3-x}$ form so-called Magneli phases, in which crystallographic shear planes with 3-D channel networks are produced that can serve as 'tunnels' for ion transport. In addition, grain boundaries in EC films may become beneficial for ion diffusion and transport.

Thermochromism is another possibility to control solar transmission through windows to create energy-efficient buildings. The most interesting thermochromic material is VO$_2$ in thin films. VO$_2$ undergoes a structural phase transition from an IR-reflecting metallic phase to a transparent insulator phase at T$_C$ = 68°C (see Fig. 8.14, showing the spectral transmittance through a thermochromic VO$_2$ film at temperatures below and above T$_C$). The thermochromic switching is highly reversible, but in VO$_2$ it is not readily applicable to practical fenestration. The critical temperature is too high for buildings, the luminous absorption is undesirably high, and the modulation of the solar transmittance is too weak. However, improvement of the optical properties of VO$_2$ has been achieved by doping with TM atoms (e.g. W, Mo), thereby decreasing T$_C$ to around room temperature, and layers of spherical nanoparticles have shown higher transmission and larger optical modulation. The combination of thermochromics with electrochromics offers an attractive direction for superfenestration, with an electrochromic device on the outer pane, a thermochromic device on the inner pane, and the two parts are thermally decoupled (Granqvist, 2014). For a comprehensive recent review of the application of thermochromic VO$_2$ for energy-efficient smart windows, with emphasis of electronic, atomic, and nano-perspectives, see (Cui *et al.*, 2018).

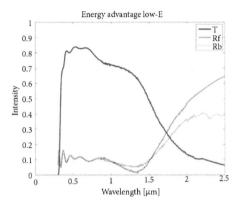

Fig. 8.15 Spectral transmittance T and reflectance from the front (coated) side (R_f) and back side (R_b) for a commercial FTO coated glass. The spectral data correspond to 71 % of the solar transmittance. Reprinted from Granqvist (2014) with permission. Copyright 2014, with permission from Elsevier.

Apart from the EC active layers, other components of EC devices are the electrodes and the ion-conducting electrolyte (see Fig. 8.12). Transparent conducting oxide films are suitable electrodes for EC devices, but they are also important elements in thin film solar cells, light emitting diodes, and display devices of various types. There exist several electrode categories (Granqvist, 2014), among which here we only mention oxide thin films. Heavily doped (meaning a few %) thin films of conducting oxides, such as In_2O_3 doped with Sn or Zn, ZnO with various dopants including Al, Ga, In, doped SnO_2, and others are typically used as transparent electrodes. Figure 8.15 displays the spectral transmittance T and reflectance R for a commercial fluorine doped SnO_2 (FTO) coated glass, which shows 71% of the solar transmittance. In order to achieve good reproducible optical and electrical properties, the thin film deposition techniques require careful deposition protocols and control; each transparent oxide appears to present particular challenges in the fabrication (Granqvist, 2011; Granqvist, 2014). An EC device has also centrally positioned an ion conducting electrolyte (Fig. 8.12). There are various options for ion-conducting electrolytes, amongst them thin solid films, polymer layers, and ionic liquids (Granqvist, 2014). They are mostly proprietary for EC products and little is known about those in commercial practise. Functionalization of polymer layers with oxide nanoparticles is an attractive option, which in addition selectively absorbs in the near-IR region via plasmon resonance absorption.

Chromogenic materials find a variety of applications which range from energy-saving smart windows in the building and glazing industry (electrochromism, thermochromism) to chemical sensor technology (chemichromism). The battery analogy of EC devices, mentioned previously, has led to the incorporation of EC materials into energy storage devices, as electrodes into batteries or supercapacitors (Yang *et al.*, 2016). Indeed, electrochromic and energy storage devices have many aspects in common: such as materials, physical and chemical operating mechanisms, and the charge and discharge properties. However, there is a basic contradiction between electrochromism and energy storage. An EC device should have a high colouration efficiency and a high switching speed, which require low charge density for high response in optical density. Conversely, batteries and supercapacitors aim for high charge densities. The combination of EC and energy storage functionalities can thus be accomplished only with moderate performance in both (Yang *et al.*, 2016).

Oxides are prevalent amongst chromogenic materials, in particular W-oxides (Deb, 2008) and Mo-oxides have found widespread attention, but more complex ternary and multi-cation systems are also gaining ground (Granqvist, 2014). Oxides in thin film morphology are used in most practical device applications, however nanoparticulate layers with tailored particle shapes have recently been shown to improve the desired optical properties. The focus in this section has been put somewhat on electrochromism, which may present a most actual contribution in the discussion of energy saving technologies via advanced fenestration in the building industry (Jelle *et al.*, 2012; Tällberg *et al.*, 2019).

8.5 Corrosion protection films

Corrosion is a natural process that converts a substrate, typically a metal, into a more stable compound such as an oxide or hydroxide by a (electro)chemical reaction with the gaseous or aqueous environment (atmosphere, electrolyte). A well-known example is the rusting of iron surfaces forming iron oxide. The corrosion results in a degradation of useful properties, such as mechanical strength, appearance, or permeability to liquids and gases. Passivation by the spontaneous formation of (*native*) ultrathin films of corrosion products, which act as a barrier layer for further reaction, provides the best means for the corrosion protection of metals. The inherent passivation properties of metal and alloy surfaces have enabled the metal-based technological progress of the modern industrial society and are also important at present for the numerous applications where metal components are in use. The passive films are oxides of the metals self-forming at the surface, and in the presence of an aqueous environment (electrolyte) they are hydroxylated at their outer part. These passive oxide films are ultrathin, in general of the order of 1–5 nm thick, well adherent to the substrate, and isolate it from the corrosive environment. The passive films are often self-healing, that is they reform when locally degraded or removed by breakdown. The self-healing capability may be poisoned in the presence of aggressive anions, which block the metal surface from reoxidation by their strong adsorption.

The passive film that blocks the transfer of cations from the metal to the outer surface/electrolyte is mostly a bilayer or multilayer structure, with an inner oxide part and an outer hydroxide part; the former is the barrier layer against cation transport, the latter an exchange layer with the outer environment (electrolyte, atmosphere) (Marcus and Maurice, 2011). More complex multilayer structures may be formed for metals with several oxidation states and for alloys. As alternatives to multilayer structures, continuous composition changes without sharp phase boundaries have also been reported (for Fe-Ni, Cr, Al, Si alloys) (Marcus and Maurice, 2011). There are two reasons behind passivity: the thermodynamics of the metal-electrolyte/atmosphere interface, i.e. the thermodynamic stability of the passivating oxide, and the slow dissolution kinetics of the oxide. Passivity breakdown may occur, in particular in the presence

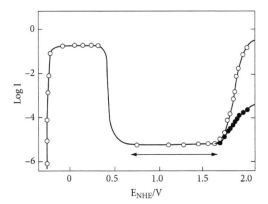

Fig. 8.16 Typical anodic polarization curves for iron in an acid solution at ambient temperature. The Fe dissolution current at higher voltages is given by the filled black symbols, the total current by the open symbols. The passive region is indicated by the horizontal arrow. Adapted from Sato and Sato (1978).

of aggressive anions (e.g. chlorides), which is typically initiated by a local breakdown event (pit growth). Precursors of passive films are ultrathin surface oxides on reconstructed metal surfaces; the grown film is mostly polycrystalline with nanoscale grain boundaries acting as defects that can initiate local passivation breakdown and localized corrosion. It has been demonstrated that grain boundaries at the nanoscale play an important role in initiating local pitting (Marcus *et al.*, 2008).

The phenomenon of passivity may be illustrated in the electrochemical environment by anodic polarization curves (current versus voltage) as shown in Fig. 8.16 for iron in sulfuric acid. On increasing the voltage from left to right, the current increases in the active region due to metal dissolution. The current then decreases and a flat plateau region is formed, which defines the passive state (from 0.5 V to 1.7 V, horizontal arrow). At higher voltages, the current increases again due to Fe dissolution and the evolution of oxygen. The general picture emerging for the passive film is sketched in Fig. 8.17: a defective oxide layer (the primary passive film) forms directly at the interface to the metal and an outer layer is precipitated via hydrolysis of cations ejected from the inner layer. The ionic/molecular transfer processes are also indicated in Fig. 8.17.

The outstanding corrosion resistance of stainless-steel has been the subject of numerous investigations over the past decades (Olsson and Landolt, 2003). Early XPS measurements of Fe-Cr-Ni containing steel have found that the oxide passive film, 1–3 nm thick, is strongly enriched in chromium, whereas the layer closest to the metal/oxide interface is enriched with Ni (Olefjord and Elfstrom, 1982). The protective action of the passive layer has thus been ascribed mainly to the stability and low solubility of the chromium oxide phase. This is confirmed by the more recent work of Diawara *et al.* (2010), shown in Fig. 8.18, in which the passivation probability as evaluated by kinetic Monte Carlo simulation is plotted versus the Cr content of the alloy. For low Cr concentrations ($< 14\%$), the metal is corroded because the Cr oxide nuclei do not completely cover the alloy surface. For high Cr concentrations ($> 16\%$),

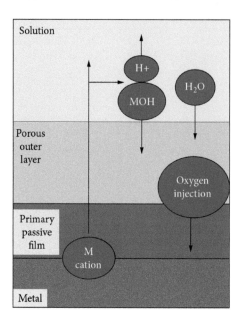

Fig. 8.17 Schematics of the formation of a bilayer passive film on metal surfaces. Molecular/ionic transfer processes are indicated.

the oxide layer covers the entire surface. Thus, the Cr oxide significance is clearly stressed.

Almost all passive films have granular polycrystalline morphology, even if grown on stainless-steel model single crystal surfaces (Marcus *et al.*, 2008; Maurice and Marcus, 2018; Ma *et al.*, 2019). The latter occurs because of the high density of nucleation centres under the corrosive environment and the inherently heterogeneous segregation character of the oxidation reactions at the alloy surfaces (see the following explanation). Also, most technical materials are polycrystals exposing a grain boundary network to the environment. Grain boundaries play an important role in local passivity breakdown as a result of enhanced grain boundary transport reactions (Marcus

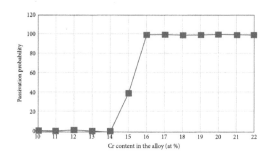

Fig. 8.18 Passivation probability versus Cr content in the Fe-Cr alloy. Reprinted from Diawara *et al.* (2010) with permission. Copyright 2010 American Chemical Society.

(a)

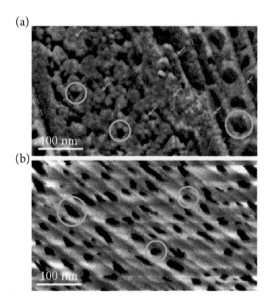

(b)

Fig. 8.19 Nanoscale morphology of the $Fe_{17}Cr_{14.5}Ni_{2.3}Mo(100)$ stainless-steel surface passivated in 0.05 M H_2SO_4(aq) as observed by STM after polarization at 0.5 V/SHE (a) and by EC-STM under polarization −0.14 V/SHE. (b) The passive oxide film has a granular morphology (arrows in (a)). Substrate terraces display depressions (circles in (a) and (b)), evidencing local protection failure caused by competing transient dissolution during passivation. Polarization in the pre-passive range in (b) causes the growth of the depressions by sustained transient dissolution. Substrate step edges are more corrosion resistant. Reprinted from Maurice and Marcus (2018) with permission from Springer. Copyright 2018.

et al., 2008). Marcus *et al.* (2008) have presented a model of localized corrosion and pit formation. Accordingly, the intergranular boundaries are less resistive to ion transfer than the defect-free oxide lattice, causing a faster metal-ion release at the surface. A film topography with depressed areas at the oxide/electrolyte and/or metal/oxide interface causes a local thinning of the oxide with metal voids underneath the oxide, which leads to a local passivity breakdown by cracking and the appearance of pits. The presence of aggressive ions (e.g. Cl^-) will aggravate the effects by forming soluble chlorine containing metal complexes and enlarged nanopits. Moreover, the strong adsorption of chlorine on the metal will hinder the repassivation and the healing (i.e. reformation) of the passive oxide layer.

Current knowledge of the nanoscale properties of passive oxide films has been brought about recently by applying a surface science approach to corrosion science (Maurice and Marcus, 2018), complemented by theoretical modelling (Diawara *et al.*, 2004; Diawara *et al.*, 2010). An UHV environment has been employed for controlled oxidation studies and surface science analytical techniques such as STM, AFM, and XPS have been applied for in-situ imaging of the nanoscale morphology and for monitoring the evolving chemical composition. The Cr enrichment in the films of stainless steel formed in the passive range results from the small dissolution rate of the Cr^{3+}-

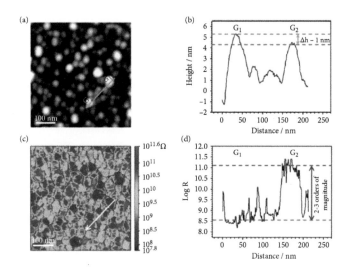

Fig. 8.20 Nanoscale morphology (a) and electrical resistance map (c) as observed by current sensing AFM of the passive oxide film formed on 316L stainless-steel exposed for 2 min to simulated pressurized water reactor environment (water at 325°C). Height (b) and resistance (d) profiles along the green and white arrows in (a) and (c), respectively. Reprinted from Massoud *et al.* (2014). Copyright 2014, with permission from Elsevier.

oxide as compared to Fe^{2+}/Fe^{3+} oxides, whereby the difference is even more effective for Cr enrichment in acid aqueous environment. The morphology of a FeCrNiMo(100) stainless-steel single crystal surface passivated in sulfuric acid aqueous solution has been investigated by in-situ electrochemical STM (EC-STM) imaging. Figure 8.19 shows topographic STM images of the $Fe_{17}Cr_{14.5}Ni_{2.3}Mo(100)$ surface after different polarizations. The terrace-step topology of the stainless steel substrate can be recognized (in particular in (b)); the covering oxide film shows a granular morphology (Fig. 8.19(a)). A grain increase from 5.3 nm to 11.5 nm has been detected between an air-formed oxide film and after passivation by potential 0.5 V/SHE (SHE = standard hydrogen electrode potential) in the middle of the passive range, while no oxide film thickness variation has been observed. It was concluded that a coalescence phenomenon accompanying the Cr enrichment of the film induced the grain size increase. The overall oxide composition of the film also changed during the passivation, with transient dissolution of the Fe oxide fraction. The substrate terraces display local depressions indicating substrate dissolution that competes transiently with oxide formation during passivation. The surface topography in Fig. 8.19(b) reveals that the transient dissolution is not homogeneous but occurs preferentially at terraces rather than at step edges. The latter is presumably due to Cr enrichment at step edges, which protects the steps. The local sites at terraces with less protection against dissolution (depressions in Fig. 8.19(b)) indicate local discontinuities in the Cr content of the oxide film.

The conductive AFM data of Fig. 8.20, recorded on a 316L stainless steel surface exposed to a simulated pressurized water reactor environment (water at 325°C), confirm the inhomogeneous character of the passive film. Not surprisingly, the passive film grows thicker at the elevated temperature than at room temperature and forms larger nanograins. The grown oxide film consists of a Cr^{3+}-rich inner barrier layer mixed with Fe^{2+}/Fe^{3+} species and an outer layer consisting of mostly Fe^{2+}/Fe^{3+} oxide. The conductive AFM data (Fig. 8.20(c,d)) map the electrical resistance of the oxide film and demonstrate that the resistivity spreads by one order of magnitude over single oxide grains and varies by 2–3 orders of magnitude at different locations. Since the outer layer of the passive film is mostly composed of Fe^{2+}/Fe^{3+} oxide of low resistivity, it has been concluded that the measured resistivity variations have to be associated with inhomogeneities of the Cr^{3+}-rich inner layer (Maurice and Marcus, 2018).

The origin of the breakdown of passivity, leading to local failure of corrosion resistance and to local corrosion and pitting, has been ascribed to the surface heterogeneity of the passive oxide layer at the microscopic level. Maurice and Marcus' group (Ma *et al.*, 2019) has investigated the mechanism causing the Cr enrichment in the surface oxide film protecting stainless steel against corrosion at nanometre length scales. These authors performed model studies of nucleation and growth of surface oxide on a single crystal austenite stainless-steel $Fe_{18}Cr_{13}Ni(100)$ surface by STM, revealing surface structure and topography under controlled oxidation conditions, and by XPS, monitoring the surface composition changes (Ma *et al.*, 2018). The XPS data confirmed that Cr^{3+}-oxide forms dominantly at the initial oxidation stages, followed

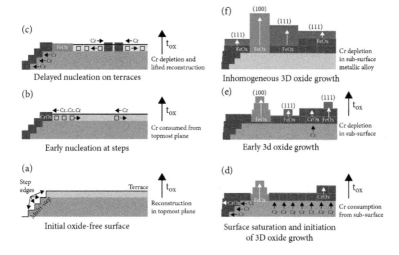

Fig. 8.21 Cross-section model of the early oxidation mechanism at the origin of the nanoscale heterogeneity of the surface oxide film on austenite stainless steel (height-to-width aspect ratio not to scale). (a) Initial oxide-free surface. (b) Early Cr oxide nucleation at the multisteps and Cr consumption with vacancy injection in topmost layer of terraces. (c) Delayed Cr and Fe oxides nucleation on terraces fully depleting Cr in the topmost plane. (d) Mixed oxide layer at surface saturation formed by Cr and Fe transport from the metallic planes below the oxide. (e) Early inhomogeneous 3-D growth of Fe oxide through the interfacial Cr oxide barrier layer or directly on the substrate. (f) Inhomogeneous 3-D growth of Fe oxides after Cr depletion of the metallic alloy below the oxide film. Reprinted from Ma *et al.* (2019).

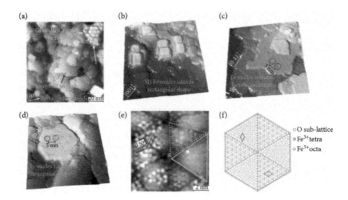

Fig. 8.22 Inhomogeneous 3-D multilayered oxide film after an oxygen exposure of 191.4 L at 250°C. (a) Topographic STM image ($100 * 100$ nm^2). The insert marked by a hexagon is an enlarged view of the granular Cr oxide interfacial layer observed locally. (b) 3-D view STM image of rectangular shapes islands ($50 * 50$ nm^2). (c) 3-D view of flat terraces ($50 * 50$ nm^2). (d) 3-D view of hexagonal terraces. (e) Atomic structure on the hexagonal terrace in (d), ($10 * 10$ nm^2). Domains and unit cells are marked. (f) Atomic model of the hexagonal structure in (e). Reprinted from Ma *et al.* (2019).

by the competitive formation of Fe^{3+} species, and eventually by the dominance of Fe^{3+}-oxide formation. Nickel did not oxidise in significant amounts. A schematic picture of the mixed oxide layer (CrO_x, FeO_x) growth on stainless steel, as established by STM and XPS studies illustrating the heterogeneous surface composition, is shown in Fig. 8.21. The nanoscale heterogeneity appears to be already generated in the early oxide nucleation phase by preferential local Cr atom supply to sites of reduced coordination (multisteps, step bunches). Cr atoms are segregated from atomic planes below the surface to grow Cr^{3+}-oxide at/near steps, but incompletely covering the surface (Fig.8.21(b,c)). Fe oxide formation is first competitive (Fig. 8.21(c,d)), but eventually becomes dominant (Fig. 8.21(e,f)). The STM images of Fig. 8.22 give an impression of the morphology of the surface after 3-D growth of the surface oxide film (after 191.1 Langmuir O_2 exposure ($1L = 1*10^{-6}$ torr O_2 sec) at 250°C), showing the multilayered growth of the crystalline oxide islands (Fig. 8.21(f)). The nanoscale heterogeneity of composition and structure of the oxide film on stainless steel is impressively documented in Fig. 8.22. Local passivity breakdown is expected in the least Cr^{3+}-enriched region of the surface oxide, self-repair of the passive layer is expected to be ineffective in those regions, where the metal is depleted in chromium.

In closing this section on passivation, it is noticed that apart from corrosion protection by the native oxide layers, the protection of metal substrates can be effected by oxide coatings applied by thin film deposition techniques. Al_2O_3 and TiO_2 as successful protective layers grown by the ALD technique have been reported (Shan *et al.*, 2008; Marin *et al.*, 2011; Abdulagatov *et al.*, 2011; Mirhashemihaghighi *et al.*, 2016).

8.6 Biological applications of oxide nanosystems

8.6.1 Biocompatible materials

The beneficial insertion of a foreign material into a living organism requires that the implantable material must exhibit natural biocompatibility to be well tolerated by the organism and to minimize immune reactions. Good biocompatibility is related to the surface reactivity of the material with oxygen-containing physiological liquids and to the formation of a protein adlayer (Kasemo, 2002), which forms the basis for cells to adhere and spread, and to a good integration of the implant in the body. Stainless steel, Ti alloys, CrCo alloys, and Ta alloys are currently used materials for osseous implants and orthopaedic, dentistry, and cardiology applications (Pacchioni, 2012; Variola *et al.*, 2009), due to their excellent mechanical strength and high corrosion resistance, which avoids the release of dangerous metal ions into the living system. The good physiological behaviour of these biocompatible metals is due to an ultrathin superficial native oxide layer, a few nm thick, which forms under physiological conditions, confers corrosion resistance, and mediates the interaction with the cell material (Anselme, 2000). The TiO_2 surface layer, for example, has a dielectric constant close to that of H_2O and carries a moderate negative charge at physiological pH, so that proteins do not become denaturated when their hydrophillic outer shell interacts with the titania surface. Also, the surface charge may attract positive ions, such as Ca^{2+}, which appears to be favourable for osseointegration.

It has been found that functionalization of the surface oxide film, in particular nanostructuring, improves protein adsorption and the capacity to assimilate implants into the human body (Anselme, 2000; Variola *et al.*, 2009). Different oxide surface topographies at the nanoscale, i.e. of the size of the biological structures, have an impact on cellular attachment or osseointegration by creation of nanoscale surface cues for biological activity. Oxidative electrochemical nanopatterning, leading to e.g. nanotubular structures of TiO_2 by anodization of Ti (Yao *et al.*, 2008), or chemical etching with a combination of strong acids and oxidants to create different nano-textures by varying the etching parameters (Ban *et al.*, 2006), have been shown to stimulate cell growth and to enhance the activity of osteoblastic cells (Variola *et al.*, 2009). Also, the density of surface hydroxyl groups is believed to influence cell activity (Lu *et al.*, 2008). From these studies, surfaces that are hydrophilic, nanorough, and porous appear to have beneficial effects on various biological phenomena. A new generation of biocompatible metals with tailored topographical nanoscale features at their oxide surfaces are thus emerging, with enhanced and selective biological responsiveness (Variola *et al.*, 2009).

While the overall results of studies with nanostructured implants are encouraging, some discrepancies in the obtained results exist (Variola and Nanci, 2011; Pacchioni, 2012). The reason is that many studies lack detailed physico-chemical characterization of the respective surfaces at the nanoscale, so that surfaces defined as nanostructured are often just inhomogeneous assemblies of 'nanofeatures' with unknown statistical distribution and poor reproducibility. The field of studies of native oxide films and of their functionalization in a biological context is thus wide open and remains a challenging endeavour to improve the accommodation of implants and lost tissue into the human body.

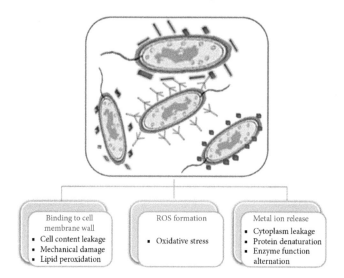

Fig. 8.23 Metal oxide nanoparticles interacting with bacteria. Molecular mechanisms contributing to the antibacterial activities of metal oxide nanoparticles. Reprinted from Stankic *et al.* (2016).

8.6.2 Toxicity, antibacterial properties and theranostic applications

Many oxide materials in nanoscale morphologies—nanoparticles (NP), nanosheets (NS)—are believed to be of low toxicity (Song *et al.*, 2014; Kalantar-zadeh *et al.*, 2016), but the cytotoxicity of some oxides towards specific cell types can also be explored for antibacterial activities (Stankic *et al.*, 2016) or theranostic (therapeutic and diagnostic) applications. Stankic *et al.* (2016) have reviewed the antibacterial and cytotoxic properties of a wide range of binary and ternary oxide nanoparticles. It has been recognized that the structure, size, and morphology of the NP play an important role in the antibacterial activities. For instance, the antibacterial efficiency increases with decreasing particle size, but the crystallographic facet orientation appears to have little effect, while increasing the lattice constant enhances the antibacterial activity. Due to their antibacterial effects and relatively low toxicity against human cells, NPs of Fe_3O_4, TiO_2, CuO, and ZnO (Reddy *et al.*, 2007) have been widely investigated as antibacterial agents in the biomedical and cosmetic industries. Several mechanisms have been proposed to explain the antibacterial action of oxide NP (see Fig. 8.23), including interaction and destruction of the bacteria cell wall, Reactive Oxygen Species (ROS) formation leading to oxidative stress and damage, and the internalization of the NPs into the bacteria cells followed by the release of toxic metal ions.

Since metal oxide NPs with different physico-chemical properties show different antibacterial mechanisms, a combination of two or more metals in oxide NP, forming ternary or multimetal oxides, may be beneficial for the elimination of diverse bacterial strains while maintaining low toxicity towards human cells. For example, ZnMgO NPs exhibit advantageous properties from both of their individual components: the high antibacteriological activity of ZnO and the low cytotoxicity of MgO (Vidic *et al.*,

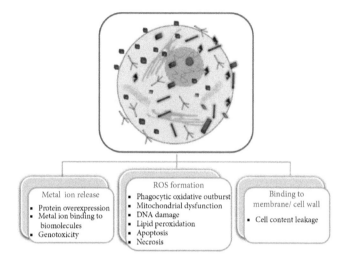

Fig. 8.24 Metal oxide nanoparticles interacting with mammallian cells. Various mechanisms may be active inducing toxicity towards mamallian cells. Reprinted from Stankic *et al.* (2016).

2013). The recognition of toxicity of oxide NPs towards mammalian cells is a necessary requirement for their safe practical antibacterial application. As with the antibacterial activities, several mechanisms for cytotoxicity of oxide NPs have been proposed (see Fig. 8.24). It appears that the most important factors determining toxicity are the inherent toxicity of the metal ions released and the induction of ROS generation, both triggered from the internalization of NP within cells. The toxicity of oxide NPs may be reduced by surface coating (Hsu *et al.*, 2014) or by doping with other metal ions (Xia *et al.*, 2011). The practical application of oxide NPs as bactericidal agents is thus possible under certain conditions and concentrations, at which low toxicity against mammalian cells has been demonstrated, e.g. for ZnO, Fe_2O_3, or Ag_2O_3 (Stankic *et al.*, 2016).

Metal oxides at the nanoscale have exciting prospects as diagnostic tools and therapeutic agents, and sometimes as both in one material (theranostic activity, i.e. therapeutic *and* diagnostic functions). The recognition of disease foci by *bioimaging* is an advanced diagnostic practise, in which metal oxides have been successfully applied. Here we mention oxide nanosystems as contrast enhancing agents in magnetic resonance imaging (MRI) and as components of fluorescence biomarker platforms, with versatile theranostic extension possibilities. Iron oxide 'nanowiskers' (a kind of nanorods, about 2 nm * 20 nm) have been reported as a positive (T_1) contrast agent for MRI (Macher *et al.*, 2015). The crystal phase of these nanowiskers has been identified as maghemite (γ-Fe_2O_3), their high surface-to-volume ratio and strong paramagnetic signal have been invoked to cause a strong positive MRI contrast enhancement (shortened longitudinal relaxation time T_1) observed both in vitro and in vivo in a rat model.

The combination of MnO_2 nanosheets with aptamers (biomolecules that bind to specific target molecules) has been designed as an activable dual fluorescence/MRI

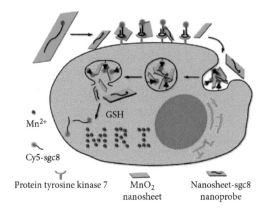

Fig. 8.25 Schematic picture of the activation mechanism of the MnO_2 nanosheet-aptamer nanoprobe for fluorescence/MRI bimodal tumour cell imaging. Reprinted from Zhao *et al.* (2014) with permission. Copyright 2014 American Chemical Society.

nanoprobe platform for tumour cell imaging (Zhao *et al.*, 2014). In this strategy, the reducible MnO_2 acts as a DNA (the aptamer) carrier, fluorescence quencher, and intracellular glutathione (GSH) activated MRI contrast agent. The activation mechanism of the MnO_2 nanosheet-aptamer nanoprobe for fluorescence/MRI tumour cell imaging is illustrated schematically in Fig. 8.25. The fluorescence of the fluorophore aptamer is quenched by the strong binding of the molecule to the MnO_2 nanosheets, as a result of resonant energy transfer to the MnO_2, where the energy is dissipated via $d-d$ electron transitions. The octahedral MnO_6 geometry in MnO_2 shields the Mn cations from the aqueous environment, reducing their influence on the longitudinal or transverse relaxation of protons responsible for the MRI contrast. In the absence of target cells, no fluorescence signalling nor MRI contrast of the nanoprobe is activated. In the presence of target cells, the aptamers bind to the targets and this weakens their adsorptive bonding to the MnO_2 surface; as a result, the fluorescence signal is recovered, and the target cells become illuminated. The aptamer attachment to the target cells facilitates the endocytosis of the MnO_2 nanosheets into the target cells, where the MnO_2 becomes degraded and reduced to Mn^{2+} by the intracellular glutathione, which is overexpressed in many tumour cells. The generated Mn^{2+} ions provide a high contrast in the MRI signalling (Park *et al.*, 2011). Due to the availability of aptamers that recognize different tumour cells, this dual-activable MnO_2 based fluorescence/MRI platform may have a wide application in cancer cell imaging (Zhao *et al.*, 2014).

Thin-layer MnO_2 nanosheets exhibit large surface areas, which can be used in a theranostic architecture for the molecular attachment of drugs in a carrier system for the controlled delivery to target cells (Chen *et al.*, 2014b). MnO_2 nanosheets feature pH-responsive degradation in a mildly acidic environment, by which the endocytosed MnO_2 are destroyed thereby delivering the drug and at the same time liberating Mn^{2+} ions, which provide an MRI active contrast agent. Tumourigenesis can generate a more acidic microenvironment in tumour tissues than in normal tissues due to the generation of lactic acid, which can activate the break-up and disintegration of the

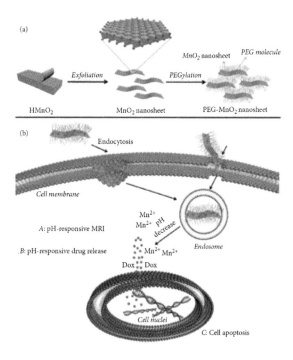

Fig. 8.26 (a) Schematic illustration of the synthetic procedure for 2-D PEG-MnO$_2$ nanosheets. (b) Theranostic function of PEG-MnO$_2$ nanosheets for intracellular pH-responsive drug delivery and MRI (PEG = polyethylene glycol). Reprinted from Chen *et al.* (2014*b*) with permission. Copyright © 2014 Wiley-VCH Verlag GmbH and Co. KGaA, Weinheim.

MnO$_2$ nanosheets and the delivery of anticancer agents, as well as the Mn^{2+} ions for tumour MRI imaging. Figure 8.26(a) sketches the fabrication of the MnO$_2$ nanosheets by exfoliation and their coating with polyethylene glycol (PEG); the latter serves to improve their stability in physiological conditions. Figure 8.26(b) illustrates the theranostic function of PEG-MnO$_2$ for intracellular pH-responsive drug delivery and MRI signalling (Chen *et al.*, 2014*b*).

Molybdenum oxide (MoO$_x$) nanosheets with a pH-dependent decomposition behaviour, strong NIR (near-infra-red) absorbance, and efficient drug loading capabilities have been proposed as a theranostic platform for efficient tumour treatment (Song *et al.*, 2016). The PEG-functionalized MoO$_x$ nanosheets show efficient accumulation in tumour cells due to the acidic environment, their high NIR absorbance offers strong contrast under photoacoustic imaging, and as photothermal agents they realize effective in vivo tumour ablation under NIR laser irradiation. The rapid clearance of PEG-MoO$_x$ from normal organs, the absence of in vivo long-term toxicity, and the favourable tumour passive retention abilities make it an attractive agent for in vivo applications. The cytotoxicity of MnO$_3$ nanoplates towards invasive breast cancer has been reported by Anh Tran *et al.* (2014). The generation of ROS in the cancer cells has been suggested to be responsible for the apoptotic cancer cell death.

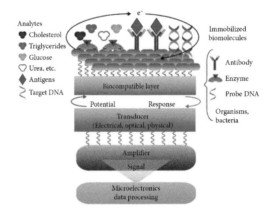

Fig. 8.27 Schematic structure and operating principle of a biosensor. Reprinted from Solanki *et al.* (2011) with permission from Nature. Copyright 2011.

The pharmacological potential of ceria nanoparticles lies in their ability to act as biological ROS scavengers, which is based on their antioxidant properties; the latter is enabled by the reversible Ce^{3+}/Ce^{4+} redox pair, which is abundantly present in ceria nanosystems (Celardo *et al.*, 2011). ROS, produced as an inevitable by-product of cell metabolism-based redox processes, can be potentially very dangerous to cells, if their concentration level rises above a critical threshold, creating oxidative stress. Oxidative stress is involved in many human diseases and the free radical concentration must be regulated or scavenged by exogenous or endogenous antioxidant systems. Nanoceria destroys ROS by reducing the prevalent superoxide (O_2^-), hydrogen peroxide (H_2O_2) and hydroxyl radicals, thus mimicking the behaviour of two antioxidant enzymes. The $3+/4+$ valence switch of ceria resembles the mechanism of redox enzymes, in that the latter use as co-factors metal atoms to catalyse the reversible redox reactions in cells. Nanoceria is considered non-toxic and apparently well tolerated by the organism, and biological studies have revealed a number of beneficial properties of ceria NPs including cell protective, neuroprotective and cardioprotective effects, improvement of cell adhesion, and anti-inflammatory properties (Celardo *et al.*, 2011; Benedetti *et al.*, 2018). However, the effects of nanoceria on cell viability appear to depend on its effect on the intracellular oxidation state: when ceria behaves as an antioxidant, it produces a pro-survival effect; but when it exerts a pro-oxidant effect, cell apoptosis may be induced. The latter may be caused by the production of hydroxyl radicals from H_2O_2 as an unwanted side effect. The discrepancy of reported results (cell pro-survival versus cell anti-survival) are difficult to rationalize, since the analysis of published studies does not establish simple correlations with cell type or particle size (Celardo *et al.*, 2011). Presumably, the complex interplay between cell features and redox properties of the nanoparticles, the latter determined by the Ce^{3+}/Ce^{4+} ratio which in turn depends on preparative and environmental aspects, is likely to determine the overall outcome.

8.6.3 Biosensing platforms

In its classical concept, a biosensor is an analytical device that transforms a biological recognition event into another signal (optical, electrical, or other physical) that can be measured and quantified. Accordingly, a biosensor consists of a biosensitive (bios-elective) layer coupled to a transducing system for signal detection (Solanki *et al.*, 2011; Shavanova *et al.*, 2016). Figure 8.27 shows the schematic structure and operating principle of a biosensor. The biological analytes are selectively immobilized on the biosensitive layer, which is fixed on the transducer, and the biological recognition reaction generates electrons which trigger an electrical, optical, electrochemical, or other physical response in the transducer, which is processed to provide the sensing information. Based on the signal detection method, biosensors can be classified as optical, electrical, electrochemical or other (piezoelectric, thermoelectric). Nanostructured metal oxides have very good prospects as promising matrices for interfacing the biological recognition layers and as effective physical or electrochemical transducer elements. In particular, 2-D oxide layers (nanosheets) have a high surface area, which produces an intense signal and high sensitivity. Today, biosensors have evolved from the classical concept of enzyme-electrode reactions to a variety of analytical methods based on bioaffinity and biocatalysis. Current trends of biosensor evolution have been reviewed by Shavanova *et al.* (2016). Here, we select a few prototypical examples to illustrate the application of thin-layer oxides in biosensing devices.

The development of highly sensitive, reliable and low-cost glucose sensors has been an important issue in the medical sciences due to the widespread occurrence of diabetes mellitis in the World population. Modern glucose sensors based on the *direct* electrochemical oxidation of glucose to gluconic acid, detected by electrochemical techniques (potentiometry, amperometry, conductometry), have been assessed by Rahman *et al.* (2010). A variety of nanostructured oxide materials — ZnO, CuO, MnO_2, TiO_2, CeO_2, amongst others — have been successfully tested as electrodes in glucose biosensors. Compared to the classical enzymatic glucose sensors, the metal oxide sensors have advantages of high sensitivity, long-term stability and of no loss of sensitivity due to denaturation of protein enzymes (Rahman *et al.*, 2010).

A different concept of glucose sensing, a luminogenetic detection system, is based

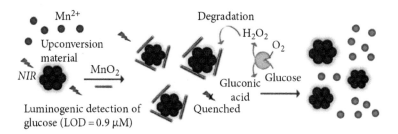

Fig. 8.28 Supramolecular self-assembly of thin-layer MnO_2 with an upconversion nanoparticle for the detection of glucose. Reprinted from He and Tian (2016) with permission. Copyright © 2016 Wiley-VCH Verlag GmbH and Co. KGaA, Weinheim.

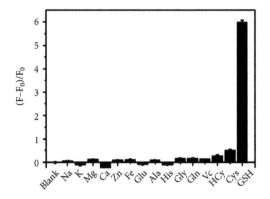

Fig. 8.29 Fluorescence responses $(F - F_0)/F_0$ of the C-nanodot-MnO$_2$ nanosheet sensing platform towards GSH and interferents (F_0: fluorescence intensity of the pristine sensor; F: fluorescence intensity of the sensor in the presence of interferents). Reprinted from Wang *et al.* (2015) with permission. Copyright 2015 by the Royal Society of Chemistry. Permission conveyed through Copyright Clearance Center, Inc.

on a fluorescence OFF-ON photoluminescence platform of MnO$_2$ in combination with fluorophores, e.g. aptamer-grafted upconversion nanoparticles (He and Tian, 2016) (see Fig. 8.28). Similar to the theranostic applications of MnO$_2$ (see Section 8.6.2), the working mechanism of this sensor relies on the degradation and reduction of MnO$_2$ nanosheets by a reducing agent, which is produced by the biochemical reaction. The fluorescence of the fluorophore upconversion (NIR→Vis) particles is quenched upon self-assembly with the MnO$_2$ nanosheets, as a result of fluorescence resonance energy transfer (FRET). The conversion of glucose by glucose oxidase generates H$_2$O$_2$, which degrades the MnO$_2$ sheets to soluble Mn^{2+} ions, thereby restoring the luminescence of the upconversion particles.

This principle of the FRET ability of oxide nanosystems and their biodegradability has also been exploited for glutathione (GSH) sensing in human blood (Wang *et al.*, 2015; Wang *et al.*, 2016). The low molecular weight thiol glutathione plays a central role in maintaining the redox balance of cells, and it is therefore important to develop biomarkers for its analysis to treat oxidative stress. Different fluorophores attached to MnO$_2$ nanosheets have been used as sensing platforms, e.g. formaldehyde resin NPs or carbon nanodots. Interestingly, these sensors show a remarkable selectivity for GSH as demonstrated in Fig. 8.29, where the fluorescence responses of the C-dot/MnO$_2$ nanosheets for GSH and for other common interferents are compared.

An electrical biosensing platform based on a field effect transistor (FET) design, employing a thin film of α-MoO$_3$ nanoflakes as a conducting channel, has been introduced by Balendhran *et al.* (2013*b*) (see Fig. 8.30), using bovine serum albumin (BSA) as a model protein. The planar α-MoO$_3$ flakes form the conduction channel with a large electron mobility and a high permittivity. Many biomolecules such as proteins are intrinsically charged and strongly adsorbed on the MoO$_3$ surface. The negatively charged BSA molecules produce a negative potential if adsorbed on the active MoO$_3$ sensing surface, which influences the field effect parameters producing a gating effect.

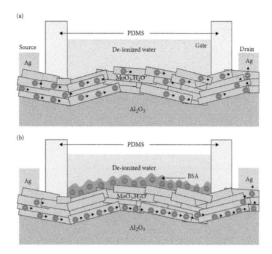

Fig. 8.30 Cross sectional schematic illustration of the MoO_3 FET-based sensor: (a) without BSA; (b) with BSA immobilized onto the active surface. Blue spheres indicate electrons, the arrowheads denote their directional mobility. BSA is negatively charged. Reprinted from Balendhran *et al.* (2013*b*) with permission. Copyright 2013 American Chemical Society.

As the MoO_3 sensing film forms an n-type FET, the negative surface charge results in a reduction of the channel conductance and thus to a detectable sensing response (Balendhran *et al.*, 2013*b*).

The widespread application of biosensing technologies in areas such as environmental monitoring, medicine, biotechnology, food analysis, agriculture and healthcare, and the wide range of biological targets makes this field very active for interdisciplinary research. Metal oxide nanosystems as transducer elements bear a great potential and may open up new horizons with functionalities for novel biodetection protocols (Shavanova *et al.*, 2016).

8.6.4 Biological responses to ferroelectric materials

Biological systems can be manipulated, moved, and displaced by electric fields. The displacement is the result of electrophoresis—the movement of charged objects under an electric field—or dielectrophoresis—the movement of neutral objects under an inhomogeneous electric field. The manipulation of other properties is based on the changes in the biological charge distribution by the electric field influence. A difficulty for the biological application of electric fields is the need for electrodes necessary to produce the electric field. Typical metal electrodes are often biologically incompatible and difficult to apply in small biosystems. Recently, the use of ferroelectric materials has been proposed to create electric fields in biological experiments (Blázquez-Castro *et al.*, 2018). Using ferroelectrics as so-called *virtual electrodes*, it is possible to generate an electric field without external circuit or external electric source. This can be realized due to the spontaneous and permanent electric polarization P_S that is inherent in these materials (see Chapter 4). In this section, the discussion will follow a recent

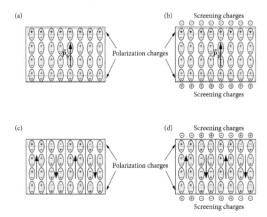

Fig. 8.31 Schematics of the cross section of a ferroelectric crystal, such as LiNbO$_3$ or LiTaO$_3$, showing the spontaneous polarization and the surface polarization and screening charges in several situations: (a) monodomain crystal with non-compensated polarization charge; (b) monodomain crystal with compensating screening charges; (c) periodically poled crystal with non-compensated screening charge; and (d) periodically poled crystal with compensating polarization charge. Reprinted from Blázquez-Castro *et al.* (2018) with the permission of AIP Publishing.

focus review by Blázquez-Castro *et al.* (2018), to which we refer for a comprehensive compilation of the published literature.

Electric fields for biological application can be generated in ferroelectric materials by ferroelectric excitation: in domain structures, by the use of the bulk photoelectric effect, and by the pyroelectric effect. The spontaneous electric polarization P$_S$ in ferroelectrics along the polar axis is a consequence of the non-centrosymmetric crystal structure, in which positive and negative charges are slightly shifted with respect to each other along this axis, as schematically indicated in Fig. 8.31(a). The spontaneous polarization gives rise to bound polarization charges, which are usually compensated by outside screening charges (Fig. 8.31(b)) to maintain charge neutrality at the crystal surface. The polarization direction can be reversed in the whole material or in a domain region by applying an electric field greater that a threshold value (the coercive field). A ferroelectric material can be monodomain (Fig. 8.31(a,b)) or polydomain (Fig. 8.31(c,d)). At the boundaries between domains (domain walls) electric fields appear (Kalinin and Bonnell, 2001).

Internal electric fields in ferroelectric materials can also be generated by bulk photovoltaic (PV) and pyroelectric (PY) effects, which have been employed in biological experiments. The PV effect, appearing in some doped ferroelectrics (e.g. particularly strong in Fe-doped LiNbO$_3$), arises from asymmetric photoexcitation of electrons from impurities due to the non-centrosymmetric crystal lattice (Fig. 8.32(a)) and produces an electric current along the polar c-axis. For biological applications of the PV effect, two geometrical configurations of the ferroelectric crystal are used, as represented in Figs. 8.32(b,c). In both cases, electric fringe fields appear outside the crystal. The pyroelectric (PY) response is related to the surface polarization and screening charges

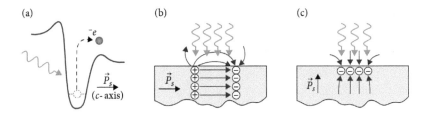

Fig. 8.32 Schematic diagram of (a) the directional electron photoexcitation in PV ferroelectric crystals such as Fe-doped LiNbO$_3$, (b) the photoexcited charges and the photovoltaic electric field when the active crystal surface is parallel to the polar axis (x-cut or y-cut), and (c) the photoexcited charges and photovoltaic electric field when the active crystal surface is perpendicular to the polar axis (z-cut). Reprinted from Blázquez-Castro *et al.* (2018) with the permission of AIP Publishing.

at the ferroelectric surfaces. After a temperature change, P$_S$ changes magnitude modifying the surface charge density and a pyroelectric field develops in the surroundings of the ferroelectric crystal. For PY application, z-cut crystals (Fig. 8.32(c)) are mostly used.

Thus, both the light-induced PV and the thermally induced PY effects create electric fields in the vicinity of ferroelectric crystals; also, electric fields can appear in the domain boundaries of polydomain ferroelectric structures. These electric fields of ferroelectric samples have been utilized as active substrates to induce trapping and patterning of biological structures, as well as cell response modulations (increased cell growth and proliferation, enhanced cell attachment, different stem cell differentiation, bactericidal activity) (Blázquez-Castro *et al.*, 2018). As an example of trapping and patterning, Fig. 8.33 shows μm-size 1-D and 2-D patterns of Himalayan cedar pollen fragments arranged on a Fe-doped LiNbO$_3$ surface, which demonstrate that it is possible to arrange microscopic biological materials with high spatial accuracy. The photoinduced electric field gratings in the LiNbO$_3$ were induced by interfering patterns of 532 nm laser light. The mechanism of trapping of these bioparticles is dielectrophoresis, since the particles are essentially neutral. The electric field is quite high and inhomogeneous near the photoexcited ferroelectric surface, which provides favourable conditions for dielectrophoretic transport. If the particles are charged, they can also be moved in the electric field and assembled according to the substrate polarity/charge pattern.

A number of tentative action mechanisms has been proposed to account for the response of biomolecules and cells to ferroelectrically generated electric fields. Whereas ferroelectric-driven electrochemistry has been formulated, its influence has been considered as a minor effect (Blázquez-Castro *et al.*, 2018). Electrokinesis (electrohydrodynamic) phenomena and several biophysical and biochemical transducing mechanisms have been proposed to mediate the biological effects, such as modulation of ionic flows into the cell, changes in plasma membrane polarization state and alteration of the distribution of membrane proteins. The intracellular Ca^{2+} ion signalling or other relevant voltage-gated ion channels (Na$^+$, K$^+$) as a mechanism for the electric field changes of cell status have been mentioned to play a crucial role in cell responses

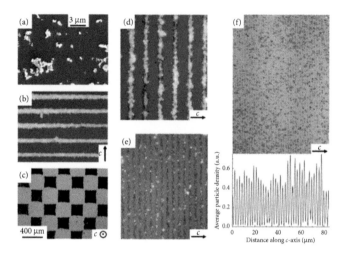

Fig. 8.33 (a) SEM image of Himalayan cedar pollen fragments. (b) Microscope image of a 65 μm period pattern of these pollen fragments obtained with sinusoidal illumination of the same spatial period. (c) Microscope image of a 2-D pollen pattern obtained after illumination with a mosaic of squares with 200 μm side. Microscope images of periodic patterns with decreasing periods: (d) 20 μm; (e) 8 μm; (f) 2 μm; at the bottom of (f) the average particle density profile along the c-axis direction is shown. Reprinted from Jubera *et al.* (2016) with the permission of AIP Publishing.

(migration, proliferation, differentiation, apoptosis). Other proposed mechanisms include the displacement of large biomolecules (proteins, other charged biopolymers) as a result of electrophoresis or electroosmosis (i.e. the movement of H_2O coupled to the movement of ions) actions. Electrokinetically-driven 'molecular crowding' in cells at ferroelectric surfaces may stimulate a variety of cell responses. 'Molecular crowding' refers to the accumulation of biomolecules in cells that may lead to local rheological changes such as increased viscosity and restructuring of water shells. Electroporation of cell membranes—the creation of membrane pores by the electric field—may cause the spontaneous movement of ions and molecules down the electrochemical gradient and subsequently necrotic cell death as a result of osmotic shock.

The ferroelectric oxides that have been mostly employed in biological studies are (Fe doped) $LiNbO_3$, $LiTaO_3$, and $BaTiO_3$ in the form of monodomain crystals, domain structures or as nanoparticles. The field of biological applications of ferroelectric materials is a very recent field of study but offers a broad new horizon of scientific possibilities. The action mechanisms that stimulate the reported biological responses to electric fields need to be further scrutinized and new experiments to pin down the biophysical nature of the observed phenomena are necessary. Envisioned applications may be foreseen in cell biology and signalling, tissue regeneration, antitumour action, and cell bioengineering.

9
Synopsis and outlook

Oxides of small size and low dimension, nanooxides in the form of thin films, nanoparticles, and other nanostructures, are fascinating materials for fundamental scientific study, have widespread use in present-day science and technology, and possess a high potential for future applications in the emerging nanotechnologies. In this book, we have attempted to address diverse topics that we believe give an impression of the nature and characteristics of nanooxides and which we feel have an input on their science and technology. In the preceding eight chapters, we present fabrication procedures and characterization methods of oxide ultra-thin films and nanoparticles, their inherent physical and chemical properties and their interaction with environmental effects. Apart from fundamental aspects, we also draw attention to prototypical examples of technological application of nanooxides in various interdisciplinary fields of physics and chemistry.

A prerequisite for the study of nanooxide systems is the synthesis of the respective samples. A variety of methods for the fabrication of oxide materials as thin films and nanoparticles has been established: from the most direct and simple method of oxide thin film formation by oxidation of bulk solids to thin film deposition methodologies and the different procedures for oxide nanoparticle generation, both in the gas phase and in liquids; this is presented in Chapter 2 Different classical theories have been applied to describe the oxidation kinetics of the outer parts of solids, mostly of a phenomenological nature, but the formation of surface oxide layers with only a few monolayers thick and particular structures, as precursor phases for the formation of oxide films with bulk structure, has only recently been recognized. To familiarize the reader with growth processes of thin films and nanoparticles, an account of the fundamental concepts of nucleation and growth is also given in Chapter 2.

The methods of study for the experimental characterization of oxide thin films and nanoparticles are manifold and can be categorized according to their information content, describing geometrical structure, chemical composition, and electronic structure as well as magnetic, vibrational, or surface chemical properties. The various experimental techniques, many of them derived from the field of surface science, are described in the first part of Chapter 3. The quantum theoretical methods to determine the atomic structure, electronic energy levels, and the total energy of oxide nanostructures at various levels of sophistication and complexity are introduced in the second part of Chapter 3. Since many oxides have complex structural and electronic

Oxide Thin Films and Nanostructures. Falko P. Netzer and Claudine Noguera, Oxford University Press (2021). © Falko P. Netzer and Claudine Noguera. DOI: 10.1093/oso/9780198834618.003.0009

behaviour, a combination of several methods is typically necessary for experimental characterization. It is also the combination of advanced experimental characterization with theoretical simulation that has been pivotal to foster the progress of oxide nanostructure research during the last two decades.

Oxide ultra-thin films, oxide nanolayers of < 10 nm thickness, display novel physical and chemical properties as compared to their bulk counterparts. Intrinsic thin film confinement effects and the reduced local environment of a large number of atoms influence the geometrical structure, the electronic band structure, and the vibrational and ferroic behaviours of oxide nanolayers. The thickness dimension of thin films is an important parameter: thick oxide films (> 100 nm) approach bulk material behaviour, but many properties show a critical film thickness limit below which novel phenomena emerge. We mention electrical behaviour, critical phenomena of ferromagnetism and ferroelectricity, defect formation, and others in Chapter 4. New structural and electronic ground state configurations, metastable in thick films, may become stable in ultra-thin films. Oxide thin films supported on diverse substrate materials (metals, oxides) are of particular relevance for a majority of technological applications and the recognition of the presence and effects of the interfaces to substrates for shaping and modifying thin film properties are important issues.

Reducing the thickness dimension to the two-dimensional limit of single-unit cell thick layers opens up the realm of two-dimensional (2-D) oxides as new materials, which is considered in Chapter 5. In view of the ongoing miniaturization trend in all fields of device technology, 2-D materials have become a very hot topic in condensed matter physics and chemistry, and 2-D oxides play a prominent role in the discovery of emergent states of matter and in prospective nanotechnology applications. The preparation of 2-D oxide samples poses particular challenges compared to other 2-D materials, which can be separated into 2-D sheets due to their layered van der Waals-type bulk solid character. 2-D oxides are realized mainly on top of a suitable support surface, which also provides the arena for most potential uses. However, progress has been made recently in the preparation of oxide nanosheets, i.e. freestanding one- to several-unit-cell thick oxide sheets, by wet chemical preparation protocols, and although reproducibly controlled fabrication is still far from perfect proof of concept experiments using 2-D oxide nanosheets in electronics, optoelectronics, biological recognition, and catalysis applications have been reported. In the field of catalysis, new paradigms have been recognized as a result of the flexibility and dynamic behaviour of 2-D oxide lattices, new adsorbate configurations due to charge transfer from metallic substrates, and the proximity to the metal-oxide interface. A new model system in the form of 'inverse catalyst', consisting of oxide nanoislands decorating a metal surface, has been successfully introduced to study the active sites at the metal-oxide interface in catalytic reaction mechanisms. Oxide nanoribbons, 2-D oxide nanostructures of limited width, are interesting objects between one- and two-dimensional, in which under-coordinated edge states may give rise to unusual phenomena of electronic and magnetic origin. Their experimental realization is still pending, but some theoretical concepts are reported in Chapter 5.

Nanoparticles (NP) constitute a state of matter that can be considered as intermediate between molecules and extended crystals. Oxide nanoparticles are influential in many different areas, from the natural environment to atmospheric chemistry or to applications in catalysis and biotechnology. The analysis of the transition between the scalable and the non-scalable size regime of oxide NPs and of their size-dependent structures and shapes is illustrative and forms the main body of Chapter 6. As with other nanostructures, the large number of under-coordinated atoms is at the basis of their particular habit and behaviour and of the existence of transitions as a function of size between polymorphic forms which are only metastable in the bulk. The interaction of oxide NP with water is of special relevance, both during their formation in the humid natural environment and in their in vitro fabrication. The dissociation of water molecules at the particle surfaces generates acid and basic sites which are critical in their catalytic behaviour. Mixed oxide NPs may occur in several types of homogeneous or phase separated forms, depending on the segregation and mixing behaviour of the respective constituents. To assist with the discussion, a brief account of the thermodynamics of solid solutions is included in Chapter 6.

Clay minerals are naturally occurring oxide nanoparticles and nanosheets of layered structure and complex chemical composition. Clays are alumino-, magneso-, or ferro-silicates in porous networks, which are formed from the alteration of rocks at the Earth surface. They have been employed in the cultural sphere of mankind since prehistoric times, in buildings, as household utilities, in human artwork and in cosmetic and pharmaceutical applications. Their complex structure and chemical compositions are investigated in Chapter 7. Clays interact strongly with water, which is important for their formation and for their swelling behaviour under the influence of humidity. Capacities for cation exchange are characteristic features of clays that originate many of their practical utilizations. A short summary of the diverse areas of clay applications in our present-day life is given in Chapter 7.

Oxide ultra-thin films and nanostructures are involved in many interdisciplinary fields of nanoscience and technology. Surface chemistry and interfacial redox reactions form a bracket in which a range of processes in applied technology contexts are discussed in Chapter 8. Heterogeneous catalysis is a mature field of application of oxide nanosystems—an account of oxide-related aspects emphasizing the surface science approach to catalysis is presented. Photocatalysis is a more recent field of catalysis, which is illustrated with the help of the prototypical example of the photochemical water splitting process involving oxide semiconductors. Solid oxide fuel cells involve complex gas-solid and solid-solid interfacial charge transfer reactions and bear great potential as portable energy sources; the basic principles are introduced in Chapter 8. In the field of solar energy materials, materials that take advantage of the properties of the solar light spectrum for energy harvesting and saving, the chromogenic functionality, which finds application in green smart window technology, is taken as a discussion topic. Corrosion protection, i.e. the passivation of materials from detrimental environmental influences by the formation of thin native oxide films, is an essential element for the utilization of metals and alloys in tools and industrial processes. The formation and nature of passivating oxide films under corrosive environment are investigated in Chapter 8. Biotechnology applications of oxide nanosystems are very active areas of actual research: biocompatibility of metallic implants fostered by oxidic surface layers, diagnostic and therapeutic functionalities of oxide NP for bioimaging diagnosis and drug delivery systems, and the use of large surface area nanosheets in biosensing platforms may be cited in this context. The response of biosystems to electric fields, using ferroelectric oxides as electric field sources, is a recent area of scientific study, which may open up new avenues of electric field induced regulations of biological processes; a brief account of this topic is given at the end of Chapter 8.

How do we envision future prospects in this field? Pondering over future developments is always difficult, it is subjective, and mainly reflects one's personal experiences, prejudices, and visions. Nonetheless, we attempt here to formulate a few ideas which we believe may be of scientific relevance in future oxide nanostructure-related research.

Oxide ultra-thin films of controlled structure and composition have been grown hitherto mainly on metal surfaces, initially because they allowed the use of STM/STS techniques on insulating compounds, and later because they represented appropriate models for so-called inverse catalysts. To a lesser extent, perovskite oxide substrates have also been used, but the diversity of other oxide structural and electronic properties has not been widely exploited. Just as an example, fundamental questions remain unsolved concerning the differences in interface bonding and charge transfer between an oxide thin film grown on an elemental metal substrate or on a metallic oxide substrate, like Fe_3O_4 or RuO_2. It is expected that oxides will be more frequently employed as substrates for oxide film growth, or to produce oxide heterostructures. In addition, growth on other substrate types could be suitable if it is intended to subsequently remove the oxide phase from the substrate, leading to the preparation of self-supported oxide layers. This may be particularly attractive for 2-D oxides, which can then be assembled using the LEGO approach into oxide heterostructures or hybrid materials of 2-D oxides with other 2-D materials. Designed functionality architectures of ultra-thin material stacks including oxides may then become possible. Improvements of the experimental handling of 2-D oxides will be imperative to fully apply the LEGO procedure to oxide materials. Further improvements in the controlled synthesis of oxide nanosheets can also be expected, where thickness, size and shape can be reproducibly tuned.

The experimental realization of oxide nanoribbons is only at its infancy and knowledge on their edge stability and size-dependent electronic structure can only be inferred from the study of the first stages of oxide growth on a substrate during which 2-D islands form. In the case of graphene, on-surface synthesis methods using coupling reactions of organic precursor molecules have enabled the controlled fabrication of carbon nanoribbons with pre-defined edge structures. With oxides, these on-surface synthesis routes have not yet been explored, suitable precursor compounds have not been identified. Whether it will be possible to apply on-surface synthesis preparation to oxide nanoribbons is unclear, but experimental scrutiny of oxide nanoribbons would certainly be a very exciting endeavour, possibly opening the way to the discovery of new protected topological states.

The controlled fabrication of complex multicomponent oxide layers, e.g. by doping or by pulsed laser deposition, is already possible and will allow the growth of complex oxide heterostructures. We expect further activities in this area and in the study of emergent phenomena at complex oxide interfaces, revealing new highly correlated 2-D magnetic or superconducting phases, as well as exotic topological phases. The doping with metal and non-metal atoms is a proved means to modify the electronic structure and chemical reactivity of a given oxide in desired ways. Segregation of dopants at the film surface or at its interface with the substrate may in some case induce the formation of a ternary compound with new specific characteristics. Doping to engineer the band structure and band edge positions for specified spectroscopic or chemical purposes or for introducing ferroic functionalities will remain an active area of oxide materials design with increasing prospects.

While most experimental technical perspectives concern the development of well-controlled synthesis protocols, there remain several important theoretical challenges to face in order to accompany/drive the research on oxide ultra-thin films and NPs. Advanced efficient electronic structure methods, such as GW, DMFT, Quantum Monte Carlo or Reduced Density-Matrix Functional Theory, designed to account for oxide spectroscopic signatures and properties of highly correlated materials, have mainly been applied to small NPs or bulk materials but are not yet able to treat supported oxide films or NPs of realistic sizes. Also, the simulation of large size systems or the description of out-of-equilibrium processes remain a challenge, but we see advances in this area thanks to machine learning methods. These methods which allow parametrizing force fields and energy landscapes speed-up computationally expensive quantum calculations. They will undoubtedly improve and be more systematically employed thanks to methodological developments and the exponential growth of computing power. A change of paradigm has recently appeared in computational materials science for the discovery of novel materials, or of new properties of existing materials, named high-throughput computing. Since it has become possible to automatically perform large sets of calculations, databases containing the calculated properties of existing and hypothetical materials have been and are being established which can be intelligently interrogated, in search of materials with desired properties. These data mining methods allow the emergence of relevant descriptors for materials design, and efficiently replace trial and error experimental methods. No doubt that all these methods will have a favourable impact on the science of oxide ultra-thin films and nanostructures.

Oxitronics, a vision of an all-oxide electronic device technology, has been formulated a few years ago (Ramirez, 2007), and a roadmap for oxide electronics has been published in 2016 and 2019 (Lorenz *et al.*, 2016; Coll *et al.*, 2019). It is expected that the utilization of oxides in ultra-thin film or 2-D form as active and passive components will remain prominent and increase in electronic and optoelectronic device design.

Oxide nanomaterials in catalysis will remain of continued interest: as components in connection with metals in metal supported catalysts (as majority support material and minority promotor agents) and as active catalysts of their own right. The large number of under-coordinated atoms in oxide NPs, which form most active reaction centres for many catalytic reactions, will ensure that oxide NP catalysts have a bright future. The search for the replacement of the expensive precious metal component in metal-oxide catalysts with cheaper materials, e.g. active oxide phases, will also spur the catalytic research of oxide materials. Programmable defect chemistry, i.e. the preparation of oxide surfaces with defined patterns of defect centres, may be a future area in oxide catalysis for achieving high reactivity and superior selectivity. In this context, oxide nanoalloys represent a promising field of research, paralleling that on metallic nanoalloys which has taken place in the last decade and has promoted advances both in growth protocols, characterization methods and theoretical simulations.

Under many circumstances, in fundamental studies or applications of oxide thin films and NPs, the presence of water appears undesirable, since its adsorption modifies all surface properties. However, contact with liquid or high-pressure gaseous water is present in all electrochemical experiments, as well as in the natural or synthetic processes of oxide formation. Much has to be learnt about hydration/dehydration processes at the surface of NPs, and about surface charging by diluted ions present in aqueous solutions. The actual hydrated state of oxide nuclei during the formation of oxide NPs in liquids is also largely unknown, because we are missing local probes to analyse it. However, by (generally) lowering the barriers through the formation of hydrated precursors or intermediates, it has a strong influence on the kinetics of nucleation, growth and dissolution. The relative thermodynamic stability of oxides, oxyhydroxides and hydroxides as a function of their degree of hydration has been established in some case in the (geo-)chemistry community, but a comprehensive knowledge of their surfaces and of the kinetics of transitions between these phases is still missing.

The interaction of oxide nanostructures with biological systems have a great potential for future applications: in diagnostic platforms, in biosensor design, or in therapeutic systems as direct medical agents or for intelligent targeted drug delivery carriers. The applications mentioned in Chapter 8 constitute merely the beginning of a new area of oxide biotechnology, where many new avenues will open.

Appendix A
The Wulff and Wulff–Kaishev constructions

This appendix illustrates the application of the Wulff and Wulff–Kaishev theorems to determine the shape of anisotropic particles at equilibrium in the simple case of tetragonal particles. Both theorems rely on the minimization of the total surface/interface energy of the particle at constant volume.

Let us consider an unsupported tetragonal particle with basal dimensions $l \times l$ and thickness e, and surface energies γ_{bas} and γ_{lat} associated to the basal and lateral faces, respectively. The total surface energy is equal to $E_S = 2l^2\gamma_{bas} + 4el\gamma_{lat}$, while the volume of the particle is simply $V = el^2$. Minimizing E_S at constant volume amounts to solving $dE_S = 0$ submitted to the constrain that $dV = 0$. The second relationship implies that $lde + 2edl = 0$, which, inserted into $dE_S = 0$ yields the value of the aspect ratio of the particle at equilibrium:

$$\frac{e}{l} = \frac{\gamma_{bas}}{\gamma_{lat}} \tag{A.1}$$

The lateral and basal dimensions are thus inversely proportional to the lateral and basal surface energies (Fig. A.1(a)). More generally, Wulff theorem states that, at equilibrium, the distance from the centre of a particle to its external facets is proportional to the surface energy of these surfaces (Zangwill, 1990).

This result can be extended to the case of a particle lying on a substrate on its basal face (Fig. A.1(b)). In that case, γ_{bas} is relevant for the external face and $\gamma_{bas} - W_{adh}$ for

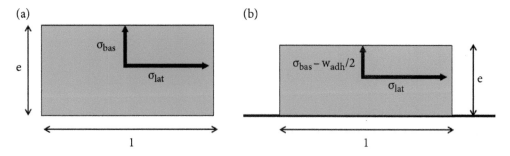

Fig. A.1 2-D representation of the Wulff construction for a) unsupported and b) supported tetragonal particles.

the one in contact with the substrate (W_{adh} the adhesion energy). The same reasoning as before yields the particle aspect ratio at equilibrium:

$$\frac{e}{l} = \frac{\gamma_{bas} - W_{adh}/2}{\gamma_{lat}} \tag{A.2}$$

If, alternatively, one of the lateral faces is in contact with the substrate, the particle equilibrium shape involves three inequivalent dimensions l, l', and e. The ratio between these lengths is then equal to:

$$\frac{e}{\gamma_{bas}} = \frac{l}{\gamma_{lat}} = \frac{l'}{\gamma_{lat} - W_{adh}/2} \tag{A.3}$$

These are simple illustrations of the Wulff–Kaishev theorem (Muller and Kern, 2000).

Appendix B
The Frenkel–Kontorova model

The original Frenkel–Kontorova model had been conceived to describe the structure of monoatomic chains interacting with a rigid substrate, under the form of a phase diagram as a function of misfit and chain-substrate interactions. It allows predicting under which conditions pseudomorphic configurations or interfacial dislocations are the most stable. In its simplest formulation, the Frenkel–Kontorova model includes two energy terms: the elastic interaction between neighbouring atoms along the chain (spring constant μ) when the bond-lengths depart from their equilibrium value b, and the interaction W with the periodic potential of the substrate (period a; W positive).

$$\mathcal{E} = \sum_n \left[\frac{\mu}{2}(x_{n+1} - x_n - b)^2 + \frac{W}{2}\left(1 - \cos\frac{2\pi x_n}{a}\right) \right] \tag{B.1}$$

Two important parameters enter the model: the misfit $\delta = (b - a)/a$ between the substrate and the chain lattices, and a length $l_0 = \sqrt{\mu a^2/2W}$ which drives the relative importance of the elastic energy and the chain-substrate interaction. As a function of the atomic displacements u_n with respect to the substrate atoms: $x_n = (n + u_n)a$, a dimensionless energy $\mathcal{E}'$ may be defined:

$$\mathcal{E}' = \frac{2\mathcal{E}}{\mu a^2} = \sum_n \left[(u_{n+1} - u_n - \delta)^2 + \frac{1}{2l_0^2}(1 - \cos 2\pi u_n) \right] \tag{B.2}$$

Such a model may be numerically solved, and its results are shown in Fig. B.1, but it is instructive to follow the derivation of Frank and van der Merwe in the case of infinite chains (Frank and van der Merwe, 1949; Frank and Van der Merwe, 1949) or Niedermeyer for finite chains (Niedermeyer, 1968) within the so-called continuum approximation, which both assume that displacements u_n are slowly functions of n, the atom index.

In this approximation, atomic displacements in an infinite chain are obtained by solving the differential equation:

$$\frac{d^2 u_n}{dn^2} = \frac{\pi}{2l_0^2}\sin 2\pi u_n \tag{B.3}$$

This is a Sine–Gordon type equation with soliton solutions, which in the present context represent the interfacial dislocations. The phase diagram may be qualitatively understood as arising from the competition between the chain-substrate interaction, which favours pseudomorphy (commensurate phase), and the elastic interactions inside

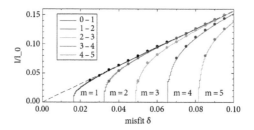

Fig. B.1 Frenkel–Kontorova phase diagram of MgO chains of 61 atoms obeying eqn. B.1, in a $(1/l_0, \delta)$ representation, with parameter values $\mu = 41.7$ eV/Å^2, and $b = 1.813$ Å determined numerically. The dots are results obtained analytically within the continuum approximation (Niedermeyer, 1968). The labels represent the number m of interfacial dislocations present in the ground state of the chain. The dashed line is the Commensurate-Incommensurate phase separation line of the infinite chain $1/l_0 = \pi\delta/2$ above which the commensurate phase is the most stable. The phase diagram is symmetric in the exchange $\delta \leftrightarrow -\delta$. The transition lines between two neighbouring dislocated phases $m-1$ and m intersect the horizontal axis (vanishing interaction with the substrate) at misfit values δ_c given by the Vernier rule $M\delta_c = m$. They have a common linear asymptote when the misfit δ and/or the interaction with the substrate increase. Reprinted from Noguera and Goniakowski (2013) with the permission of AIP Publishing.

the chain which favour a mean lattice parameter close to the natural chain value. In the commensurate phase, these two terms are of the order of $W/2$ per atom and $\mu\delta^2 a^2/2$, respectively. The commensurate phase is thus destroyed when $\mu\delta^2 a^2$ exceeds W.

Beyond this qualitative argument, the exact equation of the transition line predicted by Frank and van der Meerwe is the following:

$$\frac{1}{l_0} = \frac{\pi}{2}\delta \tag{B.4}$$

In the case of finite chains (Niedermeyer, 1968), the resolution of the continuum model is more involved. Boundary conditions at the chain ends lead to a quantization of the solutions, the quantum numbers being the number of misfit dislocations in the chain. The main difference between the phase diagram of finite and infinite chains thus lies in the partitioning of the incommensurate phase into regions corresponding to fixed numbers of dislocations (see Fig. B.1). In this approach, all physical quantities depend on only two parameters: $l_0\delta$ and N/l_0, while, in addition, the energy also possesses a term proportional to $(2N+1)/l_0$. For long chains $(2N+1 \approx 2N)$ this provides a scaling relationship between the positions of the $m - (m+1)$ transition line in chains of M_1 and M_2 atoms, respectively. If the former is defined by $\{1/l_1, \delta_1\}$ the equation of the latter is $\{M_1 l_1/M_2, \delta_1 M_1/M_2\}$. This scaling relationship is well obeyed by the numerical results.

Extension of this model for the case of binary compounds is given in Noguera and Goniakowski (2013).

Appendix C
List of acronyms

- 2-D: two-dimensional
- 2DEG: two-dimensional electron gas
- 2DHG: two-dimensional hole gas
- 3-D: three-dimensional
- AC: armchair
- AES: Auger electron spectroscopy
- AFM: atomic force microscopy
- AFM: anti-ferromagnetic
- ALD: atomic layer deposition
- ARUPS: angle-resolved ultraviolet photoemission spectroscopy
- BSE: Bethe–Salpeter equation
- CB: conduction band
- CBM: conduction band minimum
- CC: coupled cluster
- CEC: cation exchange capacity
- CI: configuration interaction
- CNT: classical nucleation theory
- CTR: crystal truncation rod
- CUS: coordinatively unsaturated sites
- CVD: chemical vapour deposition
- CVS: chemical vapour synthesis
- DFT: density functional theory
- DFTB: density functional tight binding
- DLVO: Derjaguin–Landau–Verwey–Overbeek
- DMFT: dynamical mean-field theory
- DOS: density of states
- EC: electro-chromic
- EC-STM: electrochemical STM
- EDX: energy-dispersive X-ray analysis
- EELS or ELS: electron energy loss specroscopy
- EXAFS: extended X-ray absorption fine structure
- FET: field-effect transistor
- FM: ferromagnetic

- FOM: figure of merit
- FRET: fluorescence resonance energy transfer
- FTIR: Fourier transform infra-red spectroscopy
- GGA: generalized gradient approximation
- GISAXS: grazing incidence small angle X-ray scattering
- GMR: giant magneto-resistance
- HF: Hartree–Fock
- HOMO: highest occupied molecular orbital
- HREELS: high-resolution electron energy loss spectroscopy
- HSE: Heyd–Scuseria–Ernzerhof functional
- IAP: ion activity product
- IPES: inverse photoemission spectroscopy
- IR: infra-red
- ISS: ion scattering spectroscopy
- JDOS: joint density of states
- KPAFM: Kelvin probe atomic force microscopy
- KRIPES: k-resolved inverse photoemission spectroscopy
- LDA: local density approximation
- LDOS: local density of states
- LEED: low-energy electron diffraction
- LEIS: low-energy ion scattering
- LUMO: lowest unoccupied molecular orbital
- MBE: molecular beam epitaxy
- MBPT: many body perturbation theory
- MC: Monte-Carlo
- MD: molecular dynamics
- MFM: magnetic force microscopy
- MIGS: metal-induced gap states
- ML: monolayer
- MOKE: magneto-optical Kerr effect
- MOS: metal-oxide-semiconductor device
- MOSFET: metal-oxide-semiconductor field-effect transistor
- MRI: magnetic resonance imaging
- MTJ: magnetic tunnel junction
- MUSIC: multisite complexation model
- NEXAFS: near edge X-ray absorption fine structure
- NIR: near infra-red
- NP(s): nanoparticle(s)
- NR(s): nanoribbon(s)
- NS(s): nanosheet(s)
- PBC: periodic boundary chains
- PBE: Perdew–Burke–Ernzerhof functional

- PEC: photoelectrocatalysis
- PED: photoelectron diffraction
- PEEM: photoemission electron spectroscopy
- PEG: polyethylene glycol
- PES: potential energy surface
- PHFAST: Paris Hartree–Fock atomistic simulation tool
- PL: photoluminescence
- PLD: pulsed laser deposition
- PNC: pre-nucleation clusters
- PV: photovoltaic effect
- PVD: physical vapour deposition
- PY: pyroelectric effect
- PZC: point of zero charge
- RAIRS: reflection absorption infra-red spectroscopy
- RAM: random access memory
- RHEED: reflection high-energy electron diffraction
- RHF: restricted Hartree–Fock
- ROS: reactive oxygen species
- RPA: random phase approximation
- RPES: resonant photoemission spectroscopy
- SAXS: small angle X-ray scattering
- SEM: (scanning) secondary electron microscopy
- SHE: standard hydrogen electrode
- SOFC: solid oxide fuel cell
- SPA-LEED: spot profile analysis low energy electron diffraction
- SPM: scanning probe microscopy
- SS: solid solution
- (S)TEM: (scanning) transmission electron microscopy
- STM: scanning tunnelling microscopy
- STS: scanning tunnelling spectroscopy
- SXRD: surface X-ray diffraction
- TB: tight binding
- TDDFT: time-dependent density functional theory
- TDLDA: time-dependent local density approximation
- TDS: thermal desorption spectroscopy
- TM: transition metal
- TMO: transition metal oxide
- TMR: tunnel magneto-resistance
- TPD: temperature programmed desorption
- TPRS: temperature programmed reaction spectroscopy
- UHF: unrestricted Hartree–Fock
- UHV: ultra-high-vacuum

- UPS: ultraviolet photoemission spectroscopy
- UV: ultra-violet
- VB: valence band
- VBM: valence band maximum
- WKB: Wentzel–Kramers–Brillouin
- XANES: X-ray absorption near edge structure
- XAS: X-ray absorption spectroscopy
- XMCD: X-ray magnetic circular dichroism
- XMLD: X-ray magnetic linear dichroism
- XPD: X-ray photoelectron diffraction
- XPEEM: X-ray photoemission electron microscopy
- XPS: X-ray photoelectron spectroscopy
- YSZ: yttrium stabilized zirconia
- ZZ: zigzag

References

Abdulagatov, AI, Yan, Y, Cooper, JR, Zhang, Y, Gibbs, ZM, Cavanagh, AS, Yang, RG, Lee, YC, and George, SM (2011). Al_2O_3 and TiO_2 atomic layer deposition on copper for water corrosion resistance. *ACS Applied Materials & Interfaces*, **3**(12), 4593–4601.

Acik, Muge and Chabal, Yves J (2011). Nature of graphene edges: A review. *Japanese Journal of Applied Physics*, **50**(7R), 070101.

Adamo, C, Cossi, M, and Barone, V (1999). An accurate density functional method for the study of magnetic properties: The PBE0 model. *Journal of Molecular Structure: THEOCHEM*, **493**(1-3), 145–157.

Adamson, Arthur W (1960). *Physical Chemistry of Surfaces*. Interscience Publishers, Chichester.

Agnoli, S, Sambi, M, Granozzi, G, Schoiswohl, J, Surnev, S, Netzer, FP, Ferrero, M, Ferrari, Anna Maria, and Pisani, C (2005). Experimental and theoretical study of a surface stabilized monolayer phase of nickel oxide on Pd(100). *The Journal of Physical Chemistry B*, **109**(36), 17197–17204.

Ahn, CH, Rabe, KM, and Triscone, J-M (2004). Ferroelectricity at the nanoscale: Local polarization in oxide thin films and heterostructures. *Science*, **303**(5657), 488–491.

Al-Abadleh, Hind A and Grassian, VH (2003). FT-IR study of water adsorption on aluminum oxide surfaces. *Langmuir*, **19**(2), 341–347.

Albaret, Tristan, Finocchi, Fabio, and Noguera, Claudine (1999). First principles simulations of titanium oxide clusters and surfaces. *Faraday Discussions*, **114**, 285–304.

Alders, D, Tjeng, LH, Voogt, FC, Hibma, T, Sawatzky, GA, Chen, CT, Vogel, J, Sacchi, M, and Iacobucci, S (1998). Temperature and thickness dependence of magnetic moments in NiO epitaxial films. *Physical Review B*, **57**(18), 11623.

Allan, NL, Barrera, GD, Fracchia, RM, Lavrentiev, M Yu, Taylor, MB, Todorov, IT, and Purton, JA (2001). Free energy of solid solutions and phase diagrams via quasiharmonic lattice dynamics. *Physical Review B*, **63**(9), 094203.

Allegretti, F, Parteder, G, Gragnaniello, L, Surnev, S, Netzer, FP, Barolo, A, Agnoli, S, Granozzi, G, Franchini, Cesare, and Podloucky, R (2010). Strained c(4× 2) CoO(100)-like monolayer on Pd(100): Experiment and theory. *Surface Science*, **604**(5-6), 529–534.

Alsaif, Manal MYA, Field, Matthew R, Murdoch, Billy J, Daeneke, Torben, Latham, Kay, Chrimes, Adam F, Zoolfakar, Ahmad Sabirin, Russo, Salvy P, Ou, Jian Zhen, and Kalantar-zadeh, Kourosh (2014a). Substoichiometric two-dimensional molybdenum oxide flakes: A plasmonic gas sensing platform. *Nanoscale*, **6**(21), 12780–12791.

Alsaif, Manal MYA, Latham, Kay, Field, Matthew R, Yao, David D, Medehkar,

Nikhil V, Beane, Gary A, Kaner, Richard B, Russo, Salvy P, Ou, Jian Zhen, and Kalantar-zadeh, Kourosh (2014*b*). Tunable plasmon resonances in two-dimensional molybdenum oxide nanoflakes. *Advanced Materials*, **26**(23), 3931–3937.

Altendorf, SG, Reisner, A, Tam, B, Meneghin, F, Wirth, S, and Tjeng, LH (2018). Strong modification of thin film properties due to screening across the interface. *Physical Review B*, **97**(16), 165422.

Altieri, S, Allegretti, F, Steurer, W, Surnev, S, Finazzi, Marco, Sessi, V, Brookes, NB, Valeri, Sergio, and Netzer, FP (2013). Orbital anisotropy in paramagnetic manganese oxide nanostripes. *Physical Review B*, **87**(24), 241407.

Altieri, S, Finazzi, M, Hsieh, HH, Haverkort, MW, Lin, H-J, Chen, CT, Frabboni, Stefano, Gazzadi, GC, Rota, Alberto, Valeri, Sergio et al. (2009). Image charge screening: A new approach to enhance magnetic ordering temperatures in ultrathin correlated oxide films. *Physical Review B*, **79**(17), 174431.

Altieri, S, Finazzi, Marco, Hsieh, HH, Lin, H-J, Chen, CT, Hibma, T, Valeri, S, and Sawatzky, GA (2003). Magnetic dichroism and spin structure of antiferromagnetic NiO(001) films. *Physical Review Letters*, **91**(13), 137201.

Altieri, S, Tjeng, LH, Voogt, FC, Hibma, T, and Sawatzky, GA (1999). Reduction of Coulomb and charge-transfer energies in oxide films on metals. *Physical Review B*, **59**(4), R2517.

Andersen, Tassie K, Fong, Dillon D, and Marks, Laurence D (2018). Pauling's rules for oxide surfaces. *Surface Science Reports*, **73**(5), 213–232.

Andersin, Jenni, Nevalaita, Janne, Honkala, Karoliina, and Häkkinen, Hannu (2013). The redox chemistry of gold with high-valence doped calcium oxide. *Angewandte Chemie International Edition*, **52**(5), 1424–1427.

Anh Tran, Thao, Krishnamoorthy, Karthikeyan, Song, Yeon Woo, Cho, Somi Kim, and Kim, Sang Jae (2014). Toxicity of nano molybdenum trioxide toward invasive breast cancer cells. *ACS Applied Materials & Interfaces*, **6**(4), 2980–2986.

Anisimov, Vladimir I, Aryasetiawan, Ferdi, and Lichtenstein, AI (1997). First-principles calculations of the electronic structure and spectra of strongly correlated systems: The LDA+ U method. *Journal of Physics: Condensed Matter*, **9**(4), 767.

Anisimov, Vladimir I, Zaanen, Jan, and Andersen, Ole K (1991). Band theory and Mott insulators: Hubbard *U* instead of Stoner *I*. *Physical Review B*, **44**(3), 943.

Anselme, K (2000). Osteoblast adhesion on biomaterials. *Biomaterials*, **21**(7), 667–681.

Appelo, C Anthony J and Postma, Dieke (2004). *Geochemistry, Groundwater and Pollution*. CRC Press, Boca Raton, London, New York, Washington, D.C.

Arab, Mehdi, Bougeard, Daniel, and Smirnov, Konstantin S (2003). Structure and dynamics of the interlayer water in an uncharged 2:1 clay. *Physical Chemistry Chemical Physics*, **5**(20), 4699–4707.

Arthur, John R (2002). Molecular beam epitaxy. *Surface Science*, **500**(1-3), 189–217.

Aryasetiawan, Ferdi and Gunnarsson, Olle (1998). The GW method. *Reports on Progress in Physics*, **61**(3), 237.

Aryasetiawan, F, Karlsson, Krister, Jepsen, O, and Schönberger, U (2006). Calculations of Hubbard U from first-principles. *Physical Review B*, **74**(12), 125106.

Aschauer, Ulrich and Spaldin, Nicola A (2016). Interplay between strain, defect charge state, and functionality in complex oxides. *Applied Physics Letters*, **109**(3), 031901.

Asthagiri, Aravind, Dixon, David A, Dohnálek, Zdenek, Kay, Bruce D, Rodriguez, José A, Rousseau, Roger, Stacchiola, Darío J, and Weaver, Jason F (2016). Catalytic chemistry on oxide nanostructures. In *Oxide Materials at the Two-Dimensional Limit*, pp. 251–280. Springer.

Ataca, Can, Sahin, Hasan, and Ciraci, Salim (2012). Stable, single-layer MX_2 transition-metal oxides and dichalcogenides in a honeycomb-like structure. *The Journal of Physical Chemistry C*, **116**(16), 8983–8999.

Atkinson, A (1985). Transport processes during the growth of oxide films at elevated temperature. *Reviews of Modern Physics*, **57**(2), 437.

Avena, Marcelo J, Mariscal, Marcelo M, and De Pauli, Carlos P (2003). Proton binding at clay surfaces in water. *Applied Clay Science*, **24**(1-2), 3–9.

Aykol, Muratahan and Persson, Kristin A (2018). Oxidation protection with amorphous surface oxides: Thermodynamic insights from ab initio simulations on aluminum. *ACS Applied Materials & Interfaces*, **10**(3), 3039–3045.

Backhaus-Ricoult, M, Samet, L, Trichet, M-F, Hÿtch, MJ, and Imhoff, D (2003). Interfacial chemistry in internally oxidized (Cu, Mg)-alloys. *Journal of Solid State Chemistry*, **173**(1), 172–188.

Bader, Richard FW (1991). A quantum theory of molecular structure and its applications. *Chemical Reviews*, **91**(5), 893–928.

Bae, Sukang, Kim, Hyeongkeun, Lee, Youngbin, Xu, Xiangfan, Park, Jae-Sung, Zheng, Yi, Balakrishnan, Jayakumar, Lei, Tian, Kim, Hye Ri, Song, Young Il et al. (2010). Roll-to-roll production of 30-inch graphene films for transparent electrodes. *Nature Nanotechnology*, **5**(8), 574.

Baibich, Mario Norberto, Broto, Jean Marc, Fert, Albert, Van Dau, F Nguyen, Petroff, Frédéric, Etienne, P, Creuzet, G, Friederich, A, and Chazelas, J (1988). Giant magnetoresistance of (001)Fe/(001)Cr magnetic superlattices. *Physical Review Letters*, **61**(21), 2472.

Bailey, SW (1980). Summary of recommendations of AIPEA nomenclature committee on clay minerals. *American Mineralogist*, **65**(1-2), 1–7.

Balan, Etienne, Saitta, A Marco, Mauri, Francesco, Lemaire, Céline, and Guyot, François (2002). First-principles calculation of the infrared spectrum of lizardite. *American Mineralogist*, **87**(10), 1286–1290.

Balendhran, Sivacarendran, Deng, Junkai, Ou, Jian Zhen, Walia, Sumeet, Scott, James, Tang, Jianshi, Wang, Kang L, Field, Matthew R, Russo, Salvy, Zhuiykov, Serge et al. (2013*a*). Enhanced charge carrier mobility in two-dimensional high dielectric molybdenum oxide. *Advanced Materials*, **25**(1), 109–114.

Balendhran, Sivacarendran, Walia, Sumeet, Alsaif, Manal, Nguyen, Emily P, Ou, Jian Zhen, Zhuiykov, Serge, Sriram, Sharath, Bhaskaran, Madhu, and Kalantar-zadeh, Kourosh (2013*b*). Field effect biosensing platform based on 2-D α-MoO_3. *ACS Nano*, **7**(11), 9753–9760.

Balendhran, Sivacarendran, Walia, Sumeet, Nili, Hussein, Ou, Jian Zhen, Zhuiykov, Serge, Kaner, Richard B, Sriram, Sharath, Bhaskaran, Madhu, and Kalantar-zadeh, Kourosh (2013*c*). Two-dimensional molybdenum trioxide and dichalcogenides. *Ad-*

vanced Functional Materials, **23**(32), 3952–3970.

Ban, Seiji, Iwaya, Yukari, Kono, Hiroshi, and Sato, Hideo (2006). Surface modification of titanium by etching in concentrated sulfuric acid. *Dental Materials*, **22**(12), 1115–1120.

Banfield, Jillian F. and Zhang, Hengzhong (2001). Nanoparticles in the environment. *Reviews in Mineralogy and Geochemistry*, **44**(1), 1.

Bang, Jin Ho and Suslick, Kenneth S (2010). Applications of ultrasound to the synthesis of nanostructured materials. *Advanced Materials*, **22**(10), 1039–1059.

Barcaro, Giovanni, Agnoli, Stefano, Sedona, Francesco, Rizzi, Gian Andrea, Fortunelli, Alessandro, and Granozzi, Gaetano (2009). Structure of reduced ultrathin TiO_x polar films on Pt(111). *The Journal of Physical Chemistry C*, **113**(14), 5721–5729.

Barcaro, Giovanni, Cavaliere, Emanuele, Artiglia, Luca, Sementa, Luca, Gavioli, Luca, Granozzi, Gaetano, and Fortunelli, Alessandro (2012). Building principles and structural motifs in TiO_x ultrathin films on a (111) substrate. *The Journal of Physical Chemistry C*, **116**(24), 13302–13306.

Barcaro, Giovanni, Thomas, Iorwerth Owain, and Fortunelli, Alessandro (2010). Validation of density-functional versus density-functional+ U approaches for oxide ultrathin films. *The Journal of Chemical Physics*, **132**(12), 124703.

Bard, Allen J (1979). Photoelectrochemistry and heterogeneous photo-catalysis at semiconductors. *Journal of Photochemistry*, **10**(1), 59–75.

Bartlett, Rodney J and Musiał, Monika (2007). Coupled-cluster theory in quantum chemistry. *Reviews of Modern Physics*, **79**(1), 291.

Bauer, Ernst (1958). Phänomenologische Theorie der Kristallabscheidung an Oberflächen. I. *Zeitschrift für Kristallographie-Crystalline Materials*, **110**(1-6), 372–394.

Becke, Axel D (1993). Density-functional thermochemistry. III. The role of exact exchange. *The Journal of Chemical Physics*, **98**(7), 5648–5652.

Becke, Axel D and Johnson, Erin R (2006). A simple effective potential for exchange. *The Journal of Chemical Physics*, **124**(22), 221101.

Beckel, Daniel, Bieberle-Hütter, Anja, Harvey, A, Infortuna, Anna, Muecke, Ulrich P, Prestat, M, Rupp, Jennifer LM, and Gauckler, Ludwig J (2007). Thin films for micro solid oxide fuel cells. *Journal of Power Sources*, **173**(1), 325–345.

Behler, Jörg (2011). Neural network potential-energy surfaces in chemistry: A tool for large-scale simulations. *Physical Chemistry Chemical Physics*, **13**, 17930–17955.

Bell, A T (1984). Fourier-transform infrared spectroscopy in heterogeneous catalysis. In *Chemistry and Physics of Solid Surfaces IV* (ed. R. Vanselow and R. Howe), Chapter 2. Springer.

Benedek, Nicole A, Rondinelli, James M, Djani, Hania, Ghosez, Philippe, and Lightfoot, Philip (2015). Understanding ferroelectricity in layered perovskites: New ideas and insights from theory and experiments. *Dalton Transactions*, **44**(23), 10543–10558.

Benedetti, Francesco, Amidani, Lucia, Cresi, Jacopo Stefano Pelli, Boscherini, Federico, Valeri, Sergio, D'Addato, Sergio, Nicolini, Valentina, Malavasi, Gianluca, and Luches, Paola (2018). Role of cerium oxide in bioactive glasses during catalytic dissociation of hydrogen peroxide. *Physical Chemistry Chemical Physics*, **20**(36), 23507–23514.

Benedetti, Stefania, Nilius, Niklas, and Valeri, Sergio (2015). Chromium-doped MgO thin films: Morphology, electronic structure, and segregation effects. *The Journal of Physical Chemistry C*, **119**(45), 25469–25475.

Benedetti, Stefania, Stavale, Fernando, Valeri, Sergio, Noguera, Claudine, Freund, Hans-Joachim, Goniakowski, Jacek, and Nilius, Niklas (2013). Steering the growth of metal ad-particles via interface interactions between a MgO thin film and a Mo support. *Advanced Functional Materials*, **23**(1), 75–80.

Benedetti, Stefania, Torelli, Piero, Valeri, Sergio, Benia, Hadj-Mohamed, Nilius, Niklas, and Renaud, Gilles (2008). Structure and morphology of thin MgO films on Mo(001). *Physical Review B*, **78**(19), 195411.

Benisek, Artur and Dachs, Edgar (2012). A relationship to estimate the excess entropy of mixing: Application in silicate solid solutions and binary alloys. *Journal of Alloys and Compounds*, **527**, 127–131.

Benny, Sreelekha, Grau-Crespo, Ricardo, and de Leeuw, Nora H (2009). A theoretical investigation of α-Fe_2O_3–Cr_2O_3 solid solutions. *Physical Chemistry Chemical Physics*, **11**(5), 808–815.

Bergaya, F and Lagaly, G (2013). *Handbook of Clay Science*. Volume 5. Elsevier, Amsterdam, Oxford, Cambridge.

Bergaya, F, Lagaly, G, and Vayer, M (2013). Cation and anion exchange. In *Developments in Clay Science*, Volume 5, pp. 333–359. Elsevier.

Berggren, Lars, Jonsson, Jacob C, and Niklasson, Gunnar A (2007). Optical absorption in lithiated tungsten oxide thin films: Experiment and theory. *Journal of Applied Physics*, **102**(8), 083538.

Bertel, E, Strasser, G, Netzer, FP, and Matthew, JAD (1982). Surface oxidation of ytterbium. *Surface Science*, **118**(3), 387–400.

Bhattacharya, Saswata, Levchenko, Sergey V, Ghiringhelli, Luca M, and Scheffler, Matthias (2014). Efficient ab initio schemes for finding thermodynamically stable and metastable atomic structures: Benchmark of cascade genetic algorithms. *New Journal of Physics*, **16**(12), 123016.

Bibes, Manuel, Villegas, Javier E, and Barthelemy, Agnes (2011). Ultrathin oxide films and interfaces for electronics and spintronics. *Advances in Physics*, **60**(1), 5–84.

Bickmore, Barry R, Bosbach, Dirk, Hochella, Michael F, Charlet, Laurent, and Rufe, Eric (2001). In situ atomic force microscopy study of hectorite and nontronite dissolution: Implications for phyllosilicate edge surface structures and dissolution mechanisms. *American Mineralogist*, **86**(4), 411–423.

Bickmore, Barry R, Rosso, Kevin M, Nagy, Kathryn L, Cygan, Randall T, and Tadanier, Christopher J (2003). Ab initio determination of edge surface structures for dioctahedral 2:1 phyllosilicates: Implications for acid-base reactivity. *Clays and Clay Minerals*, **51**(4), 359–371.

Biedermann, K, Gubo, M, Hammer, L, and Heinz, K (2009). Phases and phase transitions of hexagonal cobalt oxide films on Ir(100)-(1 × 1). *Journal of Physics: Condensed Matter*, **21**(18), 185003.

Bikondoa, Oier, Pang, Chi L, Ithnin, Roslinda, Muryn, Christopher A, Onishi, Hiroshi, and Thornton, Geoff (2006). Direct visualization of defect-mediated dissociation of water on $TiO_2(110)$. *Nature Materials*, **5**(3), 189.

Bilecka, Idalia and Niederberger, Markus (2010). Microwave chemistry for inorganic nanomaterials synthesis. *Nanoscale*, **2**(8), 1358–1374.

Binasch, Grünberg, Grünberg, Peter, Saurenbach, F, and Zinn, W (1989). Enhanced magnetoresistance in layered magnetic structures with antiferromagnetic interlayer exchange. *Physical Review B*, **39**(7), 4828.

Binder, Kurt (1986). Introduction: Theory and 'technical' aspects of Monte Carlo simulations. In *Monte Carlo Methods in Statistical Physics*, pp. 1–45. Springer.

Bish, DL (2013). Parallels and distinctions between clay minerals and zeolites. In *Developments in Clay Science*, Volume 5, pp. 783–800. Elsevier.

Biswas, Abhijit and Jeong, Yoon Hee (2017). Strain effect in epitaxial oxide heterostructures. In *Epitaxy* (ed. M. Zhong). Intech Open.

Blázquez-Castro, A, García-Cabañes, A, and Carrascosa, M (2018). Biological applications of ferroelectric materials. *Applied Physics Reviews*, **5**(4), 041101.

Boek, ES and Sprik, M (2003). Ab initio molecular dynamics study of the hydration of a sodium smectite clay. *The Journal of Physical Chemistry B*, **107**(14), 3251–3256.

Boffa, AB, Lin, C, Bell, AT, and Somorjai, Gabor A (1994). Lewis acidity as an explanation for oxide promotion of metals: Implications of its importance and limits for catalytic reactions. *Catalysis Letters*, **27**(3-4), 243–249.

Bond, Geoffrey C and Tahir, S Flamerz (1991). Vanadium oxide monolayer catalysts preparation, characterization and catalytic activity. *Applied Catalysis*, **71**(1), 1–31.

Borchers, JA, Carey, MJ, Erwin, RW, Majkrzak, CF, and Berkowitz, AE (1993). Spatially modulated antiferromagnetic order in CoO/NiO superlattices. *Physical Review Letters*, **70**(12), 1878.

Bordier, G and Noguera, C (1991). Electronic structure of a metal-insulator interface: Towards a theory of nonreactive adhesion. *Physical Review B*, **44**(12), 6361.

Bosbach, Dirk, Charlet, Laurent, Bickmore, Barry, and Hochella Jr, Michael F (2000). The dissolution of hectorite: In-situ, real-time observations using atomic force microscopy. *American Mineralogist*, **85**(9), 1209–1216.

Botello-Méndez, Andrés R, López-Urías, Florentino, Terrones, Mauricio, and Terrones, Humberto (2008). Magnetic behavior in zinc oxide zigzag nanoribbons. *Nano Letters*, **8**(6), 1562–1565.

Botti, Silvana, Schindlmayr, Arno, Del Sole, Rodolfo, and Reining, Lucia (2007). Time-dependent density-functional theory for extended systems. *Reports on Progress in Physics*, **70**(3), 357.

Botu, Venkatesh, Batra, Rohit, Chapman, James, and Ramprasad, Rampi (2016). Machine learning force fields: construction, validation, and outlook. *The Journal of Physical Chemistry C*, **121**(1), 511–522.

Bowen, WE, Wang, W, Cagin, E, and Phillips, JD (2008). Quantum confinement and carrier localization effects in $ZnO/Mg_xZn_{1-x}O$ wells synthesized by pulsed laser deposition. *Journal of Electronic Materials*, **37**(5), 749–754.

Bowker, Michael (2006). The surface structure of titania and the effect of reduction. *Current Opinion in Solid State and Materials Science*, **10**(3-4), 153–162.

Brack, A (2013). Clay minerals and the origin of life. In *Developments in Clay Science*, Volume 5, pp. 507–521. Elsevier.

Bradbury, Michael H and Baeyens, Bart (1997). A mechanistic description of Ni and Zn sorption on Na-montmorillonite part II: Modelling. *Journal of Contaminant Hydrology*, **27**(3-4), 223–248.

Bragg, William Lawrence (1937). *Atomic Structure of Minerals*. Cornell University Press, Ithaca, New York.

Briand, Laura E, Tkachenko, Olga P, Guraya, Monica, Gao, Xingtao, Wachs, Israel E, and Grünert, Wolfgang (2004). Surface-analytical studies of supported vanadium oxide monolayer catalysts. *The Journal of Physical Chemistry B*, **108**(15), 4823–4830.

Brigatti, Maria Franca, Galan, E, and Theng, BKG (2006). Structures and mineralogy of clay minerals. *Developments in Clay Science*, **1**, 19–86.

Brinker, C Jeffrey and Scherer, George W (2013). *Sol-Gel Science: The Physics and Chemistry of Sol-Gel Processing*. Academic Press, Boston, San Diego, New York, London, Sydney, Tokyo, Toronto.

Bromley, Stefan T (2017). Silicate nanoclusters: Understanding their cosmic relevance from bottom-up modelling. In *Clusters*, pp. 237–268. Springer.

Bromley, Stefan T, de PR Moreira, Ibério, Neyman, Konstantin M, and Illas, Francesc (2009). Approaching nanoscale oxides: Models and theoretical methods. *Chemical Society Reviews*, **38**(9), 2657–2670.

Broqvist, Peter, Kullgren, Jolla, Wolf, Matthew J, Van Duin, Adri CT, and Hermansson, Kersti (2015). ReaxFF force-field for ceria bulk, surfaces, and nanoparticles. *The Journal of Physical Chemistry C*, **119**(24), 13598–13609.

Brown, George (1982). *Crystal structures of clay minerals and their X-ray identification*. Volume 5. The Mineralogical Society of Great Britain and Ireland, Colchester, London.

Brown, ID and Altermatt, D (1985). Bond-valence parameters obtained from a systematic analysis of the inorganic crystal structure database. *Acta Crystallographica Section B: Structural Science*, **41**(4), 244–247.

Büchner, Christin, Wang, Zhu-Jun, Burson, Kristen M, Willinger, Marc-Georg, Heyde, Markus, Schlögl, Robert, and Freund, Hans-Joachim (2016). A large-area transferable wide band gap 2-D silicon dioxide layer. *ACS Nano*, **10**(8), 7982–7989.

Burgess, John (1999). *Ions in Solution: Basic Principles of Chemical Interactions*. Elsevier, Amsterdam, Oxford, Cambridge.

Burton, W-K, Cabrera, N, and Frank, FC (1951). The growth of crystals and the equilibrium structure of their surfaces. *Philosophical Transactions of the Royal Society of London. Series A, Mathematical and Physical Sciences*, **243**, 299–358.

Butler, Sheneve Z, Hollen, Shawna M, Cao, Linyou, Cui, Yi, Gupta, Jay A, Gutiérrez, Humberto R, Heinz, Tony F, Hong, Seung Sae, Huang, Jiaxing, Ismach, Ariel F et al. (2013). Progress, challenges, and opportunities in two-dimensional materials beyond graphene. *ACS Nano*, **7**(4), 2898–2926.

Cabailh, Gregory, Goniakowski, Jacek, Noguera, Claudine, Jupille, Jacques, Lazzari, Rémi, Li, Jingfeng, Lagarde, Pierre, and Trcera, Nicolas (2019). Understanding nanoscale effects in oxide/metal heteroepitaxy. *Physical Review Materials*, **3**,

046001.

Cabrera, NFMN and Mott, Nevill Francis (1949). Theory of the oxidation of metals. *Reports on Progress in Physics*, **12**(1), 163.

Caffio, M, Atrei, A, Cortigiani, B, and Rovida, G (2006). STM study of the nanostructures prepared by deposition of NiO on Ag(001). *Journal of Physics: Condensed Matter*, **18**(8), 2379.

Campbell, Charles T and Sauer, Joachim (2013). Introduction: Surface chemistry of oxides. *Chemical Reviews*, **113**(6), 3859–3862.

Campbell, Timothy, Kalia, Rajiv K, Nakano, Aiichiro, Vashishta, Priya, Ogata, Shuji, and Rodgers, Stephen (1999). Dynamics of oxidation of aluminum nanoclusters using variable charge molecular-dynamics simulations on parallel computers. *Physical Review Letters*, **82**(24), 4866.

Carrasco, Javier, Lopez, Nuria, Illas, Francesc, and Freund, H-J (2006). Bulk and surface oxygen vacancy formation and diffusion in single crystals, ultrathin films, and metal grown oxide structures. *The Journal of Chemical Physics*, **125**(7), 074711.

Carretero, MI, Gomes, CSF, and Tateo, F (2013). Clays, drugs, and human health. In *Developments in Clay Science* (ed. F. Bergaya and G. Lagaly), Volume 5, pp. 711–764. Elsevier.

Carretero, M Isabel and Pozo, Manuel (2009). Clay and non-clay minerals in the pharmaceutical industry: Part I. Excipients and medical applications. *Applied Clay Science*, **46**(1), 73–80.

Carretero, M Isabel and Pozo, Manuel (2010). Clay and non-clay minerals in the pharmaceutical and cosmetic industries Part II. Active ingredients. *Applied Clay Science*, **47**(3-4), 171–181.

Casek, Petr, Bouette-Russo, Sophie, Finocchi, Fabio, and Noguera, Claudine (2004). $SrTiO_3(001)$ thin films on MgO(001): A theoretical study. *Physical Review B*, **69**(8), 085411.

Castellarin-Cudia, C, Surnev, S, Schneider, G, Podlucky, R, Ramsey, MG, and Netzer, FP (2004). Strain-induced formation of arrays of catalytically active sites at the metal–oxide interface. *Surface Science*, **554**(2-3), L120–L126.

Catlow, CRA and Stoneham, AM (1983). Ionicity in solids. *Journal of Physics C: Solid State Physics*, **16**(22), 4321.

Catlow, C Richard A, Bromley, Stefan T, Hamad, Said, Mora-Fonz, Miguel, Sokol, Alexey A, and Woodley, Scott M (2010). Modelling nano-clusters and nucleation. *Physical Chemistry Chemical Physics*, **12**(4), 786–811.

Celardo, Ivana, Pedersen, Jens Z, Traversa, Enrico, and Ghibelli, Lina (2011). Pharmacological potential of cerium oxide nanoparticles. *Nanoscale*, **3**(4), 1411–1420.

Chadi, DJ (1979). (110) surface atomic structures of covalent and ionic semiconductors. *Physical Review B*, **19**(4), 2074.

Chakhalian, Jak, Freeland, John W, Millis, Andrew J, Panagopoulos, Christos, and Rondinelli, James M (2014). Colloquium: Emergent properties in plane view: Strong correlations at oxide interfaces. *Reviews of Modern Physics*, **86**(4), 1189.

Chambers, Scott A (2000). Epitaxial growth and properties of thin film oxides. *Surface Science Reports*, **39**(5-6), 105–180.

Chang, Fang-Ru Chou, Skipper, NT, and Sposito, Garrison (1995). Computer simulation of interlayer molecular structure in sodium montmorillonite hydrates. *Langmuir*, **11**(7), 2734–2741.

Chang, Young Jun, Kim, Choong H, Phark, S-H, Kim, YS, Yu, J, and Noh, TW (2009). Fundamental thickness limit of itinerant ferromagnetic SrRuO$_3$ thin films. *Physical Review Letters*, **103**(5), 057201.

Chen, CQ, Shi, Y, Zhang, Y St, Zhu, J, and Yan, YJ (2006). Size dependence of Young's modulus in ZnO nanowires. *Physical Review Letters*, **96**(7), 075505.

Chen, C Julian (1993). *Introduction to scanning tunneling microscopy*. Volume 4. Oxford University Press, New York.

Chen, Hanghui and Millis, Andrew (2017). Charge transfer driven emergent phenomena in oxide heterostructures. *Journal of Physics: Condensed Matter*, **29**(24), 243001.

Chen, Mingyang, Felmy, Andrew R, and Dixon, David A (2014a). Structures and stabilities of (MgO)$_n$ nanoclusters. *The Journal of Physical Chemistry A*, **118**(17), 3136–3146.

Chen, Qian, Zhu, Liyan, and Wang, Jinlan (2009). Edge-passivation induced half-metallicity of zigzag zinc oxide nanoribbons. *Applied Physics Letters*, **95**(13), 133116.

Chen, Xiaoyu, Zhou, Yong, Liu, Qi, Li, Zhengdao, Liu, Jianguo, and Zou, Zhigang (2012). Ultrathin, single-crystal WO$_3$ nanosheets by two-dimensional oriented attachment toward enhanced photocatalystic reduction of CO$_2$ into hydrocarbon fuels under visible light. *ACS Applied Materials & Interfaces*, **4**(7), 3372–3377.

Chen, Yu, Ye, Delai, Wu, Meiying, Chen, Hangrong, Zhang, Linlin, Shi, Jianlin, and Wang, Lianzhou (2014b). Break-up of two-dimensional MnO$_2$ nanosheets promotes ultrasensitive pH-triggered theranostics of cancer. *Advanced Materials*, **26**(41), 7019–7026.

Chiche, David, Digne, Mathieu, Revel, Renaud, Chanéac, Corinne, and Jolivet, Jean-Pierre (2008). Accurate determination of oxide nanoparticle size and shape based on X-ray powder pattern simulation: Application to boehmite AlOOH. *The Journal of Physical Chemistry C*, **112**(23), 8524–8533.

Chiesa, Mario, Paganini, Maria Cristina, Spoto, Giuseppe, Giamello, Elio, Di Valentin, Cristiana, Del Vitto, Annalisa, and Pacchioni, Gianfranco (2005). Single electron traps at the surface of polycrystalline MgO: Assignment of the main trapping sites. *The Journal of Physical Chemistry B*, **109**(15), 7314–7322.

Cho, Al Y and Arthur, JR (1975). Molecular beam epitaxy. *Progress in Solid State Chemistry*, **10**, 157–191.

Cho, Daeheum, Ko, Kyoung Chul, Lamiel-García, Oriol, Bromley, Stefan T, Lee, Jin Yong, and Illas, Francesc (2016). Effect of size and structure on the ground-state and excited-state electronic structure of TiO$_2$ nanoparticles. *Journal of Chemical Theory and Computation*, **12**(8), 3751–3763.

Choi, Kyoung Jin, Biegalski, Michael, Li, YL, Sharan, A, Schubert, J, Uecker, Reinhard, Reiche, P, Chen, YB, Pan, XQ, Gopalan, Venkatraman et al. (2004). Enhancement of ferroelectricity in strained BaTiO$_3$ thin films. *Science*, **306**(5698), 1005–1009.

Choi, Miri, Lin, Chungwei, Butcher, Matthew, Rodriguez, Cesar, He, Qian, Posadas,

Agham B, Borisevich, Albina Y, Zollner, Stefan, and Demkov, Alexander A (2015). Quantum confinement in transition metal oxide quantum wells. *Applied Physics Letters*, **106**(19), 192902.

Choi, Minkee, Na, Kyungsu, Kim, Jeongnam, Sakamoto, Yasuhiro, Terasaki, Osamu, and Ryoo, Ryong (2009). Stable single-unit-cell nanosheets of zeolite MFI as active and long-lived catalysts. *Nature*, **461**(7261), 246.

Choy, KL (2003). Chemical vapour deposition of coatings. *Progress in Materials Science*, **48**(2), 57–170.

Chu, YH, Zhao, T, Cruz, MP, Zhan, Q, Yang, PL, Martin, LW, Huijben, Mark, Yang, Chan-Ho, Zavaliche, F, Zheng, H et al. (2007). Ferroelectric size effects in multiferroic $BiFeO_3$ thin films. *Applied Physics Letters*, **90**(25), 252906.

Churakov, Sergey V (2006). Ab initio study of sorption on pyrophyllite: Structure and acidity of the edge sites. *The Journal of Physical Chemistry B*, **110**(9), 4135–4146.

Claeyssens, Frederik, Freeman, Colin L, Allan, Neil L, Sun, Ye, Ashfold, Michael NR, and Harding, John H (2005). Growth of ZnO thin films: Experiment and theory. *Journal of Materials Chemistry*, **15**(1), 139–148.

Clausen, Edward M and Hren, JJ (1984). The gamma to alpha transformation in thin film alumina. *MRS Online Proceedings Library Archive*, **41**, 381.

Coli, Giuliano and Bajaj, KK (2001). Excitonic transitions in ZnO/MgZnO quantum well heterostructures. *Applied Physics Letters*, **78**(19), 2861–2863.

Coll, Mariona, Fontcuberta, Josep, Althammer, M, Bibes, Manuel, Boschker, H, Calleja, Albert, Cheng, G, Cuoco, M, Dittmann, R, Dkhil, B et al. (2019). Towards oxide electronics: A roadmap. *Applied Surface Science*, **482**, 1–93.

Coluccia, Salvatore, Lavagnino, S, and Marchese, L (1988). The hydroxylated surface of MgO powders and the formation of surface sites. *Materials Chemistry and Physics*, **18**(5-6), 445–464.

Conrad, E (1996). Diffraction methods. In *Handbook of Surface Science* (ed. W. Unertl), Chapter 7. Elsevier.

Cosgrove, Terence (2010). *Colloid Science: Principles, Methods and Applications*. John Wiley & Sons, Weinheim.

Csiszar, SI, Haverkort, MW, Hu, Z, Tanaka, Arata, Hsieh, HH, Lin, H-J, Chen, CT, Hibma, T, and Tjeng, LH (2005). Controlling orbital moment and spin orientation in CoO layers by strain. *Physical Review Letters*, **95**(18), 187205.

Cui, Yuanyuan, Ke, Yujie, Liu, Chang, Chen, Zhang, Wang, Ning, Zhang, Liangmiao, Zhou, Yang, Wang, Shancheng, Gao, Yanfeng, and Long, Yi (2018). Thermochromic VO_2 for energy-efficient smart windows. *Joule*, **2**(9), 1707–1746.

Cui, Yi, Pan, Yi, Pascua, Leandro, Qiu, Hengshan, Stiehler, Christian, Kuhlenbeck, Helmut, Nilius, Niklas, and Freund, Hans-Joachim (2015*a*). Evolution of the electronic structure of CaO thin films following Mo interdiffusion at high temperature. *Physical Review B*, **91**(3), 035418.

Cui, Yi, Shao, Xiang, Prada, Stefano, Giordano, Livia, Pacchioni, Gianfranco, Freund, Hans-Joachim, and Nilius, Niklas (2014). Surface defects and their impact on the electronic structure of Mo-doped CaO films: An STM and DFT study. *Physical Chemistry Chemical Physics*, **16**(25), 12764–12772.

Cui, Yi, Stiehler, Christian, Nilius, Niklas, and Freund, Hans-Joachim (2015*b*). Probing the electronic properties and charge state of gold nanoparticles on ultrathin MgO versus thick doped CaO films. *Physical Review B*, **92**(7), 075444.

Cui, Zhi-Hao, Wu, Feng, and Jiang, Hong (2016). First-principles study of relative stability of rutile and anatase TiO_2 using the random phase approximation. *Physical Chemistry Chemical Physics*, **18**(43), 29914–29922.

Cuko, Andi, Calatayud, Monica, and Bromley, Stefan T (2018*a*). Stability of mixed-oxide titanosilicates: Dependency on size and composition from nanocluster to bulk. *Nanoscale*, **10**(2), 832–842.

Cuko, Andi, Escatllar, Antoni Macià, Calatayud, Monica, and Bromley, Stefan T (2018*b*). Properties of hydrated TiO_2 and SiO_2 nanoclusters: Dependence on size, temperature and water vapour pressure. *Nanoscale*, **10**(45), 21518–21532.

Cuko, Andi, Macià, Antoni, Calatayud, Monica, and Bromley, Stefan T (2017). Global optimisation of hydroxylated silica clusters: A cascade Monte Carlo Basin Hopping approach. *Computational and Theoretical Chemistry*, **1102**, 38–43.

Dang, Hung T, Ai, Xinyuan, Millis, Andrew J, and Marianetti, Chris A (2014). Density functional plus dynamical mean-field theory of the metal-insulator transition in early transition-metal oxides. *Physical Review B*, **90**(12), 125114.

D'Angelo, Marie, Yukawa, R, Ozawa, K, Yamamoto, S, Hirahara, T, Hasegawa, S, Silly, MG, Sirotti, F, and Matsuda, I (2012). Hydrogen-induced surface metallization of $SrTiO_3$(001). *Physical Review Letters*, **108**(11), 116802.

Davies, PK and Navrotsky, Alexandra (1983). Quantitative correlations of deviations from ideality in binary and pseudobinary solid solutions. *Journal of Solid State Chemistry*, **46**(1), 1–22.

De Siqueira, Ascenso VC, Skipper, Neal T, V, Coveney Peter, and S, Boek Edo (1997). Computer simulation evidence for enthalpy driven dehydration of smectite clays at elevated pressures and temperatures. *Molecular Physics*, **92**(1), 1–6.

Deal, Bruce E and Grove, AS (1965). General relationship for the thermal oxidation of silicon. *Journal of Applied Physics*, **36**(12), 3770–3778.

Deb, Satyen K (2008). Opportunities and challenges in science and technology of WO_3 for electrochromic and related applications. *Solar Energy Materials and Solar Cells*, **92**(2), 245–258.

Demichelis, Raffaella, Raiteri, Paolo, Gale, Julian D, Quigley, David, and Gebauer, Denis (2011). Stable prenucleation mineral clusters are liquid-like ionic polymers. *Nature Communications*, **2**, 590.

Deravi, Leila F, Swartz, Joshua D, and Wright, David W (2007). *The Biomimetic Synthesis of Metal Oxide Nanomaterials*. Wiley Online Library.

Dewhurst, David N, Yang, Yunlai, and Aplin, Andrew C (1999). Permeability and fluid flow in natural mudstones. *Geological Society, London, Special Publications*, **158**(1), 23–43.

Dholabhai, Pratik P, Pilania, Ghanshyam, Aguiar, Jeffery A, Misra, Amit, and Uberuaga, Blas P (2014). Termination chemistry-driven dislocation structure at $SrTiO_3$/MgO heterointerfaces. *Nature Communications*, **5**, 5043.

Di Valentin, Cristiana, Wang, Fenggong, and Pacchioni, Gianfranco (2013). Tungsten oxide in catalysis and photocatalysis: Hints from DFT. *Topics in Catalysis*, **56**(15-17), 1404–1419.

Diawara, Boubakar, Beh, Yves-Alain, and Marcus, Philippe (2010). Nucleation and growth of oxide layers on stainless steels (FeCr) using a virtual oxide layer model. *The Journal of Physical Chemistry C*, **114**(45), 19299–19307.

Diawara, Boubakar, Legrand, Médéric, Legendre, J-J, and Marcus, Phillipe (2004). Use of quantum chemistry results in 3-D modeling of corrosion of iron-chromium alloys. *Journal of the Electrochemical Society*, **151**(3), B172–B178.

Dick Jr, BG and Overhauser, AW (1958). Theory of the dielectric constants of alkali halide crystals. *Physical Review*, **112**(1), 90.

Didier, Fabrice and Jupille, Jacques (1994). Layer-by-layer growth mode of silver on magnesium oxide (100). *Surface Science*, **307**, 587–590.

Diebold, Ulrike (2003). The surface science of titanium dioxide. *Surface Science Reports*, **48**(5-8), 53–229.

Diodati, Stefano, Dolcet, Paolo, Casarin, Maurizio, and Gross, Silvia (2015). Pursuing the crystallization of mono-and polymetallic nanosized crystalline inorganic compounds by low-temperature wet-chemistry and colloidal routes. *Chemical Reviews*, **115**(20), 11449–11502.

Dion, M, Rydberg, H, Schröder, E, Langreth, DC, and Lundqvist, BI (2004). Van der Waals density functional for general geometries. *Physical Review Letters*, **92**, 246401.

Djenadic, Ruzica and Winterer, Markus (2012, 07). Chemical vapor synthesis of nanocrystalline oxides. In *Nanoparticles from the Gasphase: Formation, Structure, Properties, NanoScience and Technology*, Volume 79, pp. 49–76. Springer.

Dose, Volker (1983). Ultraviolet bremsstrahlung spectroscopy. *Progress in Surface Science*, **13**(3), 225–283.

Doudin, N, Kuhness, D, Blatnik, M, Barcaro, G, Negreiros, FR, Sementa, L, Fortunelli, A, Surnev, S, and Netzer, FP (2016). Nanoscale domain structure and defects in a 2-D WO_3 layer on Pd(100). *The Journal of Physical Chemistry C*, **120**(50), 28682–28693.

Drits, VA (2003). Structural and chemical heterogeneity of layer silicates and clay minerals. *Clay Minerals*, **38**(4), 403–432.

D'Souza, Lawrence and Richards, Ryan (2006, 07). Synthesis of metal-oxide nanoparticles: Liquid-solid transformations. In *Synthesis, Properties, and Applications of Oxide Nanomaterials*, pp. 81 – 117. Wiley.

Du, Yingge, Li, Guoqiang, Peterson, Erik W, Zhou, Jing, Zhang, Xin, Mu, Rentao, Dohnálek, Zdenek, Bowden, Mark, Lyubinetsky, Igor, and Chambers, Scott A (2016). Iso-oriented monolayer α-MoO_3(010) films epitaxially grown on $SrTiO_3$(001). *Nanoscale*, **8**(5), 3119–3124.

Duan, Haohong, Wang, Dingsheng, and Li, Yadong (2015). Green chemistry for nanoparticle synthesis. *Chemical Society Reviews*, **44**(16), 5778–5792.

Dudarev, SL, Botton, GA, Savrasov, SY, Humphreys, CJ, and Sutton, AP (1998). Electron-energy-loss spectra and the structural stability of nickel oxide: An LSDA+U study. *Physical Review B*, **57**, 1505–1509.

Dudarev, SL, Liechtenstein, AI, Castell, MR, Briggs, GAD, and Sutton, AP (1997). Surface states on NiO(100) and the origin of the contrast reversal in atomically resolved scanning tunneling microscope images. *Physical Review B*, **56**(8), 4900.

Duffy, DM and Stoneham, AM (1983). Conductivity and 'negative U' for ionic grain boundaries. *Journal of Physics C: Solid State Physics*, **16**(21), 4087.

Dugourd, Philippe, Hudgins, Robert R, and Jarrold, Martin F (1997). High-resolution ion mobility studies of sodium chloride nanocrystals. *Chemical Physics Letters*, **267**(1-2), 186–192.

Dulub, Olga, Hebenstreit, Wilhelm, and Diebold, Ulrike (2000). Imaging cluster surfaces with atomic resolution: The strong metal-support interaction state of Pt supported on TiO_2(110). *Physical Review Letters*, **84**(16), 3646.

Dumas, Angela, Martin, François, Ferrage, Eric, Micoud, Pierre, Le Roux, Christophe, and Petit, Sabine (2013). Synthetic talc advances: Coming closer to nature, added value, and industrial requirements. *Applied Clay Science*, **85**, 8–18.

Dutta, Sudipta and Pati, Swapan K (2010). Novel properties of graphene nanoribbons: A review. *Journal of Materials Chemistry*, **20**(38), 8207–8223.

Duval, Y, Mielczarski, JA, Pokrovsky, OS, Mielczarski, E, and Ehrhardt, JJ (2002). Evidence of the existence of three types of species at the quartz-aqueous solution interface at pH 0-10: XPS surface group quantification and surface complexation modeling. *The Journal of Physical Chemistry B*, **106**(11), 2937–2945.

Ebensperger, Christina, Gubo, Matthias, Meyer, Wolfgang, Hammer, Lutz, and Heinz, Klaus (2010). Substrate-induced structural modulation of a CoO(111) bilayer on Ir(100). *Physical Review B*, **81**(23), 235405.

Eberl, DD (1984). Clay mineral formation and transformation in rocks and soils. *Philosophical Transactions of the Royal Society of London. Series A, Mathematical and Physical Sciences*, **311**(1517), 241–257.

Ebrahimi, Davoud, Pellenq, Roland J-M, and Whittle, Andrew J (2012). Nanoscale elastic properties of montmorillonite upon water adsorption. *Langmuir*, **28**(49), 16855–16863.

Echt, O, Sattler, K, and Recknagel, E (1981). Magic numbers for sphere packings: Experimental verification in free xenon clusters. *Physical Review Letters*, **47**(16), 1121.

Eck, S, Castellarin-Cudia, C, Surnev, S, Ramsey, MG, and Netzer, FP (2002). Growth and thermal properties of ultrathin cerium oxide layers on Rh(111). *Surface Science*, **520**(3), 173–185.

Ederer, Claude and Spaldin, Nicola A (2005). Effect of epitaxial strain on the spontaneous polarization of thin film ferroelectrics. *Physical Review Letters*, **95**(25), 257601.

Ehlmann, Bethany L, Berger, Gilles, Mangold, Nicolas, Michalski, Joseph R, Catling, David C, Ruff, Steven W, Chassefière, Eric, Niles, Paul B, Chevrier, Vincent, and Poulet, Francois (2013). Geochemical consequences of widespread clay mineral formation in Mars ancient crust. *Space Science Reviews*, **174**(1-4), 329–364.

Engel, Eberhard and Dreizler, Reiner (2013). *Density Functional Theory*. Springer Verlag, Berlin, Heidelberg, New York, London, Paris, Tokyo.

Enterkin, James A, Subramanian, Arun K, Russell, Bruce C, Castell, Martin R, Poeppelmeier, Kenneth R, and Marks, Laurence D (2010). A homologous series of structures on the surface of $SrTiO_3$(110). *Nature Materials*, **9**(3), 245.

Ertl, Gerhard and Küppers, Jürgen (1985). *Low Energy Electrons and Surface Chemistry*. Wiley-VCH Verlagsgesellschaft, Weinheim.

Escher, Susanne GET, Lazauskas, Tomas, Zwijnenburg, Martijn A, and Woodley, Scott M (2018). Synthesis target structures for alkaline earth oxide clusters. *Inorganics*, **6**(1), 29.

Eyring, Henry (1935). The activated complex and the absolute rate of chemical reactions. *Chemical Reviews*, **17**(1), 65–77.

Farrow, MR, Chow, Y, and Woodley, SM (2014). Structure prediction of nanoclusters; A direct or a pre-screened search on the DFT energy landscape? *Physical Chemistry Chemical Physics*, **16**(39), 21119–21134.

Feng, Wei, Wang, Xiaona, Zhang, Jia, Wang, Lifeng, Zheng, Wei, Hu, PingAn, Cao, Wenwu, and Yang, Bin (2014). Synthesis of two-dimensional β-Ga_2O_3 nanosheets for high-performance solar blind photodetectors. *Journal of Materials Chemistry C*, **2**(17), 3254–3259.

Fernando, Amendra, Weerawardene, KL Dimuthu M, Karimova, Natalia V, and Aikens, Christine M (2015). Quantum mechanical studies of large metal, metal oxide, and metal chalcogenide nanoparticles and clusters. *Chemical Reviews*, **115**(12), 6112–6216.

Ferrando, Riccardo, Jellinek, Julius, and Johnston, Roy L (2008). Nanoalloys: From theory to applications of alloy clusters and nanoparticles. *Chemical Reviews*, **108**(3), 845–910.

Ferrari, Anna M, Casassa, Silvia, and Pisani, Cesare (2005*a*). Electronic structure and morphology of MgO submonolayers at the Ag(001) surface: An ab initio model study. *Physical Review B*, **71**(15), 155404.

Ferrari, Anna Maria, Casassa, S, Pisani, Cesare, Altieri, S, Rota, Alberto, and Valeri, Sergio (2005*b*). Polar and non-polar domain borders in MgO ultrathin films on Ag(001). *Surface Science*, **588**(1-3), 160–166.

Ferrari, Anna Maria, Roetti, Carla, and Pisani, Cesare (2007). Water dissociation at MgO sub-monolayers on silver: A periodic model study. *Physical Chemistry Chemical Physics*, **9**(19), 2350–2354.

Fester, Jakob, Bajdich, Michal, Walton, Alex S, Sun, Zhaozong, Plessow, Philipp N, Vojvodic, Aleksandra, and Lauritsen, Jeppe V (2017). Comparative analysis of cobalt oxide nanoisland stability and edge structures on three related noble metal surfaces: Au(111), Pt(111) and Ag(111). *Topics in Catalysis*, **60**(6-7), 503–512.

Feya, Oleg D, Wang, Qinggao, Lepeshkin, Sergey V, Baturin, Vladimir S, Uspenskii, Yurii A, and Oganov, Artem R (2018). Tetrahedral honeycomb surface reconstructions of quartz, cristobalite and stishovite. *Scientific Reports*, **8**(1), 11947.

Finazzi, Marco and Altieri, Salvatore (2003). Magnetic dipolar anisotropy in strained antiferromagnetic films. *Physical Review B*, **68**(5), 054420.

Finnis, Mike (2003). *Interatomic Forces in Condensed Matter*. Volume 1. Oxford University Press, Oxford.

Fondell, Mattis, Jacobsson, T Jesper, Boman, Mats, and Edvinsson, Tomas (2014). Optical quantum confinement in low dimensional hematite. *Journal of Materials Chemistry A*, **2**(10), 3352–3363.

Fong, DD, Kolpak, AM, Eastman, JA, Streiffer, SK, Fuoss, PH, Stephenson, GB,

Thompson, Carol, Kim, DM, Choi, Kyoung Jin, Eom, CB et al. (2006). Stabilization of monodomain polarization in ultrathin $PbTiO_3$ films. *Physical Review Letters*, **96**(12), 127601.

Fong, Dillon D, Stephenson, G Brian, Streiffer, Stephen K, Eastman, Jeffrey A, Auciello, Orlando, Fuoss, Paul H, and Thompson, Carol (2004). Ferroelectricity in ultrathin perovskite films. *Science*, **304**(5677), 1650–1653.

Förster, Daniel F, Klinkhammer, Jürgen, Busse, Carsten, Altendorf, Simone G, Michely, Thomas, Hu, Zhiwei, Chin, Yi-Ying, Tjeng, LH, Coraux, Johann, and Bourgault, Daniel (2011). Epitaxial europium oxide on Ni(100) with single-crystal quality. *Physical Review B*, **83**(4), 045424.

Franchini, Cesare, Li, F, Surnev, S, Podloucky, R, Allegretti, F, and Netzer, FP (2012). Tailor-made ultrathin manganese oxide nanostripes: 'Magic widths' on $Pd(11n)$ terraces. *Journal of Physics: Condensed Matter*, **24**(4), 042001.

Franchini, Cesare, Podloucky, R, Allegretti, F, Li, F, Parteder, G, Surnev, S, and Netzer, FP (2009a). Structural and vibrational properties of two-dimensional Mn_xO_y layers on Pd(100): Experiments and density functional theory calculations. *Physical Review B*, **79**(3), 035420.

Franchini, Cesare, Zabloudil, J, Podloucky, R, Allegretti, F, Li, F, Surnev, S, and Netzer, FP (2009b). Interplay between magnetic, electronic, and vibrational effects in monolayer Mn_3O_4 grown on Pd(100). *The Journal of Chemical Physics*, **130**(12), 124707.

Franchy, René (2000). Growth of thin, crystalline oxide, nitride and oxynitride films on metal and metal alloy surfaces. *Surface Science Reports*, **38**(6-8), 195–294.

Frank, FC and van der Merwe, Jan H (1949). One-dimensional dislocations. I. static theory. *Proceedings of the Royal Society of London. Series A. Mathematical and Physical Sciences*, **198**(1053), 205–216.

Frank, Frederick Charles and Van der Merwe, JH (1949). One-dimensional dislocations. II. misfitting monolayers and oriented overgrowth. *Proceedings of the Royal Society of London. Series A. Mathematical and Physical Sciences*, **198**(1053), 216–225.

Freeman, Colin L, Claeyssens, Frederik, Allan, Neil L, and Harding, John H (2006). Graphitic nanofilms as precursors to wurtzite films: Theory. *Physical Review Letters*, **96**(6), 066102.

Freitag, A, Staemmler, V, Cappus, D, Ventrice Jr, CA, Al Shamery, K, Kuhlenbeck, H, and Freund, H-J (1993). Electronic surface state of NiO(100). *Chemical Physics Letters*, **210**(1-3), 10–14.

Freund, Hans-Joachim (2007). Metal-supported ultrathin oxide film systems as designable catalysts and catalyst supports. *Surface Science*, **601**(6), 1438–1442.

Freund, Hans-Joachim (2011). Oxide films as catalytic materials and models of real catalysts. In *Oxide Ultrathin Films: Science and Technology* (ed. G. Pacchioni and S. Valeri), pp. 145–179. Wiley Online Library.

Freund, Hans-Joachim and Pacchioni, Gianfranco (2008). Oxide ultra-thin films on metals: New materials for the design of supported metal catalysts. *Chemical Society Reviews*, **37**(10), 2224–2242.

Freysoldt, Christoph, Rinke, Patrick, and Scheffler, Matthias (2007). Ultrathin oxides:

Bulk-oxide-like model surfaces or unique films? *Physical Review Letters*, **99**(8), 086101.

Fritz, B, Clement, A, Amal, Y, and Noguera, C (2009). Simulation of the nucleation and growth of simple clay minerals in weathering processes: The NANOKIN code. *Geochimica et Cosmochimica Acta*, **73**(5), 1340–1358.

Froitzheim, H, Ibach, H, and Mills, DL (1975). Surface optical constants of silicon and germanium derived from electron-energy-loss spectroscopy. *Physical Review B*, **11**(12), 4980.

Fu, Qiang, Li, Wei-Xue, Yao, Yunxi, Liu, Hongyang, Su, Hai-Yan, Ma, Ding, Gu, Xiang-Kui, Chen, Limin, Wang, Zhen, Zhang, Hui et al. (2010). Interface-confined ferrous centers for catalytic oxidation. *Science*, **328**(5982), 1141–1144.

Fujishima, Akira and Honda, Kenichi (1972). Electrochemical photolysis of water at a semiconductor electrode. *Nature*, **238**(5358), 37–38.

Fukumura, T, Yamada, Y, Toyosaki, H, Hasegawa, T, Koinuma, H, and Kawasaki, M (2004). Exploration of oxide-based diluted magnetic semiconductors toward transparent spintronics. *Applied Surface Science*, **223**(1-3), 62–67.

Furche, Filipp and Ahlrichs, Reinhart (2002). Adiabatic time-dependent density functional methods for excited state properties. *The Journal of Chemical Physics*, **117**(16), 7433–7447.

Galán, E (2006). Genesis of clay minerals. *Developments in Clay Science*, **1**, 1129–1162.

Gallagher, Marc C, Fyfield, Margaret S, Bumm, Lloyd A, Cowin, James P, and Joyce, Stephen A (2003). Structure of ultrathin MgO films on Mo(001). *Thin Solid Films*, **445**(1), 90–95.

Galloway, HC, Benitez, JJ, and Salmeron, M (1994). Growth of FeO_x on Pt(111) studied by scanning tunneling microscopy. *Journal of Vacuum Science & Technology A: Vacuum, Surfaces, and Films*, **12**(4), 2302–2307.

Galloway, HC, Sautet, P, and Salmeron, M (1996). Structure and contrast in scanning tunneling microscopy of oxides: FeO monolayer on Pt(111). *Physical Review B*, **54**(16), R11145.

Ganduglia-Pirovano, M Veronica, Hofmann, Alexander, and Sauer, Joachim (2007). Oxygen vacancies in transition metal and rare earth oxides: Current state of understanding and remaining challenges. *Surface Science Reports*, **62**(6), 219–270.

Gao, Peng, Zhang, Zhangyuan, Li, Mingqiang, Ishikawa, Ryo, Feng, Bin, Liu, Heng-Jui, Huang, Yen-Lin, Shibata, Naoya, Ma, Xiumei, Chen, Shulin et al. (2017). Possible absence of critical thickness and size effect in ultrathin perovskite ferroelectric films. *Nature Communications*, **8**, 15549.

Gao, W, Fujikawa, Y, Saiki, K, and Koma, A (1993). Surface phonons of LiBr/Si(100) epitaxial layers by high resolution electron energy loss spectroscopy. *Solid State Communications*, **87**(11), 1013–1015.

Garcia-Matres, E, Stüßer, N, Hofmann, M, and Reehuis, M (2003). Magnetic phases in $Mn_{1-x}Fe_xWO_4$ studied by neutron powder diffraction. *The European Physical Journal B-Condensed Matter and Complex Systems*, **32**(1), 35–42.

Garoufalis, Christos, Barnasas, Alexandros, Stamatelatos, Alkeos, Karoutsos, Vagelis, Grammatikopoulos, Spyridon, Poulopoulos, Panagiotis, and Baskoutas, Sotirios

(2018). A study of quantum confinement effects in ultrathin NiO films performed by experiment and theory. *Materials*, **11**(6), 949.

Garoufalis, CS, Poulopoulos, P, Bouropoulos, N, Barnasas, A, and Baskoutas, S (2017). Growth and optical properties of Fe_2O_3 thin films: A study of quantum confinement effects by experiment and theory. *Physica E: Low-dimensional Systems and Nanostructures*, **89**, 67–71.

Gautier, M, Duraud, JP, Van, L Pham, and Guittet, MJ (1991). Modifications of α-Al_2O_3 (0001) surfaces induced by thermal treatments or ion bombardment. *Surface Science*, **250**(1-3), 71–80.

Gautier-Soyer, M, Jollet, F, and Noguera, C (1996). Influence of surface relaxation on the electronic states of the α-Al_2O_3(0001) surface: A self-consistent tight-binding approach. *Surface Science*, **352**, 755–759.

Gebauer, Denis, Kellermeier, Matthias, Gale, Julian D, Bergström, Lennart, and Cölfen, Helmut (2014). Pre-nucleation clusters as solute precursors in crystallisation. *Chemical Society Reviews*, **43**(7), 2348–2371.

Gebauer, Denis, Völkel, Antje, and Cölfen, Helmut (2008). Stable prenucleation calcium carbonate clusters. *Science*, **322**(5909), 1819–1822.

Geerlings, Paul, De Proft, F, and Langenaeker, W (2003). Conceptual density functional theory. *Chemical Reviews*, **103**(5), 1793–1874.

Geiger, Charles A (2001). Thermodynamic mixing properties of binary oxide and silicate solid solutions determined by direct measurements: The role of strain. *Solid Solutions in Silicate and Oxide Systems of Geological Importance. EMU Notes in Mineralogy*, **3**, 71.

Geim, Andre K and Grigorieva, Irina V (2013). Van der Waals heterostructures. *Nature*, **499**(7459), 419.

Georges, Antoine, Kotliar, Gabriel, Krauth, Werner, and Rozenberg, Marcelo J (1996). Dynamical mean-field theory of strongly correlated fermion systems and the limit of infinite dimensions. *Reviews of Modern Physics*, **68**(1), 13.

Gerson, AR, Cookson, DJ, and Prince, KC (2009). Synchrotron-based techniques. In *Handbook of Surface and Interface Analysis* (ed. J. Riviere and S. Myrha), Chapter 7. CRC Press.

Geysermans, P, Finocchi, F, Goniakowski, Jacek, Hacquart, R, and Jupille, J (2009). Combination of (100), (110) and (111) facets in MgO crystals shapes from dry to wet environment. *Physical Chemistry Chemical Physics*, **11**(13), 2228–2233.

Geysermans, Pascale and Noguera, Claudine (2009). Advances in atomistic simulations of mineral surfaces. *Journal of Materials Chemistry*, **19**(42), 7807–7821.

Ghosez, Ph and Rabe, KM (2000). Microscopic model of ferroelectricity in stress-free $PbTiO_3$ ultrathin films. *Applied Physics Letters*, **76**(19), 2767–2769.

Ghosh, SP, Das, KC, Tripathy, N, Bose, G, Kim, DH, Lee, TI, Myoung, JM, and Kar, JP (2016). Ultraviolet photodetection characteristics of zinc oxide thin films and nanostructures. In *IOP Conference Series: Materials Science and Engineering*, Volume 115, p. 012035. IOP Publishing.

Ghosh Chaudhuri, Rajib and Paria, Santanu (2011). Core/shell nanoparticles: Classes, properties, synthesis mechanisms, characterization, and applications. *Chemical Reviews*, **112**(4), 2373–2433.

Gibbs, J Willard (1876a). On the equilibrium of heterogeneous substances. *Transactions of the Connecticut Academy of Arts and Sciences*, **3**, 108–248.

Gibbs, J Willard (1876b). On the equilibrium of heterogeneous substances. *Transactions of the Connecticut Academy of Arts and Sciences*, **3**, 343–524.

Giessibl, Franz J (2003). Advances in atomic force microscopy. *Reviews of Modern Physics*, **75**(3), 949.

Gionco, Chiara, Battiato, Alfio, Vittone, Ettore, Paganini, Maria Cristina, and Giamello, Elio (2013a). Structural and spectroscopic properties of high temperature prepared ZrO_2–TiO_2 mixed oxides. *Journal of Solid State Chemistry*, **201**, 222–228.

Gionco, Chiara, Paganini, Maria Cristina, Agnoli, Stefano, Reeder, Askia Enrico, and Giamello, Elio (2013b). Structural and spectroscopic characterization of CeO_2–TiO_2 mixed oxides. *Journal of Materials Chemistry A*, **1**(36), 10918–10926.

Gionco, Chiara, Paganini, Maria C, Giamello, Elio, Burgess, Robertson, Di Valentin, Cristiana, and Pacchioni, Gianfranco (2014). Cerium-doped zirconium dioxide, a visible-light-sensitive photoactive material of third generation. *The Journal of Physical Chemistry Letters*, **5**(3), 447–451.

Giordano, Livia, Goniakowski, Jacek, and Pacchioni, Gianfranco (2003). Properties of MgO(100) ultrathin layers on Pd(100): Influence of the metal support. *Physical Review B*, **67**(4), 045410.

Giordano, Livia, Goniakowski, Jacek, and Suzanne, Jean (1998). Partial dissociation of water molecules in the (3× 2) water monolayer deposited on the MgO(100) surface. *Physical Review Letters*, **81**(6), 1271.

Giordano, Livia, Lewandowski, Mikolaj, Groot, IMN, Sun, Y-N, Goniakowski, Jacek, Noguera, Claudine, Shaikhutdinov, Shamil, Pacchioni, Gianfranco, and Freund, H-J (2010). Oxygen-induced transformations of an FeO(111) film on Pt(111): A combined DFT and STM study. *The Journal of Physical Chemistry C*, **114**(49), 21504–21509.

Giordano, Livia and Pacchioni, Gianfranco (2011). Oxide films at the nanoscale: New structures, new functions, and new materials. *Accounts of Chemical Research*, **44**(11), 1244–1252.

Giordano, Livia, Pacchioni, Gianfranco, Goniakowski, Jacek, Nilius, Niklas, Rienks, Emile DL, and Freund, Hans-Joachim (2007). Interplay between structural, magnetic, and electronic properties in a FeO/Pt(111) ultrathin film. *Physical Review B*, **76**(7), 075416.

Giordano, Livia, Pacchioni, Gianfranco, Noguera, Claudine, and Goniakowski, Jacek (2013). Spectroscopic evidences of charge transfer phenomena and stabilization of unusual phases at iron oxide monolayers grown on Pt(111). *Topics in Catalysis*, **56**(12), 1074–1081.

Glass, Colin W, Oganov, Artem R, and Hansen, Nikolaus (2006). USPEX—evolutionary crystal structure prediction. *Computer Physics Communications*, **175**(11-12), 713–720.

Godby, RW, Schlüter, M, and Sham, LJ (1988). Self-energy operators and exchange-correlation potentials in semiconductors. *Physical Review B*, **37**(17), 10159.

Goian, Veronica, Schumann, Florian O, and Widdra, Wolf (2018). Growth and lattice dynamics of ultrathin BaO films on Pt(001). *Journal of Physics: Condensed*

Matter, **30**(9), 095001.

Goldstein, Joseph I, Newbury, Dale E, Michael, Joseph R, Ritchie, Nicholas WM, Scott, John Henry J, and Joy, David C (2017). *Scanning Electron Microscopy and X-ray Microanalysis*. Springer, Berlin, Heidelberg, New York, Barcelona, Hong Kong, London, Milan, Paris, Tokyo.

Gong, Cheng, Li, Lin, Li, Zhenglu, Ji, Huiwen, Stern, Alex, Xia, Yang, Cao, Ting, Bao, Wei, Wang, Chenzhe, Wang, Yuan et al. (2017). Discovery of intrinsic ferromagnetism in two-dimensional van der Waals crystals. *Nature*, **546**(7657), 265.

Gong, Cheng and Zhang, Xiang (2019). Two-dimensional magnetic crystals and emergent heterostructure devices. *Science*, **363**(6428), eaav4450.

Gong, Xue-Qing and Selloni, Annabella (2005). Reactivity of anatase TiO_2 nanoparticles: The role of the minority (001) surface. *The Journal of Physical Chemistry B*, **109**(42), 19560–19562.

Goniakowski, Jacek, Finocchi, Fabio, and Noguera, Claudine (2007a). Polarity of oxide surfaces and nanostructures. *Reports on Progress in Physics*, **71**(1), 016501.

Goniakowski, Jacek, Giordano, Livia, and Noguera, Claudine (2010). Polarity of ultrathin MgO(111) films deposited on a metal substrate. *Physical Review B*, **81**(20), 205404.

Goniakowski, J, Giordano, L, and Noguera, Claudine (2013). Polarity compensation in low-dimensional oxide nanostructures: The case of metal-supported MgO nanoribbons. *Physical Review B*, **87**(3), 035405.

Goniakowski, Jacek and Noguera, Claudine (1995). Relaxation and rumpling mechanisms on oxide surfaces. *Surface Science*, **323**(1-2), 129–141.

Goniakowski, Jacek and Noguera, Claudine (1999). Characteristics of Pd deposition on the MgO(111) surface. *Physical Review B*, **60**(23), 16120.

Goniakowski, Jacek and Noguera, Claudine (2002). Microscopic mechanisms of stabilization of polar oxide surfaces: Transition metals on the MgO(111) surface. *Physical Review B*, **66**(8), 085417.

Goniakowski, Jacek and Noguera, Claudine (2004). Electronic states and Schottky barrier height at metal/MgO(100) interfaces. *Interface Science*, **12**(1), 93–103.

Goniakowski, Jacek and Noguera, Claudine (2009). Polarization and rumpling in oxide monolayers deposited on metallic substrates. *Physical Review B*, **79**(15), 155433.

Goniakowski, J and Noguera, Claudine (2011). Polarity at the nanoscale. *Physical Review B*, **83**(11), 115413.

Goniakowski, Jacek and Noguera, Claudine (2014). Conditions for electronic reconstruction at stoichiometric polar/polar interfaces. *Journal of Physics: Condensed Matter*, **26**(48), 485010.

Goniakowski, Jacek and Noguera, Claudine (2016). Insulating oxide surfaces and nanostructures. *Comptes Rendus Physique*, **17**(3-4), 471–480.

Goniakowski, Jacek and Noguera, Claudine (2018). Intrinsic properties of pure and mixed monolayer oxides in the honeycomb structure: M_2O_3 and $MM'O_3$ (M, M'= Ti, V, Cr, Fe). *The Journal of Physical Chemistry C*, **123**(13), 7898–7910.

Goniakowski, Jacek and Noguera, Claudine (2019). Properties of M_2O_3/Au(111) honeycomb monolayers (M = Sc, Ti, V, Cr, Mn, Fe, Co, Ni). *The Journal of*

Physical Chemistry C, **123**, 9272–9281.

Goniakowski, Jacek, Noguera, Claudine, and Giordano, Livia (2004). Using polarity for engineering oxide nanostructures: Structural phase diagram in free and supported MgO(111) ultrathin films. *Physical Review Letters*, **93**(21), 215702.

Goniakowski, Jacek, Noguera, Claudine, and Giordano, Livia (2007*b*). Prediction of uncompensated polarity in ultrathin films. *Physical Review Letters*, **98**(20), 205701.

Goniakowski, Jacek, Noguera, Claudine, Giordano, Livia, and Pacchioni, Gianfranco (2009). Adsorption of metal adatoms on FeO(111) and MgO(111) monolayers: Effects of charge state of adsorbate on rumpling of supported oxide film. *Physical Review B*, **80**(12), 125403.

Goodenough, John B (1963). *Magnetism and Chemical Bond*. Volume 1. Interscience Publishers, Chichester.

Goodenough, John B (1971). Metallic oxides. *Progress in Solid State Chemistry*, **5**, 145–399.

Goodenough, John B and Huang, Yun-Hui (2007). Alternative anode materials for solid oxide fuel cells. *Journal of Power Sources*, **173**(1), 1–10.

Gordon, Tamar, Perlstein, Benny, Houbara, Ofir, Felner, Israel, Banin, Ehud, and Margel, Shlomo (2011). Synthesis and characterization of zinc/iron oxide composite nanoparticles and their antibacterial properties. *Colloids and Surfaces A: Physicochemical and Engineering Aspects*, **374**(1-3), 1–8.

Gragnaniello, L, Agnoli, S, Parteder, G, Barolo, A, Bondino, F, Allegretti, F, Surnev, S, Granozzi, G, and Netzer, FP (2010). Cobalt oxide nanolayers on Pd(100): The thickness-dependent structural evolution. *Surface Science*, **604**(21-22), 2002–2011.

Gragnaniello, L, Barcaro, G, Sementa, L, Allegretti, F, Parteder, G, Surnev, S, Steurer, W, Fortunelli, A, and Netzer, FP (2011). The two-dimensional cobalt oxide (9 × 2) phase on Pd(100). *The Journal of Chemical Physics*, **134**(18), 184706.

Gragnaniello, Luca, Ma, Teng, Barcaro, Giovanni, Sementa, Luca, Negreiros, Fabio R, Fortunelli, Alessandro, Surnev, Svetlozar, and Netzer, Falko P (2012). Ordered arrays of size-selected oxide nanoparticles. *Physical Review Letters*, **108**(19), 195507.

Grahame, David C (1947). The electrical double layer and the theory of electrocapillarity. *Chemical Reviews*, **41**(3), 441–501.

Granqvist, Claes G (1995). *Handbook of Inorganic Electrochromic Materials*. Elsevier, Amsterdam, Oxford, Cambridge.

Granqvist, Claes G (2003). Solar energy materials. *Advanced Materials*, **15**(21), 1789–1803.

Granqvist, Claes-Göran (2011). Transparent conducting and chromogenic oxide films as solar energy materials. *Oxide Ultrathin Films: Science and Technology*, 221–238.

Granqvist, Claes G (2012). Oxide electrochromics: An introduction to devices and materials. *Solar Energy Materials and Solar Cells*, **99**, 1–13.

Granqvist, Claes G (2014). Electrochromics for smart windows: Oxide-based thin films and devices. *Thin Solid Films*, **564**, 1–38.

Greene, Joseph E (2017). Tracing the recorded history of thin-film sputter deposition: From the 1800s to 2017. *Journal of Vacuum Science & Technology A: Vacuum, Surfaces, and Films*, **35**(5), 05C204.

Greiner, Mark T, Chai, Lily, Helander, Michael G, Tang, Wing-Man, and Lu, Zheng-

Hong (2012). Transition metal oxide work functions: The influence of cation oxidation state and oxygen vacancies. *Advanced Functional Materials*, **22**(21), 4557–4568.

Grimme, Stefan (2006). Semiempirical GGA-type density functional constructed with a long-range dispersion correction. *Journal of Computational Chemistry*, **27**(15), 1787–1799.

Grinter, David C and Thornton, Geoff (2011). Characterization tools of ultrathin oxide films. In *Oxide Ultrathin Films: Science and Technology*, pp. 27–46. Wiley Online Library.

Gross, Eberhard KU and Runge, Erich (1986). *Many-Particle Theory*. Teubner, Stuttgart.

Gross, E. K. U. and Burke, K. (2006). Basics of time-dependent density functional theory. *Lecture Notes in Physics*, **706**, 1–17.

Guggenheim, EA (1937). The theoretical basis of Raoult/s law. *Transactions of the Faraday Society*, **33**, 151–156.

Güller, Francisco, Llois, Ana Maria, Goniakowski, J, and Noguera, Claudine (2013). Polarity effects in unsupported polar nanoribbons. *Physical Review B*, **87**(20), 205423.

Guo, Q, Xu, C, and Goodman, DW (1998). Ultrathin films of NiO on MgO(100): Studies of the oxide–oxide interface. *Langmuir*, **14**(6), 1371–1374.

Guo, Yuzheng, Clark, Stewart J, and Robertson, John (2012). Electronic and magnetic properties of Ti_2O_3, Cr_2O_3, and Fe_2O_3 calculated by the screened exchange hybrid density functional. *Journal of Physics: Condensed Matter*, **24**(32), 325504.

Hacquart, Romain and Jupille, Jacques (2007). Hydrated MgO smoke crystals from cubes to octahedra. *Chemical Physics Letters*, **439**(1-3), 91–94.

Haeni, JH, Irvin, P, Chang, W, Uecker, Reinhard, Reiche, P, Li, YL, Choudhury, S, Tian, W, Hawley, ME, Craigo, B et al. (2004). Room-temperature ferroelectricity in strained $SrTiO_3$. *Nature*, **430**(7001), 758.

Haghiri-Gosnet, AM, Arnal, T, Soulimane, R, Koubaa, M, and Renard, JP (2004). Spintronics: Perspectives for the half-metallic oxides. *Physica Status Solidi (A)*, **201**(7), 1392–1397.

Hallil, A, Tétot, R, Berthier, F, Braems, I, and Creuze, J (2006). Use of a variable-charge interatomic potential for atomistic simulations of bulk, oxygen vacancies, and surfaces of rutile TiO_2. *Physical Review B*, **73**(16), 165406.

Han, Xiguang, Zhou, Xi, Jiang, Yaqi, and Xie, Zhaoxiong (2012). The preparation of spiral ZnO nanostructures by top-down wet-chemical etching and their related properties. *Journal of Materials Chemistry*, **22**(21), 10924–10928.

Hanke, W and Sham, LJ (1979). Many-particle effects in the optical excitations of a semiconductor. *Physical Review Letters*, **43**(5), 387.

Hanlon, Damien, Backes, Claudia, Higgins, Thomas M, Hughes, Marguerite, O'Neill, Arlene, King, Paul, McEvoy, Niall, Duesberg, Georg S, Mendoza Sanchez, Beatriz, Pettersson, Henrik et al. (2014). Production of molybdenum trioxide nanosheets by liquid exfoliation and their application in high-performance supercapacitors. *Chemistry of Materials*, **26**(4), 1751–1763.

Hao, Weiduo, Flynn, Shannon L, Alessi, Daniel S, and Konhauser, Kurt O (2018). Change of the point of zero net proton charge (pHPZNPC) of clay minerals with ionic strength. *Chemical Geology*, **493**, 458–467.

Harl, Judith and Kresse, Georg (2009). Accurate bulk properties from approximate many-body techniques. *Physical Review Letters*, **103**(5), 056401.

Harrison, Walter A (2012). *Electronic Structure and the Properties of Solids: The Physics of the Chemical Bond*. Dover Publications, Inc., New York.

Hartman, P (1973). Structure and morphology. *Crystal Growth: An Introduction*, **1**, 367–402.

Hartman, Piet and Perdok, WG (1955). On the relations between structure and morphology of crystals. I. *Acta Crystallographica*, **8**(1), 49–52.

Hävecker, Michael, Wrabetz, Sabine, Kröhnert, Jutta, Csepei, Lenard-Istvan, d'Alnoncourt, Raoul Naumann, Kolen'ko, Yury V, Girgsdies, Frank, Schlögl, Robert, and Trunschke, Annette (2012). Surface chemistry of phase-pure M1 MoVTeNb oxide during operation in selective oxidation of propane to acrylic acid. *Journal of Catalysis*, **285**(1), 48–60.

Havelia, Sarthak, Wang, Shanling, Balasubramaniam, KR, Schultz, Andrew M, Rohrer, Gregory S, and Salvador, Paul A (2013). Combinatorial substrate epitaxy: A new approach to growth of complex metastable compounds. *CrystEngComm*, **15**(27), 5434–5441.

Hayun, Shmuel, Shvareva, Tatiana Y, and Navrotsky, Alexandra (2011). Nanoceria– energetics of surfaces, interfaces and water adsorption. *Journal of the American Ceramic Society*, **94**(11), 3992–3999.

He, Tao, Pan, Fengchun, Xi, Zexiao, Zhang, Xuejuan, Zhang, Hongyu, Wang, Zhenhai, Zhao, Mingwen, Yan, Shishen, and Xia, Yueyuan (2010). First-principles study of titania nanoribbons: Formation, energetics, and electronic properties. *The Journal of Physical Chemistry C*, **114**(20), 9234–9238.

He, Xiao-Peng and Tian, He (2016). Photoluminescence architectures for disease diagnosis: From graphene to thin-layer transition metal dichalcogenides and oxides. *Small*, **12**(2), 144–160.

Hedin, L (1965). New method for calculating 1-particle Green's function with application to electron-gas problem. *Physical Review*, **139**(3A), A796+.

Hedin, Lars and Lundqvist, Stig (1970). Effects of electron-electron and electron-phonon interactions on the one-electron states of solids. In *Solid State Physics*, Volume 23. Elsevier.

Henderson, Michael A (2002). The interaction of water with solid surfaces: Fundamental aspects revisited. *Surface Science Reports*, **46**(1-8), 1–308.

Henrich, Victor E and Cox, Paul Anthony (1996). *The Surface Science of Metal Oxides*. Cambridge University Press, Cambridge, New York, Melbourne.

Henry, Marc, Jolivet, Jean Pierre, and Livage, Jacques (1992). Aqueous chemistry of metal cations: Hydrolysis, condensation and complexation. In *Chemistry, Spectroscopy and Applications of Sol-Gel Glasses*, pp. 153–206. Springer.

Herklotz, A, Rus, SF, Kc, S, Cooper, VR, Huon, A, Guo, E-J, and Ward, TZ (2017). Symmetry driven control of optical properties in WO_3 films. *APL Materials*, **5**(6), 066106.

Heyd, Jochen, Scuseria, Gustavo E, and Ernzerhof, Matthias (2003). Hybrid functionals based on a screened Coulomb potential. *The Journal of Chemical*

Physics, **118**(18), 8207–8215.

Hiemstra, T, De Wit, JCM, and Van Riemsdijk, WH (1989*a*). Multisite proton adsorption modeling at the solid/solution interface of (hydr)oxides: A new approach: II. Application to various important (hydr)oxides. *Journal of Colloid and Interface Science*, **133**(1), 105 – 117.

Hiemstra, T, Van Riemsdijk, WH, and Bolt, GH (1989*b*). Multisite proton adsorption modeling at the solid/solution interface of (hydr)oxides: A new approach: I. Model description and evaluation of intrinsic reaction constants. *Journal of Colloid and Interface Science*, **133**(1), 91–104.

Hiemstra, Tjisse and Van Riemsdijk, Willem H (1996). A surface structural approach to ion adsorption: The charge distribution (CD) model. *Journal of Colloid and Interface Science*, **179**(2), 488–508.

Hiley, Craig I and Walton, Richard I (2016). Controlling the crystallisation of oxide materials by solvothermal chemistry: Tuning composition, substitution and morphology of functional solids. *CrystEngComm*, **18**(40), 7656–7670.

Hisatomi, Takashi, Kubota, Jun, and Domen, Kazunari (2014). Recent advances in semiconductors for photocatalytic and photoelectrochemical water splitting. *Chemical Society Reviews*, **43**(22), 7520–7535.

Hisatomi, Takashi, Minegishi, Tsutomu, and Domen, Kazunari (2012). Kinetic assessment and numerical modeling of photocatalytic water splitting toward efficient solar hydrogen production. *Bulletin of the Chemical Society of Japan*, **85**(6), 647–655.

Hohenberg, Pierre and Kohn, Walter (1964). Inhomogeneous electron gas. *Physical Review*, **136**(3B), B864.

Honkala, Karoliina (2014). Tailoring oxide properties: An impact on adsorption characteristics of molecules and metals. *Surface Science Reports*, **69**(4), 366–388.

Hotta, Y, Susaki, T, and Hwang, HY (2007). Polar discontinuity doping of the $LaVO_3/SrTiO_3$ interface. *Physical Review Letters*, **99**(23), 236805.

Hou, Wan-Guo and Song, Shu-E (2004). Intrinsic surface reaction equilibrium constants of structurally charged amphoteric hydrotalcite-like compounds. *Journal of Colloid and Interface Science*, **269**(2), 381–387.

Hsu, Alexander, Liu, Fangzhou, Leung, Yu Hang, Ma, Angel PY, Djurišić, Aleksandra B, Leung, Frederick CC, Chan, Wai Kin, and Lee, Hung Kay (2014). Is the effect of surface modifying molecules on antibacterial activity universal for a given material? *Nanoscale*, **6**(17), 10323–10331.

Hu, Jun, Xiao, X-D, Ogletree, DF, and Salmeron, M (1995). Imaging the condensation and evaporation of molecularly thin films of water with nanometer resolution. *Science*, **268**(5208), 267–269.

Hu, Xiao Liang and Michaelides, Angelos (2010). The kaolinite (001) polar basal plane. *Surface Science*, **604**(2), 111–117.

Hu, Zhimi, Xiao, Xu, Jin, Huanyu, Li, Tianqi, Chen, Ming, Liang, Zhun, Guo, Zhengfeng, Li, Jia, Wan, Jun, Huang, Liang et al. (2017). Rapid mass production of two-dimensional metal oxides and hydroxides via the molten salts method. *Nature Communications*, **8**, 15630.

Huang, Bevin, Clark, Genevieve, Navarro-Moratalla, Efrén, Klein, Dahlia R, Cheng, Ran, Seyler, Kyle L, Zhong, Ding, Schmidgall, Emma, McGuire, Michael A, Cobden,

David H et al. (2017). Layer-dependent ferromagnetism in a van der Waals crystal down to the monolayer limit. *Nature*, **546**(7657), 270.

Huang, Liangliang, Gubbins, Keith E, Li, Licheng, and Lu, Xiaohua (2014). Water on titanium dioxide surface: A revisiting by reactive molecular dynamics simulations. *Langmuir*, **30**(49), 14832–14840.

Hubbard, J (1963). Electron correlations in narrow energy bands. *Proceedings of the Royal Society of London A: Mathematical, Physical and Engineering Sciences*, **276**(1365), 238–257.

Hubbard, J (1964). Electron correlations in narrow energy bands. II. The degenerate band case. *Proceedings of the Royal Society of London A: Mathematical, Physical and Engineering Sciences*, **277**(1369), 237–259.

Huczko, Appl (2000). Template-based synthesis of nanomaterials. *Applied Physics A*, **70**(4), 365–376.

Hüfner, Stephan (2013). *Photoelectron Spectroscopy: Principles and Applications*. Springer Science & Business Media, Berlin, Heidelberg, New York, Barcelona, Hong Kong, London, Milan, Paris, Tokyo.

Hwang, Harold Y, Iwasa, Yoh, Kawasaki, Masashi, Keimer, Bernhard, Nagaosa, Naoto, and Tokura, Yoshinori (2012). Emergent phenomena at oxide interfaces. *Nature Materials*, **11**(2), 103.

Hybertsen, Mark S and Louie, Steven G (1986). Electron correlation in semiconductors and insulators: Band gaps and quasiparticle energies. *Physical Review B*, **34**(8), 5390.

Hybertsen, Mark S and Louie, Steven G (1988). Model dielectric matrices for quasiparticle self-energy calculations. *Physical Review B*, **37**(5), 2733.

Ibach, Harald and Mills, Douglas L (2013). *Electron Energy Loss Spectroscopy and Surface Vibrations*. Academic Press, Boston, San Diego, New York, London, Sydney, Tokyo, Toronto.

Idriss, Hicham and Barteau, Mark A (2000). Active sites on oxides: From single crystals to catalysts. In *Advances in Catalysis*, Volume 45, pp. 261–331. Elsevier.

Imhoff, D, Laurent, S, Colliex, C, and Backhaus-Ricoult, M (1999). Determination of the characteristic interfacial electronic states of {111} Cu-MgO interfaces by ELNES. *The European Physical Journal-Applied Physics*, **5**(1), 9–18.

Ingle, NJC and Elfimov, IS (2008). Influence of epitaxial strain on the ferromagnetic semiconductor EuO: First-principles calculations. *Physical Review B*, **77**(12), 121202.

Ishigami, K, Yoshimatsu, K, Toyota, D, Takizawa, M, Yoshida, T, Shibata, Goro, Harano, T, Takahashi, Y, Kadono, T, Verma, VK et al. (2015). Thickness-dependent magnetic properties and strain-induced orbital magnetic moment in $SrRuO_3$ thin films. *Physical Review B*, **92**(6), 064402.

Ishihara, Tatsumi (2012). Oxide ultrathin films for solid oxide fuel cells. In *Oxide Ultrathin Films*, pp. 201–220. Wiley-VCH.

Israelachvili, Jacob N (2011). *Intermolecular and Surface Forces*. Academic Press, Boston, San Diego, New York, London, Sydney, Tokyo, Toronto.

Jacobson, Allan J (2009). Materials for solid oxide fuel cells. *Chemistry of Materials*, **22**(3), 660–674.

Janáky, C, Rajeshwar, K, De Tacconi, NR, Chanmanee, W, and Huda, MN (2013). Tungsten-based oxide semiconductors for solar hydrogen generation. *Catalysis Today*, **199**, 53–64.

Jelle, Bjørn Petter, Hynd, Andrew, Gustavsen, Arild, Arasteh, Dariush, Goudey, Howdy, and Hart, Robert (2012). Fenestration of today and tomorrow: A state-of-the-art review and future research opportunities. *Solar Energy Materials and Solar Cells*, **96**, 1–28.

Jerratsch, Jan Frederik, Nilius, Niklas, Freund, Hans-Joachim, Martinez, Umberto, Giordano, Livia, and Pacchioni, Gianfranco (2009). Lithium incorporation into a silica thin film: Scanning tunneling microscopy and density functional theory. *Physical Review B*, **80**(24), 245423.

Jiang, Zhi-Yuan, Kuang, Qin, Xie, Zhao-Xiong, and Zheng, Lan-Sun (2010). Syntheses and properties of micro/nanostructured crystallites with high-energy surfaces. *Advanced Functional Materials*, **20**(21), 3634–3645.

Johnson, Alex C, Lai, Bo-Kuai, Xiong, Hui, and Ramanathan, Shriram (2009). An experimental investigation into micro-fabricated solid oxide fuel cells with ultra-thin $La_{0.6}Sr_{0.4}Co_{0.8}Fe_{0.2}O_{0.3}$ cathodes and yttria-doped zirconia electrolyte films. *Journal of Power Sources*, **186**(2), 252–260.

Johnson, Richard W, Hultqvist, Adam, and Bent, Stacey F (2014). A brief review of atomic layer deposition: From fundamentals to applications. *Materials Today*, **17**(5), 236–246.

Johnston, Roy L (2003). Evolving better nanoparticles: Genetic algorithms for optimising cluster geometries. *Dalton Transactions*, **22**, 4193–4207.

Jolivet, Jean-Pierre (2019). *Metal Oxide Nanostructures Chemistry: Synthesis from Aqueous Solutions*. Oxford University Press, New York.

Jolivet, Jean-Pierre, Froidefond, Cédric, Pottier, Agnès, Chanéac, Corinne, Cassaignon, Sophie, Tronc, Elisabeth, and Euzen, Patrick (2004). Size tailoring of oxide nanoparticles by precipitation in aqueous medium. A semi-quantitative modelling. *Journal of Materials Chemistry*, **14**(21), 3281–3288.

Jones, RO and Gunnarsson, O (1989). The density functional formalism, its applications and prospects. *Reviews of Modern Physics*, **61**, 689–746.

Jubera, Mariano, Elvira, I, García-Cabañes, A, Bella, JL, and Carrascosa, Mercedes (2016). Trapping and patterning of biological objects using photovoltaic tweezers. *Applied Physics Letters*, **108**(2), 023703.

Julliere, Michel (1975). Tunneling between ferromagnetic films. *Physics Letters A*, **54**(3), 225–226.

Junquera, Javier and Ghosez, Philippe (2003). Critical thickness for ferroelectricity in perovskite ultrathin films. *Nature*, **422**(6931), 506.

Kalantar-zadeh, Kourosh, Ou, Jian Zhen, Daeneke, Torben, Mitchell, Arnan, Sasaki, Takayoshi, and Fuhrer, Michael S (2016). Two dimensional and layered transition metal oxides. *Applied Materials Today*, **5**, 73–89.

Kalantar-Zadeh, Kourosh, Tang, Jianshi, Wang, Minsheng, Wang, Kang L, Shailos, Alexandros, Galatsis, Kosmas, Kojima, Robert, Strong, Veronica, Lech, Andrew, Wlodarski, Wojtek et al. (2010). Synthesis of nanometre-thick MoO_3 sheets. *Nanoscale*, **2**(3), 429–433.

Kalantar-zadeh, Kourosh, Vijayaraghavan, Aravind, Ham, Moon-Ho, Zheng, Haidong, Breedon, Michael, and Strano, Michael S (2010). Synthesis of atomically thin WO_3 sheets from hydrated tungsten trioxide. *Chemistry of Materials*, **22**(19), 5660–5666.

Kalinin, Sergei V and Bonnell, Dawn A (2001). Local potential and polarization screening on ferroelectric surfaces. *Physical Review B*, **63**(12), 125411.

Kan, Erjun, Li, Ming, Hu, Shuanglin, Xiao, Chuanyun, Xiang, Hongjun, and Deng, Kaiming (2013). Two-dimensional hexagonal transition-metal oxide for spintronics. *The Journal of Physical Chemistry Letters*, **4**(7), 1120–1125.

Kasemo, Bengt (2002). Biological surface science. *Surface Science*, **500**(1-3), 656–677.

Kawakami, Hiroshi and Yoshida, Satohiro (1985). Quantum-chemical studies of alumina. Part 2.—Lewis acidity. *Journal of the Chemical Society, Faraday Transactions 2: Molecular and Chemical Physics*, **81**(7), 1129–1137.

Kelly, Peter J and Arnell, R Derek (2000). Magnetron sputtering: A review of recent developments and applications. *Vacuum*, **56**(3), 159–172.

Kevan, Stephen Douglas (1992). *Angle-Resolved Photoemission: Theory and Current Applications*. Volume 74. Elsevier, Amsterdam, London, New York, Tokyo.

Khorshidi, Alireza and Peterson, Andrew A (2016). Amp: A modular approach to machine learning in atomistic simulations. *Computer Physics Communications*, **207**, 310–324.

Kiguchi, Manabu, Entani, Shiro, Saiki, Koichiro, Goto, Takayuki, and Koma, Atsushi (2003). Atomic and electronic structure of an unreconstructed polar MgO(111) thin film on Ag(111). *Physical Review B*, **68**(11), 115402.

Kiguchi, M, Goto, T, Saiki, K, Sasaki, T, Iwasawa, Y, and Koma, A (2002). Atomic and electronic structures of MgO/Ag(001) heterointerface. *Surface Science*, **512**(1-2), 97–106.

Kim, Bongjae, Liu, Peitao, Tomczak, Jan M, and Franchini, Cesare (2018). Strain-induced tuning of the electronic Coulomb interaction in 3d transition metal oxide perovskites. *Physical Review B*, **98**(7), 075130.

Kim, YJ, Gao, Y, and Chambers, SA (1997*a*). Core-level X-ray photoelectron spectra and X-ray photoelectron diffraction of RuO_2(110) grown by molecular beam epitaxy on TiO_2(110). *Applied Surface Science*, **120**(3-4), 250–260.

Kim, YD, Stultz, J, and Goodman, DW (2002). Dissociation of water on MgO(100). *The Journal of Physical Chemistry B*, **106**(7), 1515–1517.

Kim, YJ, Westphal, C, Ynzunza, RX, Galloway, HC, Salmeron, M, Van Hove, MA, and Fadley, CS (1997*b*). Interlayer interactions in epitaxial oxide growth: FeO on Pt(111). *Physical Review B*, **55**(20), R13448.

Kleibeuker, JE, Zhong, Z, Nishikawa, H, Gabel, J, Müller, A, Pfaff, F, Sing, M, Held, K, Claessen, R, Koster, Gertjan et al. (2014). Electronic reconstruction at the isopolar $LaTiO_3$/$LaFeO_3$ interface: An X-ray photoemission and density-functional theory study. *Physical Review Letters*, **113**(23), 237402.

Klimeš, J, Bowler, David R, and Michaelides, Angelos (2011). Van der Waals density functionals applied to solids. *Physical Review B*, **83**, 195131.

Klimeš, J and Michaelides, A (2012). Perspective: Advances and challenges in treating van der Waals dispersion forces in density functional theory. *The Journal of Chemical Physics*, **137**(12), 120901.

Knözinger, Eric, Jacob, Karl-Heinz, and Hofmann, Peter (1993*a*). Adsorption of hydrogen on highly dispersed MgO. *Journal of the Chemical Society, Faraday Transactions*, **89**(7), 1101–1107.

Knözinger, Erich, Jacob, Karl-Heinz, Singh, Surjit, and Hofmann, Peter (1993*b*). Hydroxyl groups as IR active surface probes on MgO crystallites. *Surface Science*, **290**(3), 388–402.

Knözinger, H and Ratnasamy, Ph (1978). Catalytic aluminas: Surface models and characterization of surface sites. *Catalysis Reviews Science and Engineering*, **17**(1), 31–70.

Kohn, Walter and Sham, Lu Jeu (1965). Self-consistent equations including exchange and correlation effects. *Physical Review*, **140**(4A), A1133.

Kohn, Walter and Sherrill, C. David (2014). Editorial: Reflections on fifty years of density functional theory. *The Journal of Chemical Physics*, **140**(18), 18A201.

Köksal, Okan, Baidya, Santu, and Pentcheva, Rossitza (2018). Confinement-driven electronic and topological phases in corundum-derived 3d-oxide honeycomb lattices. *Physical Review B*, **97**(3), 035126.

Kolsbjerg, EL, Peterson, AA, and Hammer, B (2018). Neural-network-enhanced evolutionary algorithm applied to supported metal nanoparticles. *Physical Review B*, **97**(19), 195424.

Koningsberger, DC and Prins, R (1988). *X-ray Absorption: Principles, Applications, Techniques of EXAFS, SEXAFS, and XANES*. Volume 92. John Wiley and Sons, New York.

Korotcenkov, Ghenadii (2018). *Metal Oxides in Energy Technologies*. Elsevier, Amsterdam, Oxford, Cambridge.

Kosmulski, Marek (2018). The pH dependent surface charging and points of zero charge. VII. Update. *Advances in Colloid and Interface Science*, **251**, 115–138.

Kostov, KL, Schumann, FO, Polzin, S, Sander, D, and Widdra, W (2016). NiO growth on Ag(001): A layer-by-layer vibrational study. *Physical Review B*, **94**(7), 075438.

Kou, Liangzhi, Li, Chun, Zhang, Zhuhua, and Guo, Wanlin (2009). Electric-field-and hydrogen-passivation-induced band modulations in armchair ZnO nanoribbons. *The Journal of Physical Chemistry C*, **114**(2), 1326–1330.

Kou, Liangzhi, Li, Chun, Zhang, Zhuhua, and Guo, Wanlin (2010). Tuning magnetism in zigzag ZnO nanoribbons by transverse electric fields. *ACS Nano*, **4**(4), 2124–2128.

Kresse, Georg, Schmid, Michael, Napetschnig, Evelyn, Shishkin, Maxim, Köhler, Lukas, and Varga, Peter (2005). Structure of the ultrathin aluminum oxide film on NiAl (110). *Science*, **308**(5727), 1440–1442.

Kudo, Akihiko and Miseki, Yugo (2009). Heterogeneous photocatalyst materials for water splitting. *Chemical Society Reviews*, **38**(1), 253–278.

Kuhness, D, Pomp, S, Mankad, V, Barcaro, G, Sementa, L, Fortunelli, A, Netzer, FP, and Surnev, S (2016). Two-dimensional iron oxide bi-and trilayer structures on Pd(100). *Surface Science*, **645**, 13–22.

Kundu, Asish K, Barman, Sukanta, and Menon, Krishnakumar SR (2018). Evolution of surface antiferromagnetic Néel temperature with film coverage in ultrathin MnO films on Ag(001). *Journal of Magnetism and Magnetic Materials*, **466**, 186–191.

Kung, Harold H (1989). *Transition Metal Oxides: Surface Chemistry and Catalysis*. Volume 45. Elsevier, Amsterdam, Oxford, Cambridge.

Lagaly, G and Malberg, R (1990). Disaggregation of alkylammonium montmorillonites in organic solvents. *Colloids and Surfaces*, **49**, 11–27.

Lagaly, G and Ziesmer, S (2003). Colloid chemistry of clay minerals: The coagulation of montmorillonite dispersions. *Advances in Colloid and Interface Science*, **100**, 105–128.

Lambin, Ph, Senet, Patrick, and Lucas, AA (1991). Validity of the dielectric approximation in describing electron-energy-loss spectra of surface and interface phonons in thin films of ionic crystals. *Physical Review B*, **44**(12), 6416.

Lamiel-Garcia, Oriol, Cuko, Andi, Calatayud, Monica, Illas, Francesc, and Bromley, Stefan T (2017). Predicting size-dependent emergence of crystallinity in nanomaterials: Titania nanoclusters versus nanocrystals. *Nanoscale*, **9**(3), 1049–1058.

Le, H-LT, Goniakowski, Jacek, and Noguera, C (2018). Properties of mixed transition metal oxides: MM'O$_3$ in corundum-type structures (M, M'= Al, Ti, V, Cr, and Fe). *Physical Review Materials*, **2**(8), 085001.

Le, H-LT, Goniakowski, Jacek, and Noguera, C (2019). (0001) interfaces between M$_2$O$_3$ corundum oxides (M= Al, Ti, V, Cr, Fe). *Surface Science*, **679**, 17–23.

Lei, Lijun, Wu, Zhiwei, Liu, Huan, Qin, Zhangfeng, Chen, Chengmeng, Luo, Li, Wang, Guofu, Fan, Weibin, and Wang, Jianguo (2018). A facile method for the synthesis of graphene-like 2-D metal oxides and their excellent catalytic application in the hydrogenation of nitroarenes. *Journal of Materials Chemistry A*, **6**(21), 9948–9961.

Leisenberger, FP, Surnev, S, Koller, G, Ramsey, MG, and Netzer, FP (2000). Probing the metal sites of a vanadium oxide–Pd(111) 'inverse catalyst': Adsorption of CO. *Surface Science*, **444**(1-3), 211–220.

Levin, Igor and Brandon, David (1998). Metastable alumina polymorphs: Crystal structures and transition sequences. *Journal of the American Ceramic Society*, **81**(8), 1995–2012.

Li, F, Parteder, G, Allegretti, F, Franchini, Cesare, Podloucky, R, Surnev, S, and Netzer, FP (2009a). Two-dimensional manganese oxide nanolayers on Pd(100): The surface phase diagram. *Journal of Physics: Condensed Matter*, **21**(13), 134008.

Li, Ju, Shan, Zhiwei, and Ma, Evan (2014a). Elastic strain engineering for unprecedented materials properties. *MRS Bulletin*, **39**(2), 108–114.

Li, Q, Liang, JH, Luo, YM, Ding, Z, Gu, T, Hu, Z, Hua, CY, Lin, H-J, Pi, TW, Kang, SP et al. (2016). Antiferromagnetic proximity effect in epitaxial CoO/NiO/MgO(001) systems. *Scientific Reports*, **6**, 22355.

Li, Rong and Cheng, Longjiu (2012). Structural determination of (Al$_2$O$_3$)$_n$ (n= 1–7) clusters based on density functional calculation. *Computational and Theoretical Chemistry*, **996**, 125–131.

Li, Rengui, Han, Hongxian, Zhang, Fuxiang, Wang, Donge, and Li, Can (2014b). Highly efficient photocatalysts constructed by rational assembly of dual-cocatalysts separately on different facets of BiVO$_4$. *Energy & Environmental Science*, **7**(4), 1369–1376.

Li, Wei, He, Da, Hu, Guoxiang, Li, Xiang, Banerjee, Gourab, Li, Jingyi, Lee, Shin Hee, Dong, Qi, Gao, Tianyue, Brudvig, Gary W et al. (2018). Selective CO production by photoelectrochemical methane oxidation on TiO_2. *ACS Central Science*, **4**(5), 631–637.

Li, Xuesong, Cai, Weiwei, An, Jinho, Kim, Seyoung, Nah, Junghyo, Yang, Dongxing, Piner, Richard, Velamakanni, Aruna, Jung, Inhwa, Tutuc, Emanuel et al. (2009*b*). Large-area synthesis of high-quality and uniform graphene films on copper foils. *Science*, **324**(5932), 1312–1314.

Li, XD, Chen, TP, Liu, P, Liu, Y, and Leong, KC (2013). Effects of free electrons and quantum confinement in ultrathin ZnO films: A comparison between undoped and Al-doped ZnO. *Optics Express*, **21**(12), 14131–14138.

Li, Yueliang, Yu, Rong, Zhou, Huihua, Cheng, Zhiying, Wang, Xiaohui, Li, Longtu, and Zhu, Jing (2015). Direct observation of thickness dependence of ferroelectricity in freestanding $BaTiO_3$ thin film. *Journal of the American Ceramic Society*, **98**(9), 2710–2712.

Li, Zhenjun, Zhang, Zhenrong, Kim, Yu Kwon, Smith, R Scott, Netzer, Falko, Kay, Bruce D, Rousseau, Roger, and Dohnalek, Zdenek (2011). Growth of ordered ultrathin tungsten oxide films on Pt(111). *The Journal of Physical Chemistry C*, **115**(13), 5773–5783.

Liang, Lin, Li, Kun, Xiao, Chong, Fan, Shaojuan, Liu, Jiao, Zhang, Wenshuai, Xu, Wenhui, Tong, Wei, Liao, Jiaying, Zhou, Yingying et al. (2015). Vacancy associates-rich ultrathin nanosheets for high performance and flexible nonvolatile memory device. *Journal of the American Chemical Society*, **137**(8), 3102–3108.

Liang, XiaoPing, Tan, FeiHu, Wei, Feng, and Du, Jun (2019). Research progress of all solid-state thin film lithium battery. In *IOP Conference Series: Earth and Environmental Science*, Volume 218, p. 012138. IOP Publishing.

Lichtensteiger, C, Zubko, P, Stengel, M, Aguado-Puente, P, Triscone, J-M, Ghosez, Philippe, and Junquera, J (2012). Ferroelectricity in ultrathin film capacitors. *Oxide Ultrathin Films: Science and Technology*, 265–308.

Lichtenstein, AI and Katsnelson, MI (1998). Ab initio calculations of quasiparticle band structure in correlated systems: LDA++ approach. *Physical Review B*, **57**(12), 6884.

Lide, David R et al. (2012). *CRC Handbook of Chemistry and Physics*. CRC Press, Boca Raton, London, New York, Washington, D.C.

Lin, Yongjing, Yuan, Guangbi, Liu, Rui, Zhou, Sa, Sheehan, Stafford W, and Wang, Dunwei (2011). Semiconductor nanostructure-based photoelectrochemical water splitting: A brief review. *Chemical Physics Letters*, **507**(4-6), 209–215.

Ling, Sanliang, Watkins, Matthew B, and Shluger, Alexander L (2013). Effects of oxide roughness at metal oxide interface: MgO on Ag(001). *The Journal of Physical Chemistry C*, **117**(10), 5075–5083.

Lorenz, Manuel, Rao, MS Ramachandra, Venkatesan, Thirumalai, Fortunato, Elvira, Barquinha, Pedro, Branquinho, Rita, Salgueiro, Daniela, Martins, Rodrigo, Carlos, Emanuel, Liu, Ao et al. (2016). The 2016 oxide electronic materials and oxide interfaces roadmap. *Journal of Physics D: Applied Physics*, **49**(43), 433001.

Lu, Chengliang, Hu, Weijin, Tian, Yufeng, and Wu, Tom (2015). Multiferroic oxide thin films and heterostructures. *Applied Physics Reviews*, **2**(2), 021304.

Lu, Xiong, Wang, Yingbo, Yang, Xiudong, Zhang, Qiyi, Zhao, Zhanfeng, Weng, Lu-Tao, and Leng, Yang (2008). Spectroscopic analysis of titanium surface functional groups under various surface modification and their behaviors in vitro and in vivo. *Journal of Biomedical Materials Research Part A*, **84**(2), 523–534.

Lucas, AA (1968). Phonon modes of an ionic crystal slab. *The Journal of Chemical Physics*, **48**(7), 3156–3168.

Luches, Paola, Giordano, Livia, Grillo, Vincenzo, Gazzadi, Gian Carlo, Prada, Stefano, Campanini, Marco, Bertoni, Giovanni, Magen, Cesar, Pagliuca, Federico, Pacchioni, Gianfranco et al. (2015). Atomic scale structure and reduction of cerium oxide at the interface with platinum. *Advanced Materials Interfaces*, **2**(18), 1500375.

Lundgren, Edvin, Mikkelsen, Anders, Andersen, Jesper N, Kresse, Georg, Schmid, Michael, and Varga, Peter (2006). Surface oxides on close-packed surfaces of late transition metals. *Journal of Physics: Condensed Matter*, **18**(30), R481.

Lyakhov, Andriy O, Oganov, Artem R, Stokes, Harold T, and Zhu, Qiang (2013). New developments in evolutionary structure prediction algorithm USPEX. *Computer Physics Communications*, **184**(4), 1172–1182.

Ma, Li, Wiame, Frédéric, Maurice, Vincent, and Marcus, Philippe (2018). New insight on early oxidation stages of austenitic stainless steel from in situ XPS analysis on single-crystalline $Fe_{18}Cr_{13}Ni$. *Corrosion Science*, **140**, 205–216.

Ma, Li, Wiame, Frédéric, Maurice, Vincent, and Marcus, Philippe (2019). Origin of nanoscale heterogeneity in the surface oxide film protecting stainless steel against corrosion. *npj Materials Degradation*, **3**(1), 1–9.

Ma, Renzhi and Sasaki, Takayoshi (2014). Two-dimensional oxide and hydroxide nanosheets: Controllable high-quality exfoliation, molecular assembly, and exploration of functionality. *Accounts of Chemical Research*, **48**(1), 136–143.

Macdonald, Digby D (1999). Passivity–the key to our metals-based civilization. *Pure and Applied Chemistry*, **71**(6), 951–978.

Macher, Thomas, Totenhagen, John, Sherwood, Jennifer, Qin, Ying, Gurler, Demet, Bolding, Mark S, and Bao, Yuping (2015). Ultrathin iron oxide nanowhiskers as positive contrast agents for magnetic resonance imaging. *Advanced Functional Materials*, **25**(3), 490–494.

Mackrodt, WC (1988). Atomistic simulation of oxide surfaces. *Physics and Chemistry of Minerals*, **15**(3), 228–237.

Mackrodt, WC, Davey, RJ, Black, SN, and Docherty, R (1987). The morphology of α-Al_2O_3 and α-Fe_2O_3: The importance of surface relaxation. *Journal of Crystal Growth*, **80**(2), 441–446.

Maeda, Kazuhiko (2013). Z-scheme water splitting using two different semiconductor photocatalysts. *ACS Catalysis*, **3**(7), 1486–1503.

Maier, J (2005). Nanoionics: Ion transport and electrochemical storage in confined systems. *Nature Materials*, **4**(11), 805.

Mak, Kin Fai, Lee, Changgu, Hone, James, Shan, Jie, and Heinz, Tony F (2010). Atomically thin MoS_2: A new direct-gap semiconductor. *Physical Review Letters*, **105**(13), 136805.

Maksymovych, Peter, Huijben, Mark, Pan, Minghu, Jesse, Stephen, Balke, Nina,

Chu, Ying-Hao, Chang, Hye Jung, Borisevich, Albina Y, Baddorf, Arthur P, Rijnders, Guus et al. (2012). Ultrathin limit and dead-layer effects in local polarization switching of BiFeO$_3$. *Physical Review B*, **85**(1), 014119.

Manna, PK and Yusuf, SM (2014). Two interface effects: Exchange bias and magnetic proximity. *Physics Reports*, **535**(2), 61–99.

Mannhart, Jochen and Schlom, DG (2010). Oxide interfaces—an opportunity for electronics. *Science*, **327**(5973), 1607–1611.

Marcus, Philippe and Maurice, Vincent (2011). Oxide passive films and corrosion protection. *Oxide Ultrathin Films: Science and Technology*, 119–144.

Marcus, P, Maurice, V, and Strehblow, H-H (2008). Localized corrosion (pitting): A model of passivity breakdown including the role of the oxide layer nanostructure. *Corrosion Science*, **50**(9), 2698–2704.

Marin, E, Lanzutti, A, Guzman, L, and Fedrizzi, L (2011). Corrosion protection of AISI 316 stainless steel by ALD alumina/titania nanometric coatings. *Journal of Coatings Technology and Research*, **8**(5), 655.

Markov, Ivan V (1995). *Crystal Growth for Beginners: Fundamentals of Nucleation, Crystal Growth, and Epitaxy*. World Scientific, New Jersey, London, Singapore.

Marques, Miguel AL, Ullrich, Carsten A, Nogueira, Fernando, Rubio, Angel, Burke, Kieron, and Gross, Eberhard KU (2006). *Time-dependent density functional theory*. Volume 706. Springer Science & Business Media, Berlin, Heidelberg, New York, Barcelona, Hong Kong, London, Milan, Paris, Tokyo.

Marques, Miguel AL, Vidal, Julien, Oliveira, Micael JT, Reining, Lucia, and Botti, Silvana (2011). Density-based mixing parameter for hybrid functionals. *Physical Review B*, **83**(3), 035119.

Mars, P and Van Krevelen, Do W (1954). Oxidations carried out by means of vanadium oxide catalysts. *Chemical Engineering Science*, **3**, 41–59.

Martin, Richard M, Reining, Lucia, and Ceperley, David M (2016). *Interacting Electrons*. Cambridge University Press, Cambridge, New York, Melbourne, Madrid, Cape Town, Singapore, São Paulo.

Massoud, Hisham Z, Plummer, James D, and Irene, Eugene A (1985*a*). Thermal oxidation of silicon in dry oxygen accurate determination of the kinetic rate constants. *Journal of the Electrochemical Society*, **132**(7), 1745–1753.

Massoud, Hisham Z, Plummer, James D, and Irene, Eugene A (1985*b*). Thermal oxidation of silicon in dry oxygen growth-rate enhancement in the thin regime I. Experimental results. *Journal of the Electrochemical Society*, **132**(11), 2685–2693.

Massoud, Hisham Z, Plummer, James D, and Irene, Eugene A (1985*c*). Thermal oxidation of silicon in dry oxygen: Growth-rate enhancement in the thin regime II. Physical mechanisms. *Journal of The Electrochemical Society*, **132**(11), 2693–2700.

Massoud, Toni, Maurice, Vincent, Klein, Lorena H, Seyeux, Antoine, and Marcus, Philippe (2014). Nanostructure and local properties of oxide layers grown on stainless steel in simulated pressurized water reactor environment. *Corrosion Science*, **84**, 198–203.

Matsui, Keitaro, Pradhan, Bhabendra K, Kyotani, Takashi, and Tomita, Akira (2001). Formation of nickel oxide nanoribbons in the cavity of carbon nanotubes. *The Journal of Physical Chemistry B*, **105**(24), 5682–5688.

Matsumoto, Yuji, Murakami, Makoto, Shono, Tomoji, Hasegawa, Tetsuya, Fukumura, Tomoteru, Kawasaki, Masashi, Ahmet, Parhat, Chikyow, Toyohiro, Koshihara, Shin-ya, and Koinuma, Hideomi (2001). Room-temperature ferromagnetism in transparent transition metal-doped titanium dioxide. *Science*, **291**(5505), 854–856.

Matxain, Jose M, Fowler, Joseph E, and Ugalde, Jesus M (2000). Small clusters of II-VI materials: Zn_iO_i , i= 1–9. *Physical Review A*, **62**(5), 053201.

Maurice, Vincent and Marcus, Philippe (2018). Current developments of nanoscale insight into corrosion protection by passive oxide films. *Current Opinion in Solid State and Materials Science*, **22**(4), 156–167.

Mazo, Mikhail A, Manevitch, Leonid I, Gusarova, Elena B, Shamaev, Mikhail Yu, Berlin, Alexander A, Balabaev, Nikolay K, and Rutledge, Gregory C (2008). Molecular dynamics simulation of thermomechanical properties of montmorillonite crystal. 1. isolated clay nanoplate. *The Journal of Physical Chemistry B*, **112**(10), 2964–2969.

McCabe, RW and Adams, JM (2013). Clay minerals as catalysts. In *Developments in Clay Science*, Volume 5, pp. 491–538. Elsevier.

McFarland, Eric W and Metiu, Horia (2013). Catalysis by doped oxides. *Chemical Reviews*, **113**(6), 4391–4427.

McGlashan, Maxwell Len (2007). *Chemical thermodynamics*. Volume 2. Royal Society of Chemistry, Bristol.

Meldgaard, Søren A, Kolsbjerg, Esben L, and Hammer, Bjørk (2018). Machine learning enhanced global optimization by clustering local environments to enable bundled atomic energies. *The Journal of Chemical Physics*, **149**(13), 134104.

Mermin, N David and Wagner, Herbert (1966). Absence of ferromagnetism or antiferromagnetism in one-or two-dimensional isotropic Heisenberg models. *Physical Review Letters*, **17**(22), 1133.

Mesri, Gholamreza and Olson, Roy E (1971). Mechanisms controlling the permeability of clays. *Clays and Clay minerals*, **19**(3), 151–158.

Metiu, Horia, Chrétien, Steeve, Hu, Zhenpeng, Li, Bo, and Sun, XiaoYing (2012). Chemistry of Lewis acid–base pairs on oxide surfaces. *The Journal of Physical Chemistry C*, **116**(19), 10439–10450.

Meunier, Alain (2005). *Clays*. Springer Science & Business Media, Berlin, Heidelberg, New York, Barcelona, Hong Kong, London, Milan, Paris, Tokyo.

Meunier, A (2006). Why are clay minerals small? *Clay Minerals*, **41**(2), 551–566.

Meyer, B and Vanderbilt, David (2001). Ab initio study of $BaTiO_3$ and $PbTiO_3$ surfaces in external electric fields. *Physical Review B*, **63**(20), 205426.

Michot, Laurent J, Bihannic, Isabelle, Porsch, Katharina, Maddi, Solange, Baravian, Christophe, Mougel, Julien, and Levitz, Pierre (2004). Phase diagrams of Wyoming Na-montmorillonite clay. influence of particle anisotropy. *Langmuir*, **20**(25), 10829–10837.

Migani, Annapaola, Vayssilov, Georgi N, Bromley, Stefan T, Illas, Francesc, and Neyman, Konstantin M (2010). Greatly facilitated oxygen vacancy formation in ceria nanocrystallites. *Chemical Communications*, **46**(32), 5936–5938.

Militzer, B, Wenk, H-R, Stackhouse, S, and Stixrude, L (2011). First-principles calculation of the elastic moduli of sheet silicates and their application to shale anisotropy. *American Mineralogist*, **96**(1), 125–137.

Millot, Georges (2013). *Geology of Clays: Weathering – Sedimentology – Geochemistry*. Springer Science & Business Media, Berlin, Heidelberg, New York, Barcelona, Hong Kong, London, Milan, Paris, Tokyo.

Mirhashemihaghighi, Shadi, Światowska, Jolanta, Maurice, Vincent, Seyeux, Antoine, Zanna, Sandrine, Salmi, Emma, Ritala, Mikko, and Marcus, Philippe (2016). Corrosion protection of aluminium by ultra-thin atomic layer deposited alumina coatings. *Corrosion Science*, **106**, 16–24.

Molina-Mendoza, Aday J, Lado, Jose L, Island, Joshua O, Niño, Miguel Angel, Aballe, Lucía, Foerster, Michael, Bruno, Flavio Y, López-Moreno, Alejandro, Vaquero-Garzon, Luis, van der Zant, Herre SJ et al. (2016). Centimeter-scale synthesis of ultrathin layered MoO_3 by van der Waals epitaxy. *Chemistry of Materials*, **28**(11), 4042–4051.

Monch, Winfried (1990). On the physics of metal-semiconductor interfaces. *Reports on Progress in Physics*, **53**(3), 221.

Morgan, Benjamin J (2009). Preferential stability of the d-BCT phase in ZnO thin films. *Physical Review B*, **80**(17), 174105.

Muller, DA, Shashkov, DA, Benedek, R, Yang, LH, Silcox, J, and Seidman, David N (1998). Atomic scale observations of metal-induced gap states at {222} MgO/Cu interfaces. *Physical Review Letters*, **80**(21), 4741.

Müller, Martina, Miao, Guo-Xing, and Moodera, Jagadeesh S (2009). Thickness dependence of ferromagnetic-and metal-insulator transition in thin EuO films. *Journal of Applied Physics*, **105**(7), 07C917.

Muller, P and Kern, R (2000). Equilibrium nano-shape changes induced by epitaxial stress (generalised Wulf-Kaishew theorem). *Surface Science*, **457**(1-2), 229–253.

Mulliken, Robert S (1955). Electronic population analysis on LCAO–MO molecular wave functions. i. *The Journal of Chemical Physics*, **23**(10), 1833–1840.

Myhr, OR and Grong, O (2000). Modelling of non-isothermal transformations in alloys containing a particle distribution. *Acta Materialia*, **48**(7), 1605–1615.

Myrach, Philipp, Nilius, Niklas, Levchenko, Sergey V, Gonchar, Anastasia, Risse, Thomas, Dinse, Klaus-Peter, Boatner, Lynn A, Frandsen, Wiebke, Horn, Raimund, Freund, Hans-Joachim et al. (2010). Temperature-dependent morphology, magnetic and optical properties of Li-Doped MgO. *ChemCatChem*, **2**(7), 854–862.

Nagarajan, Valanoor, Junquera, J, He, JQ, Jia, CL, Waser, R, Lee, K, Kim, YK, Baik, Sunggi, Zhao, Tong, Ramesh, Ramamoorthy et al. (2006). Scaling of structure and electrical properties in ultrathin epitaxial ferroelectric heterostructures. *Journal of Applied Physics*, **100**(5), 051609.

Nagy, A. (1998). Density functional. theory and application to atoms and molecules. *Physics Reports*, **298**(1), 1 – 79.

Nam, Ki Min, Kim, Yong-Il, Jo, Younghun, Lee, Seung Mi, Kim, Bog G, Choi, Ran, Choi, Sang-Il, Song, Hyunjoon, and Park, Joon T (2012). New crystal structure: Synthesis and characterization of hexagonal wurtzite MnO. *Journal of the American Chemical Society*, **134**(20), 8392–8395.

Nam, Ki Min, Seo, Won Seok, Song, Hyunjoon, and Park, Joon Taik (2017). Non-native transition metal monoxide nanostructures: Unique physicochemical properties and phase transformations of CoO, MnO and ZnO. *NPG Asia Materials*, **9**(3), e364.

Navrotsky, Alexandra (2003). Energetics of nanoparticle oxides: Interplay between surface energy and polymorphism. *Geochemical Transactions*, **4**(1), 34.

Navrotsky, Alexandra (2014). Progress and new directions in calorimetry: A 2014 perspective. *Journal of the American Ceramic Society*, **97**(11), 3349–3359.

Negreiros, FR, Obermüller, Th, Blatnik, Matthias, Mohammadi, Malihe, Fortunelli, Alessandro, Netzer, Falko P, and Surnev, Svetlozar (2019). Ultrathin WO_3 bilayer on Ag(100): A model for the structure of 2-D WO_3 nanosheets. *The Journal of Physical Chemistry C*, **123**(45), 27584–27593.

Neogi, Soumya Ganguly and Chaudhury, Pinaki (2014). Structure, spectroscopy and electronic properties of neutral lattice-like $(MgO)_n$ clusters: A study based on a blending of DFT with stochastic algorithms inspired by natural processes. *Structural Chemistry*, **25**(4), 1229–1244.

Netzer, Falko P (1988). Electron energy loss spectroscopy in reflection geometry. In *X-Ray Spectroscopy in Atomic and Solid State Physics*, pp. 335–365. Springer.

Netzer, Falko P (2010). 'Small and beautiful'–the novel structures and phases of nano-oxides. *Surface Science*, **604**(5-6), 485–489.

Netzer, Falko P, Allegretti, Francesco, and Surnev, Svetlozar (2010). Low-dimensional oxide nanostructures on metals: Hybrid systems with novel properties. *Journal of Vacuum Science & Technology B, Nanotechnology and Microelectronics: Materials, Processing, Measurement, and Phenomena*, **28**(1), 1–16.

Netzer, Falko P and Fortunelli, Alessandro (2016). *Oxide materials at the two-dimensional limit*. Volume 234. Springer, Switzerland.

Netzer, Falko P and Surnev, Svetlozar (2016). Structure concepts in two-dimensional oxide materials. In *Oxide Materials at the Two-Dimensional Limit*, pp. 1–38. Springer.

Neukermans, Sven, Janssens, Ewald, Silverans, RE, and Lievens, Peter (2007). Magic numbers for shells of electrons and shells of atoms in binary clusters. *The Chemical Physics of Solid Surfaces*, **12**, 271–297.

Neuzil, CE (2019). Permeability of clays and shales. *Annual Review of Earth and Planetary Sciences*, **47**, 247–273.

Niedermeyer, R (1968). Formation of anisotropic islands in layer growth on isotropic monocrystalline substrates. *Thin Films*, **1**(1), 25.

Nielsen, Arne E (1964). *Kinetics of precipitation*. Volume 18. Pergamon, Oxford, New York.

Niklasson, Gunnar A and Granqvist, Claes G (2007). Electrochromics for smart windows: Thin films of tungsten oxide and nickel oxide, and devices based on these. *Journal of Materials Chemistry*, **17**(2), 127–156.

Nilius, Niklas (2009). Properties of oxide thin films and their adsorption behavior studied by scanning tunneling microscopy and conductance spectroscopy. *Surface Science Reports*, **64**(12), 595–659.

Nilius, Niklas (2015). Exploring routes to tailor the physical and chemical properties of oxides via doping: An STM study. *Journal of Physics: Condensed Matter*, **27**(30), 303001.

Nilius, Niklas and Freund, Hans-Joachim (2015). Activating nonreducible oxides via doping. *Accounts of Chemical Research*, **48**(5), 1532–1539.

Noguera, Claudine (1996). *Physics and chemistry at oxide surfaces*. Cambridge University Press, Cambridge, New York, Melbourne, Madrid, Cape Town, Singapore, São Paulo.

Noguera, Claudine (2000). Polar oxide surfaces. *Journal of Physics: Condensed Matter*, **12**(31), R367.

Noguera, C and Fritz, B (2017). Solid solution/exsolution. In *Encyclopedia of Geochemistry: A Comprehensive Reference Source on the Chemistry of the Earth* (ed. W. M. White), pp. S:1–8. Springer International Publishing, Cham.

Noguera, Claudine, Godet, J, and Goniakowski, Jacek (2010). MgO/metal interfaces at low coverage: An order N, semiempirical Hartree-Fock simulation. *Physical Review B*, **81**(15), 155409.

Noguera, Claudine and Goniakowski, Jacek (2008). Polarity in oxide ultrathin films. *Journal of Physics: Condensed Matter*, **20**(26), 264003.

Noguera, Claudine and Goniakowski, Jacek (2012). Polarity in oxide nano-objects. *Chemical Reviews*, **113**(6), 4073–4105.

Noguera, Claudine and Goniakowski, J (2013). Structural phase diagrams of supported oxide nanowires from extended Frenkel-Kontorova models of diatomic chains. *The Journal of Chemical Physics*, **139**(8), 084703.

Noguera, Claudine and Mackrodt, WC (2006). A theoretical study of magnetic phase transitions in ultra-thin films: Application to NiO. *Surface Science*, **600**(4), 861–872.

Noguera, Claudine, Pojani, Ariana, Casek, Petr, and Finocchi, Fabio (2002). Electron redistribution in low-dimensional oxide structures. *Surface Science*, **507**, 245–255.

Novoselov, Kostya S, Geim, Andre K, Morozov, Sergei V, Jiang, D, Zhang, Y, Dubonos, Sergey V, Grigorieva, Irina V, and Firsov, Alexandr A (2004). Electric field effect in atomically thin carbon films. *Science*, **306**(5696), 666–669.

Obermüller, Thomas (2015). *Growth of transition metal oxides in 2-D layers*. University of Graz.

Obermüller, Thomas, Doudin, Nassar, Kuhness, David, Surnev, Svetlozar, and Netzer, Falko P (2017). Ultrathin oxide films: Epitaxy at the two-dimensional limit. *Journal of Materials Research*, **32**(21), 3924–3935.

Odelius, M, Bernasconi, M, and Parrinello, M (1997). Two dimensional ice adsorbed on mica surface. *Physical Review Letters*, **78**(14), 2855.

Ogale, Satishchandra B (2010). Dilute doping, defects, and ferromagnetism in metal oxide systems. *Advanced Materials*, **22**(29), 3125–3155.

Oh, WS, Xu, C, Kim, DY, and Goodman, DW (1997). Preparation and characterization of epitaxial titanium oxide films on Mo(100). *Journal of Vacuum Science & Technology A: Vacuum, Surfaces, and Films*, **15**(3), 1710–1716.

Ohtomo, Akira and Tsukazaki, Atsushi (2005). Pulsed laser deposition of thin films and superlattices based on ZnO. *Semiconductor Science and Technology*, **20**(4), S1.

Olbrich, Reinhard, Murgida, Gustavo E, Ferrari, Valeria, Barth, Clemens, Llois, Ana M, Reichling, Michael, and Ganduglia-Pirovano, M Verónica (2017). Surface stabilizes ceria in unexpected stoichiometry. *The Journal of Physical Chemistry*

C, **121**(12), 6844–6851.

Olefjord, Ingemar and Elfstrom, Bengt-Olof (1982). The composition of the surface during passivation of stainless steels. *Corrosion*, **38**(1), 46–52.

Olsson, C-OA and Landolt, D (2003). Passive films on stainless steels—chemistry, structure and growth. *Electrochimica Acta*, **48**(9), 1093–1104.

Omomo, Yoshitomo, Sasaki, Takayoshi, Wang, Lianzhou, and Watanabe, Mamoru (2003). Redoxable nanosheet crystallites of MnO_2 derived via delamination of a layered manganese oxide. *Journal of the American Chemical Society*, **125**(12), 3568–3575.

Onida, Giovanni, Reining, Lucia, and Rubio, Angel (2002). Electronic excitations: Density-functional versus many-body Green's function approaches. *Reviews of Modern Physics*, **74**, 601–659.

Osada, M, Ebina, Y, Fukuda, K, Ono, K, Takada, K, Yamaura, K, Takayama-Muromachi, E, and Sasaki, T (2006*a*). Ferromagnetism in two-dimensional $Ti_{0.8}Co_{0.2}O_2$ nanosheets. *Physical Review B*, **73**(15), 153301.

Osada, Minoru, Ebina, Yasuo, Takada, Kazunori, and Sasaki, Takayoshi (2006*b*). Gigantic magneto–optical effects in multilayer assemblies of two-dimensional titania nanosheets. *Advanced Materials*, **18**(3), 295–299.

Osada, Minoru and Sasaki, Takayoshi (2009). Exfoliated oxide nanosheets: New solution to nanoelectronics. *Journal of Materials Chemistry*, **19**(17), 2503–2511.

Osada, Minoru and Sasaki, Takayoshi (2012). Two-dimensional dielectric nanosheets: Novel nanoelectronics from nanocrystal building blocks. *Advanced Materials*, **24**(2), 210–228.

Ostlund, Neil S and Szabo, Attila (1996). *Modern Quantum Chemistry: Introduction to Advanced Electronic Structure Theory*. Dover Publications, Inc., New York.

Outhwaite, Christopher W, Bhuiyan, Lutful B, and Levine, Samuel (1980). Theory of the electric double layer using a modified Poisson–Boltzman equation. *Journal of the Chemical Society, Faraday Transactions 2: Molecular and Chemical Physics*, **76**, 1388–1408.

Over, Herbert, Kim, Young Dae, Seitsonen, AP, Wendt, Stefan, Lundgren, Edvin, Schmid, Michael, Varga, Peter, Morgante, Alberto, and Ertl, G (2000). Atomic-scale structure and catalytic reactivity of the RuO_2(110) surface. *Science*, **287**(5457), 1474–1476.

Pacchioni, Gianfranco (2012). Two-dimensional oxides: Multifunctional materials for advanced technologies. *Chemistry–A European Journal*, **18**(33), 10144–10158.

Pacchioni, Gianfranco (2016). Role of structural flexibility on the physical and chemical properties of metal-supported oxide ultrathin films. In *Oxide Materials at the Two-Dimensional Limit*, pp. 91–118. Springer.

Pacchioni, Gianfranco and Freund, Hajo (2012). Electron transfer at oxide surfaces. The MgO paradigm: From defects to ultrathin films. *Chemical Reviews*, **113**(6), 4035–4072.

Pacchioni, Gianfranco and Freund, Hans-Joachim (2018). Controlling the charge state of supported nanoparticles in catalysis: Lessons from model systems. *Chemical Society Reviews*, **47**(22), 8474–8502.

Pacchioni, Gianfranco and Valeri, Sergio (2012). *Oxide Ultrathin Films: Science and*

Technology. John Wiley & Sons, Weinheim.

Paier, Joachim (2016). Hybrid density functionals applied to complex solid catalysts: Successes, limitations, and prospects. *Catalysis Letters*, **146**(5), 861–885.

Paineau, Erwan, Bihannic, Isabelle, Baravian, Christophe, Philippe, Adrian-Marie, Davidson, Patrick, Levitz, Pierre, Funari, Sérgio S, Rochas, Cyrille, and Michot, Laurent J (2011). Aqueous suspensions of natural swelling clay minerals. 1. Structure and electrostatic interactions. *Langmuir*, **27**(9), 5562–5573.

Pan, Yi, Benedetti, Stefania, Noguera, Claudine, Giordano, Livia, Goniakowski, Jacek, and Nilius, Niklas (2012). Compensating edge polarity: A means to alter the growth orientation of MgO nanostructures on Au(111). *The Journal of Physical Chemistry C*, **116**(20), 11126–11132.

Parbhakar, K, Lewandowski, J, and Dao, LH (1995). Simulation model for Ostwald ripening in liquids. *Journal of Colloid and Interface Science*, **174**(1), 142–147.

Park, Joonho, Lee, Yeageun, Chang, Ikwhang, Cho, Gu Young, Ji, Sanghoon, Lee, Wonyoung, and Cha, Suk Won (2016). Atomic layer deposition of yttria-stabilized zirconia thin films for enhanced reactivity and stability of solid oxide fuel cells. *Energy*, **116**, 170–176.

Park, Mihyun, Lee, Nohyun, Choi, Seung Hong, An, Kwangjin, Yu, Seung-Ho, Kim, Jeong Hyun, Kwon, Seung-Hae, Kim, Dokyoon, Kim, Hyoungsu, Baek, Sung-Il et al. (2011). Large-scale synthesis of ultrathin manganese oxide nanoplates and their applications to T1 MRI contrast agents. *Chemistry of Materials*, **23**(14), 3318–3324.

Parks, George A (1965). The isoelectric points of solid oxides, solid hydroxides, and aqueous hydroxo complex systems. *Chemical Reviews*, **65**(2), 177–198.

Parr, RG and Weitao, Y (1989). *Density-Functional Theory of Atoms and Molecules*. International Series of Monographs on Chemistry. Oxford University Press, Oxford.

Patil, KC (2008). *Chemistry of Nanocrystalline Oxide Materials: Combustion Synthesis, Properties and Applications*. World Scientific, New Jersey, London, Singapore, Beijing, Shanghai, Hong Kong, Taipei, Chennai.

Pauling, Linus (1929). The principles determining the structure of complex ionic crystals. *Journal of the American Chemical Society*, **51**(4), 1010–1026.

Peden, Charles HF and Goodman, D Wayne (1986). Kinetics of carbon monoxide oxidation over ruthenium (0001). *The Journal of Physical Chemistry*, **90**(7), 1360–1365.

Perdew, JP and Zunger, Alex (1981). Self-interaction correction to density-functional approximations for many-electron systems. *Physical Review B*, **23**, 5048–5079.

Perdew, John P, Burke, Kieron, and Ernzerhof, Matthias (1996a). Generalized Gradient Approximation made simple. *Physical Review Letters*, **77**, 3865–3868.

Perdew, John P, Burke, Kieron, and Wang, Yue (1996b). Generalized gradient approximation for the exchange-correlation hole of a many-electron system. *Physical Review B*, **54**, 16533–16539.

Perdew, John P, Chevary, JA, Vosko, SH, Jackson, Koblar A, Pederson, Mark R, Singh, DJ, and Fiolhais, Carlos (1992). Atoms, molecules, solids, and surfaces: Applications of the generalized gradient approximation for exchange and correlation. *Physical Review B*, **46**, 6671–6687.

Perdew, John P, Parr, Robert G, Levy, Mel, and Balduz Jr, Jose L (1982). Density-functional theory for fractional particle number: Derivative discontinuities of the energy. *Physical Review Letters*, **49**(23), 1691.

Perednis, Dainius and Gauckler, Ludwig J (2005). Thin film deposition using spray pyrolysis. *Journal of Electroceramics*, **14**(2), 103–111.

Phillips, JC (1970). Ionicity of the chemical bond in crystals. *Reviews of Modern Physics*, **42**(3), 317.

Pisani, C, Dovesi, R, and Roetti, C (1992). *Hartree-Fock Ab Initio Treatment of Crystalline Systems*. Volume 48. Springer Verlag, Berlin, Heidelberg, New York, London, Paris, Tokyo.

Plummer, EW and Eberhardt, W (1982). Angle-resolved photoemission as a tool for the study of surfaces. *Advances in Chemical Physics*, **49**, 533–656.

Pojani, A, Finocchi, F, Goniakowski, J, and Noguera, C (1997). A theoretical study of the stability and electronic structure of the polar (111) face of MgO. *Surface Science*, **387**(1-3), 354–370.

Pomp, S, Kuhness, D, Barcaro, G, Sementa, L, Mankad, V, Fortunelli, A, Sterrer, M, Netzer, FP, and Surnev, S (2016). Two-dimensional iron tungstate: A ternary oxide layer with honeycomb geometry. *The Journal of Physical Chemistry C*, **120**(14), 7629–7638.

Pongsai, SB (2006). Computational study on thermodynamics of mixing and phase behaviour for CoO/FeO and CoO/MnO solid solutions. *Journal of Molecular Structure: THEOCHEM*, **761**(1-3), 171–175.

Pople, John A and Segal, Gerald A (1965). Approximate self-consistent molecular orbital theory. II. calculations with Complete Neglect of Differential Overlap. *The Journal of Chemical Physics*, **43**(10), S136–S151.

Pople, John A and Segal, Gerald A (1966). Approximate self-consistent molecular orbital theory. III. CNDO results for AB_2 and AB_3 systems. *The Journal of Chemical Physics*, **44**(9), 3289–3296.

Pottier, Agnes, Cassaignon, Sophie, Chanéac, Corinne, Villain, Françoise, Tronc, Elisabeth, and Jolivet, Jean-Pierre (2003). Size tailoring of TiO_2 anatase nanoparticles in aqueous medium and synthesis of nanocomposites. Characterization by Raman spectroscopy. *Journal of Materials Chemistry*, **13**(4), 877–882.

Poulopoulos, Panagiotis, Baskoutas, Sotirios, Pappas, Spiridon D, Garoufalis, Christos S, Droulias, Sotirios A, Zamani, Atieh, and Kapaklis, Vassilios (2011). Intense quantum confinement effects in Cu_2O thin films. *The Journal of Physical Chemistry C*, **115**(30), 14839–14843.

Prada, Stefano, Giordano, Livia, and Pacchioni, Gianfranco (2012). Li, Al, and Ni substitutional doping in MgO ultrathin films on metals: Work function tuning via charge compensation. *The Journal of Physical Chemistry C*, **116**(9), 5781–5786.

Prada, Stefano, Giordano, Livia, Pacchioni, Gianfranco, and Goniakowski, Jacek (2016). Theoretical description of metal/oxide interfacial properties: The case of MgO/Ag(001). *Applied Surface Science*, **390**, 578–582.

Prada, Stefano, Martinez, Umberto, and Pacchioni, Gianfranco (2008). Work function changes induced by deposition of ultrathin dielectric films on metals: A theoretical analysis. *Physical Review B*, **78**(23), 235423.

Prada, Stefano, Rosa, Massimo, Giordano, Livia, Di Valentin, Cristiana, and Pacchioni, Gianfranco (2011). Density functional theory study of TiO_2/Ag interfaces and their role in memristor devices. *Physical Review B*, **83**(24), 245314.

Pravarthana, D, Trassin, Morgan, Haw Chu, Jiun, Lacotte, M, David, A, Ramesh, R, Salvador, PA, and Prellier, Wilfrid (2014). $BiFeO_3/La_{0.7}Sr_{0.3}MnO_3$ heterostructures deposited on spark plasma sintered $LaAlO_3$ substrates. *Applied Physics Letters*, **104**(8), 082914.

Press, WH, Teukolsky, SA, Vetterling, WT, and Flannery, BP (1992)). *Numerical Recipies in fortran*. Cambridge University Press, Cambridge.

Prinz, Günther M, Gerber, Timm, Lorke, Axel, and Müller, Martina (2016). Quantum confinement in EuO heterostructures. *Applied Physics Letters*, **109**(20), 202401.

Protheroe, AR, Steinbrunn, A, and Gallon, TE (1983). The electron energy loss spectra of some alkaline earth oxides. *Surface Science*, **126**(1-3), 534–542.

Pulay, Peter (1969). Ab initio calculation of force constants and equilibrium geometries in polyatomic molecules: I. Theory. *Molecular Physics*, **17**(2), 197–204.

Purton, JohnáA, Allan, NeiláL, and Blundy, JonáD (1997). Impurity cations in the bulk and the {001} surface of muscovite: An atomistic simulation study. *Journal of Materials Chemistry*, **7**(9), 1947–1951.

Purton, JA, Allan, NL, Lavrentiev, M Yu, Todorov, IT, and Freeman, CL (2006). Computer simulation of mineral solid solutions. *Chemical Geology*, **225**(3-4), 176–188.

Purton, John A, Parker, Stephen C, and Allan, Neil L (2013). Monte Carlo simulation and free energies of mixed oxide nanoparticles. *Physical Chemistry Chemical Physics*, **15**(17), 6219–6225.

Qin, Z-H, Lewandowski, Mikolaj, Sun, Y-N, Shaikhutdinov, Shamil, and Freund, H-J (2008). Encapsulation of Pt nanoparticles as a result of strong metal- support interaction with Fe_3O_4(111). *The Journal of Physical Chemistry C*, **112**(27), 10209–10213.

Radha, AV, Bomati-Miguel, Oscar, Ushakov, Sergey V, Navrotsky, Alexandra, and Tartaj, Pedro (2009). Surface enthalpy, enthalpy of water adsorption, and phase stability in nanocrystalline monoclinic zirconia. *Journal of the American Ceramic Society*, **92**(1), 133–140.

Rahman, Md, Ahammad, AJ, Jin, Joon-Hyung, Ahn, Sang Jung, Lee, Jae-Joon et al. (2010). A comprehensive review of glucose biosensors based on nanostructured metal-oxides. *Sensors*, **10**(5), 4855–4886.

Ramirez, Arthur P (2007). Oxide electronics emerge. *Science*, **315**(5817), 1377–1378.

Ranade, MR, Navrotsky, A, Zhang, HZ, Banfield, JF, Elder, SH, Zaban, A, Borse, PH, Kulkarni, SK, Doran, GS, and Whitfield, HJ (2002). Energetics of nanocrystalline TiO_2. *Proceedings of the National Academy of Sciences*, **99**(suppl 2), 6476–6481.

Ratke, Lorenz and Voorhees, Peter W (2002). *Growth and Coarsening: Ostwald Ripening in Material Processing*. Springer Science & Business Media, Berlin, Heidelberg, New York, Barcelona, Hong Kong, London, Milan, Paris, Tokyo.

Raybaud, P, Digne, M, Iftimie, R, Wellens, W, Euzen, P, and Toulhoat, H (2001). Morphology and surface properties of boehmite (γ-AlOOH): A density functional theory study. *Journal of Catalysis*, **201**(2), 236–246.

Rayleigh, Lord (1885). On waves propagated along the plane surface of an elastic solid. *Proceedings of the London Mathematical Society*, **17**, 4–11.

Reddy, Kongara M, Feris, Kevin, Bell, Jason, Wingett, Denise G, Hanley, Cory, and Punnoose, Alex (2007). Selective toxicity of zinc oxide nanoparticles to prokaryotic and eukaryotic systems. *Applied Physics Letters*, **90**(21), 213902.

Refson, K, Wogelius, RA, Fraser, DG, Payne, MC, Lee, MH, and Milman, V (1995). Water chemisorption and reconstruction of the MgO surface. *Physical Review B*, **52**(15), 10823.

Reina, Alfonso, Jia, Xiaoting, Ho, John, Nezich, Daniel, Son, Hyungbin, Bulovic, Vladimir, Dresselhaus, Mildred S, and Kong, Jing (2008). Large area, few-layer graphene films on arbitrary substrates by chemical vapor deposition. *Nano Letters*, **9**(1), 30–35.

Reining, Lucia (2018). The GW approximation: Content, successes and limitations. *Wiley Interdisciplinary Reviews: Computational Molecular Science*, **8**(3), e1344.

Reining, Lucia, Onida, Giovanni, and Godby, RW (1997). Elimination of unoccupied-state summations in ab initio self-energy calculations for large supercells. *Physical Review B*, **56**(8), R4301.

Renaud, G, Lazzari, R, Revenant, C, Barbier, A, Noblet, M, Ulrich, O, Leroy, F, Jupille, J, Borensztein, Y, Henry, CR et al. (2003). Real-time monitoring of growing nanoparticles. *Science*, **300**(5624), 1416–1419.

Reuter, Karsten, Stampfl, Catherine, Ganduglia-Pirovano, M Verónica, and Scheffler, Matthias (2002). Atomistic description of oxide formation on metal surfaces: The example of ruthenium. *Chemical Physics Letters*, **352**(5-6), 311–317.

Reyren, N, Thiel, S, Caviglia, AD, Kourkoutis, LF, Hammerl, G, Richter, C, Schneider, CW, Kopp, T, Rüetschi, A-S, Jaccard, D, Gabay, M, Muller, DA, J-M, Triscone, and J, Mannhart (2007). Superconducting interfaces between insulating oxides. *Science*, **317**(5842), 1196–1199.

Richter, NA, Stavale, Fernando, Levchenko, Sergey V, Nilius, Niklas, Freund, H-J, and Scheffler, Matthias (2015). Defect complexes in Li-doped MgO. *Physical Review B*, **91**(19), 195305.

Ritter, M, Ranke, W, and Weiss, W (1998). Growth and structure of ultrathin FeO films on Pt(111) studied by STM and LEED. *Physical Review B*, **57**(12), 7240.

Rivière, John C (1990). *Surface analytical techniques*. Oxford Science Publications, Oxford.

Riviere, John C and Myhra, Sverre (2012). *Characterization of Nanostructures*. CRC Press, Boca Raton, London, New York, Washington, D.C.

Roberts, Christopher and Johnston, Roy L (2001). Investigation of the structures of MgO clusters using a genetic algorithm. *Physical Chemistry Chemical Physics*, **3**(22), 5024–5034.

Robinson, IK, Waskiewicz, WK, Tung, RT, and Bohr, Jakob (1986). Ordering at Si(111)/a-Si and Si(111)/SiO$_2$ interfaces. *Physical Review Letters*, **57**(21), 2714.

Rödel, TC, Bareille, C, Fortuna, F, Baumier, C, Bertran, F, Le Fèvre, P, Gabay, M, Cubelos, O Hijano, Rozenberg, MJ, Maroutian, T et al. (2014). Orientational tuning of the Fermi sea of confined electrons at the SrTiO$_3$ (110) and (111) surfaces. *Physical Review Applied*, **1**(5), 051002.

Rödl, C, Fuchs, F, Furthmüller, J, and Bechstedt, F (2009). Quasiparticle band structures of the antiferromagnetic transition-metal oxides MnO, FeO, CoO, and NiO. *Physical Review B*, **79**, 235114.

Rodriguez, José A, Graciani, Jesús, Evans, Jaime, Park, Joon B, Yang, Fan, Stacchiola, Dario, Senanayake, Sanjaya D, Ma, Shuguo, Pérez, Manuel, Liu, Ping et al. (2009). Water-gas shift reaction on a highly active inverse $CeO_x/Cu(111)$ catalyst: Unique role of ceria nanoparticles. *Angewandte Chemie International Edition*, **48**(43), 8047–8050.

Ronning, C, Shang, NG, Gerhards, I, Hofsäss, H, and Seibt, M (2005). Nucleation mechanism of the seed of tetrapod ZnO nanostructures. *Journal of Applied Physics*, **98**(3), 034307.

Roozeboom, F, Fransen, T, Mars, P, and Gellings, PJ (1979). Vanadium oxide monolayer catalysts. I. Preparation, characterization, and thermal stability. *Zeitschrift für Anorganische und Allgemeine Chemie*, **449**(1), 25–40.

Rousseau, Roger, Dixon, David A, Kay, Bruce D, and Dohnalek, Zdenek (2014). Dehydration, dehydrogenation, and condensation of alcohols on supported oxide catalysts based on cyclic $(WO_3)_3$ and $(MoO_3)_3$ clusters. *Chemical Society Reviews*, **43**(22), 7664–7680.

Roy, Poulomi, Berger, Steffen, and Schmuki, Patrik (2011). TiO_2 nanotubes: Synthesis and applications. *Angewandte Chemie International Edition*, **50**(13), 2904–2939.

Rücker, H, Molinari, Elisa, and Lugli, P (1992). Microscopic calculation of the electron-phonon interaction in quantum wells. *Physical Review B*, **45**(12), 6747.

Rui, Xianhong, Lu, Ziyang, Yu, Hong, Yang, Dan, Hng, Huey Hoon, Lim, Tuti Mariana, and Yan, Qingyu (2013). Ultrathin V_2O_5 nanosheet cathodes: Realizing ultrafast reversible lithium storage. *Nanoscale*, **5**(2), 556–560.

Ruiz Puigdollers, Antonio, Schlexer, Philomena, Tosoni, Sergio, and Pacchioni, Gianfranco (2017). Increasing oxide reducibility: The role of metal/oxide interfaces in the formation of oxygen vacancies. *ACS Catalysis*, **7**(10), 6493–6513.

Runge, Erich and Gross, E. K. U. (1984). Density-functional theory for time-dependent systems. *Physical Review Letters*, **52**, 997–1000.

Russel, William Bailey, Russel, WB, Saville, Dudley A, and Schowalter, William Raymond (1991). *Colloidal Dispersions*. Cambridge University Press, Cambridge, New York, Melbourne.

Sachert, Steffen, Polzin, Sebastian, Kostov, Krassimir, and Widdra, Wolf (2010). Thickness dependent vibrational and electronic properties of MnO(100) thin films grown on Pt(111). *Physical Review B*, **81**(19), 195424.

Sadat Nabi, Hasan and Pentcheva, Rossitza (2011, Jun). Energetic stability and magnetic coupling in $(Cr_{1-x}Fe_x)_2O_3$: Evidence for a ferrimagnetic ilmenite-type superlattice from first principles. *Physical Review B*, **83**, 214424.

Sai, Na, Kolpak, Alexie M, and Rappe, Andrew M (2005). Ferroelectricity in ultrathin perovskite films. *Physical Review B*, **72**(2), 020101.

Salluzzo, M, Cezar, JC, Brookes, NB, Bisogni, V, De Luca, GM, Richter, Christoph, Thiel, Stefan, Mannhart, Jochen, Huijben, M, Brinkman, A et al. (2009). Orbital reconstruction and the two-dimensional electron gas at the $LaAlO_3/SrTiO_3$ interface. *Physical Review Letters*, **102**(16), 166804.

Sanchez-Dominguez, Margarita, Boutonnet, Magali, and Solans, Conxita (2009). A novel approach to metal and metal oxide nanoparticle synthesis: The oil-in-water microemulsion reaction method. *Journal of Nanoparticle Research*, **11**(7), 1823.

Sandberg, Robert B, Hansen, Martin H, Nørskov, Jens K, Abild-Pedersen, Frank, and Bajdich, Michal (2018). Strongly modified scaling of CO hydrogenation in metal supported TiO nanostripes. *ACS Catalysis*, **8**(11), 10555–10563.

Santander-Syro, AF, Copie, O, Kondo, T, Fortuna, F, Pailhes, S, Weht, R, Qiu, XG, Bertran, F, Nicolaou, A, Taleb-Ibrahimi, A et al. (2011). Two-dimensional electron gas with universal subbands at the surface of $SrTiO_3$. *Nature*, **469**(7329), 189.

Sasaki, Takayoshi and Watanabe, Mamoru (1998). Osmotic swelling to exfoliation. Exceptionally high degrees of hydration of a layered titanate. *Journal of the American Chemical Society*, **120**(19), 4682–4689.

Sato, Hisako, Ono, Kanta, Johnston, Cliff T, and Yamagishi, Akihiko (2005). First-principles studies on the elastic constants of a 1:1 layered kaolinite mineral. *American Mineralogist*, **90**(11-12), 1824–1826.

Sato, N and Sato, N (1978). *Passivity of Metals*. The Electrochemical Soc., Princeton, NJ.

Savio, L, Celasco, E, Vattuone, L, Rocca, M, and Senet, P (2003). MgO/Ag(100): Confined vibrational modes in the limit of ultrathin films. *Physical Review B*, **67**(7), 075420.

Sayle, Dean C, Gay, DH, Rohl, AH, Catlow, CRA, Harding, JA, Perri, MA, and Nortier, P (1996). Computer modelling of V_2O_5: Surface structures, crystal morphology and ethene sorption. *Journal of Materials Chemistry*, **6**(4), 653–660.

Sayle, Thi XT, Parker, Stephen C, and Sayle, Dean C (2004). Shape of CeO_2 nanoparticles using simulated amorphisation and recrystallisation. *Chemical Communications*, **21**, 2438–2439.

Schintke, Silvia, Messerli, Stéphane, Pivetta, Marina, Patthey, François, Libioulle, Laurent, Stengel, Massimiliano, De Vita, Alessandro, and Schneider, Wolf-Dieter (2001). Insulator at the ultrathin limit: MgO on Ag(001). *Physical Review Letters*, **87**(27), 276801.

Schlögl, Robert (2015). Heterogeneous catalysis. *Angewandte Chemie International Edition*, **54**(11), 3465–3520.

Schmid, Michael, Kresse, G, Buchsbaum, Andreas, Napetschnig, Evelyn, Gritschneder, S, Reichling, M, and Varga, Peter (2007). Nanotemplate with holes: Ultrathin alumina on Ni_3Al (111). *Physical Review Letters*, **99**(19), 196104.

Schoiswohl, J, Eck, S, Ramsey, MG, Andersen, Jesper N, Surnev, S, and Netzer, FP (2005*a*). Vanadium oxide nanostructures on Rh(111): Promotion effect of CO adsorption and oxidation. *Surface Science*, **580**(1-3), 122–136.

Schoiswohl, J, Sock, M, Chen, Q, Thornton, G, Kresse, G, Ramsey, MG, Surnev, S, and Netzer, FP (2007). Metal supported oxide nanostructures: Model systems for advanced catalysis. *Topics in Catalysis*, **46**(1-2), 137–149.

Schoiswohl, J, Sock, M, Eck, S, Surnev, S, Ramsey, MG, Netzer, FP, and Kresse, G (2004). Atomic-level growth study of vanadium oxide nanostructures on Rh(111). *Physical Review B*, **69**(15), 155403.

Schoiswohl, J, Surnev, S, and Netzer, FP (2005*b*). Reactions on inverse model catalyst surfaces: Atomic views by STM. *Topics in Catalysis*, **36**(1-4), 91–105.

Schoiswohl, J, Surnev, S, Netzer, FP, and Kresse, G (2006). Vanadium oxide nanostructures: From zero- to three-dimensional. *Journal of Physics: Condensed Matter*, **18**(4), R1.

Schoonheydt, Robert A, Johnston, Cliff T, and Bergaya, Faiza (2018). *Surface and Interface Chemistry of Clay Minerals*. Elsevier, Amsterdam, Oxford, Cambridge.

Schwoebel, Richard L and Shipsey, Edward J (1966). Step motion on crystal surfaces. *Journal of Applied Physics*, **37**(10), 3682–3686.

Seah, M Pl and Dench, WA (1979). Quantitative electron spectroscopy of surfaces: A standard data base for electron inelastic mean free paths in solids. *Surface and Interface Analysis*, **1**(1), 2–11.

Sebastian, Ina, Bertrams, Thomas, Meinel, Klaus, and Neddermeyer, Henning (1999). Scanning tunnelling microscopy on the growth and structure of NiO(100) and CoO(100) thin films. *Faraday Discussions*, **114**, 129–140.

Sedona, Francesco, Granozzi, Gaetano, Barcaro, Giovanni, and Fortunelli, Alessandro (2008). Defect evolution in oxide nanophases: The case of a zigzag-like TiO_x phase on Pt(111). *Physical Review B*, **77**(11), 115417.

Sedona, Francesco, Rizzi, Gian Andrea, Agnoli, Stefano, Llabrés i Xamena, Francesc X, Papageorgiou, Anthoula, Ostermann, Dieter, Sambi, Mauro, Finetti, Paola, Schierbaum, Klaus, and Granozzi, Gaetano (2005). Ultrathin TiO_x films on Pt(111): a LEED, XPS, and STM investigation. *The Journal of Physical Chemistry B*, **109**(51), 24411–24426.

Seifert, G (2007). Tight-binding density functional theory: An approximate Kohn-Sham DFT scheme. *The Journal of Physical Chemistry A*, **111**(26), 5609–5613.

Seixas, L, Rodin, AS, Carvalho, A, and Neto, AH Castro (2016). Multiferroic two-dimensional materials. *Physical Review Letters*, **116**(20), 206803.

Sekizaki, S, Osada, M, and Nagashio, K (2017). Molecularly-thin anatase field-effect transistors fabricated through the solid state transformation of titania nanosheets. *Nanoscale*, **9**(19), 6471–6477.

Sementa, Luca, Barcaro, Giovanni, Negreiros, Fabio R, Thomas, Iorwerth O, Netzer, Falko P, Ferrari, Anna Maria, and Fortunelli, Alessandro (2012). Work function of oxide ultrathin films on the Ag(100) surface. *Journal of Chemical Theory and Computation*, **8**(2), 629–638.

Senet, Patrick, Lambin, Ph, and Lucas, AA (1995). Standing-wave optical phonons confined in ultrathin overlayers of ionic materials. *Physical Review Letters*, **74**(4), 570.

Setter, N, Damjanovic, D, Eng, L, Fox, G, Gevorgian, Spartak, Hong, S, Kingon, A, Kohlstedt, H, Park, NY, Stephenson, GB et al. (2006). Ferroelectric thin films: Review of materials, properties, and applications. *Journal of Applied Physics*, **100**(5), 051606.

Shabbir, Babar, Nadeem, Muhammad, Dai, Zhigao, Fuhrer, Michael S, Xue, Qi-Kun, Wang, Xiaolin, and Bao, Qiaoliang (2018). Long range intrinsic ferromagnetism in two dimensional materials and dissipationless future technologies. *Applied Physics Reviews*, **5**(4), 041105.

Shahrokhi, Masoud and Leonard, Céline (2016). Quasi-particle energies and optical excitations of wurtzite BeO and its nanosheet. *Journal of Alloys and Compounds*, **682**, 254–262.

Shaikhutdinov, Sh, Ritter, Michael, and Weiss, Werner (2000). Hexagonal heterolayers on a square lattice: A combined STM and LEED study of FeO(111) on Pt(100). *Physical Review B*, **62**(11), 7535.

Shan, CX, Hou, Xianghui, and Choy, Kwang-Leong (2008). Corrosion resistance of TiO_2 films grown on stainless steel by atomic layer deposition. *Surface and Coatings Technology*, **202**(11), 2399–2402.

Shannon, RD (1976). Revised effective ionic radii and systematic studies of interatomic distances in halides and chalcogenides. *Acta Cristallographica A*, **32**, 751–767.

Shao, Xiang, Myrach, Philipp, Nilius, Niklas, and Freund, Hans-Joachim (2011a). Growth and morphology of calcium-oxide films grown on Mo(001). *The Journal of Physical Chemistry C*, **115**(17), 8784–8789.

Shao, Xiang, Nilius, Niklas, Myrach, Philipp, Freund, Hans-Joachim, Martinez, Umberto, Prada, Stefano, Giordano, Livia, and Pacchioni, Gianfranco (2011b). Strain-induced formation of ultrathin mixed-oxide films. *Physical Review B*, **83**(24), 245407.

Shao, Xiang, Prada, Stefano, Giordano, Livia, Pacchioni, Gianfranco, Nilius, Niklas, and Freund, Hans-Joachim (2011c). Tailoring the shape of metal ad-particles by doping the oxide support. *Angewandte Chemie International Edition*, **50**(48), 11525–11527.

Shavanova, Kateryna, Bakakina, Yulia, Burkova, Inna, Shtepliuk, Ivan, Viter, Roman, Ubelis, Arnolds, Beni, Valerio, Starodub, Nickolaj, Yakimova, Rositsa, and Khranovskyy, Volodymyr (2016). Application of 2-D non-graphene materials and 2-D oxide nanostructures for biosensing technology. *Sensors*, **16**(2), 223.

Shluger, Alexander L, Sushko, Peter V, and Kantorovich, Lev N (1999). Spectroscopy of low-coordinated surface sites: Theoretical study of MgO. *Physical Review B*, **59**(3), 2417.

Si, H and Pan, BC (2010). Strain-induced semiconducting-metallic transition for ZnO zigzag nanoribbons. *Journal of Applied Physics*, **107**(9), 094313.

Si, Liang, Zhong, Zhicheng, Tomczak, Jan M, and Held, Karsten (2015). Route to room-temperature ferromagnetic ultrathin $SrRuO_3$ films. *Physical Review B*, **92**(4), 041108.

Sierka, Marek, Todorova, Tanya K., Sauer, Joachim, Kaya, Sarp, Stacchiola, Dario, Weissenrieder, Jonas, Shaikhutdinov, Shamil, and Freund, Hans-Joachim (2007). Oxygen adsorption on Mo(112) surface studied by ab initio genetic algorithm and experiment. *The Journal of Chemical Physics*, **126**(23), 234710.

Sobhan, MA, Kivaisi, RT, Stjerna, B, and Granqvist, CG (1996). Thermochromism of sputter deposited $W_xV_{1-x}O_2$ films. *Solar Energy Materials and Solar Cells*, **44**(4), 451–455.

Söhnel, Otakar (1982). Electrolyte crystal-aqueous solution interfacial tensions from crystallization data. *Journal of Crystal Growth*, **57**(1), 101–108.

Solanki, Pratima R, Kaushik, Ajeet, Agrawal, Ved V, and Malhotra, Bansi D (2011). Nanostructured metal oxide-based biosensors. *NPG Asia Materials*, **3**(1), 17.

Somorjai, Gabor A and Li, Yimin (2010). *Introduction to surface chemistry and catalysis*. John Wiley & Sons, Weinheim.

Song, Guosheng, Hao, Jiali, Liang, Chao, Liu, Teng, Gao, Min, Cheng, Liang, Hu, Junqing, and Liu, Zhuang (2016). Degradable molybdenum oxide nanosheets with rapid clearance and efficient tumor homing capabilities as a therapeutic nanoplatform. *Angewandte Chemie International Edition*, **55**(6), 2122–2126.

Song, Jinhui, Zhang, Yan, Xu, Chen, Wu, Wenzuo, and Wang, Zhong Lin (2011). Polar charges induced electric hysteresis of ZnO nano/microwire for fast data storage. *Nano Letters*, **11**(7), 2829–2834.

Song, Sha-Sha, Xia, Bao-Yu, Chen, Jie, Yang, Jiang, Shen, Xiu, Fan, Sai-Jun, Guo, Mei-li, Sun, Yuan-Ming, and Zhang, Xiao-Dong (2014). Two dimensional TiO_2 nanosheets: In vivo toxicity investigation. *RSC Advances*, **4**(80), 42598–42603.

Spence, John CH (2013). *High-resolution electron microscopy*. Oxford University Press, Oxford.

Spencer, Michelle JS (2012). Gas sensing applications of 1-D-nanostructured zinc oxide: Insights from density functional theory calculations. *Progress in Materials Science*, **57**(3), 437–486.

Sponza, Lorenzo, Goniakowski, Jacek, and Noguera, Claudine (2016). Confinement effects in ultrathin ZnO polymorph films: Electronic and optical properties. *Physical Review B*, **93**(19), 195435.

Sposito, Garrison (1998). On points of zero charge. *Environmental Science & Technology*, **32**(19), 2815–2819.

Sposito, Garrison and Prost, Rene (1982). Structure of water adsorbed on smectites. *Chemical Reviews*, **82**(6), 553–573.

Stankic, Slavica, Cottura, Maeva, Demaille, Dominique, Noguera, Claudine, and Jupille, Jacques (2011). Nucleation and growth concepts applied to the formation of a stoichiometric compound in a gas phase: The case of MgO smoke. *Journal of Crystal Growth*, **329**(1), 52–56.

Stankic, Slavica, Müller, Markus, Diwald, Oliver, Sterrer, Martin, Knözinger, Erich, and Bernardi, Johannes (2005). Size-dependent optical properties of MgO nanocubes. *Angewandte Chemie International Edition*, **44**(31), 4917–4920.

Stankic, Slavica, Sternig, Andreas, Finocchi, Fabio, Bernardi, Johannes, and Diwald, Oliver (2010). Zinc oxide scaffolds on MgO nanocubes. *Nanotechnology*, **21**(35), 355603.

Stankic, Slavica, Suman, Sneha, Haque, Francia, and Vidic, Jasmina (2016). Pure and multi metal oxide nanoparticles: Synthesis, antibacterial and cytotoxic properties. *Journal of Nanobiotechnology*, **14**(1), 73.

Stavale, Fernando, Nilius, Niklas, and Freund, Hans-Joachim (2012). Cathodoluminescence of near-surface centres in Cr-doped MgO(001) thin films probed by scanning tunnelling microscopy. *New Journal of Physics*, **14**(3), 033006.

Stengel, Massimiliano and Vanderbilt, David (2009). Berry-phase theory of polar discontinuities at oxide-oxide interfaces. *Physical Review B*, **80**(24), 241103.

Stengel, Massimiliano, Vanderbilt, David, and Spaldin, Nicola A (2009). First-principles modeling of ferroelectric capacitors via constrained displacement field calculations. *Physical Review B*, **80**(22), 224110.

Steurer, W, Allegretti, F, Surnev, S, Barcaro, G, Sementa, L, Negreiros, F, Fortunelli, A, and Netzer, FP (2011). Metamorphosis of ultrathin Ni oxide nanostructures on Ag(100). *Physical Review B*, **84**(11), 115446.

Steurer, Wolfram, Surnev, Svetlozar, Fortunelli, Alessandro, and Netzer, Falko P (2012). Scanning tunneling microscopy imaging of NiO(100)(1 × 1) islands embedded in Ag(100). *Surface Science*, **606**(9-10), 803–807.

Stillinger, Frank H and Weber, Thomas A (1985). Computer simulation of local order in condensed phases of silicon. *Physical Review B*, **31**(8), 5262.

Stöhr, Joachim (2013). *NEXAFS Spectroscopy*. Volume 25. Springer Science & Business Media, Berlin, Heidelberg, New York, Barcelona, Hong Kong, London, Milan, Paris, Tokyo.

Stöhr, Joachim and Siegmann, Hans Christoph (2007). *Magnetism: From Fundamentals to Nanoscale Dynamics*. Volume 152. Springer Science & Business Media, Berlin, Heidelberg, New York, Barcelona, Hong Kong, London, Milan, Paris, Tokyo.

Street, SC, Xu, C, and Goodman, DW (1997). The physical and chemical properties of ultrathin oxide films. *Annual Review of Physical Chemistry*, **48**(1), 43–68.

Streiffer, SK, Eastman, JA, Fong, DD, Thompson, Carol, Munkholm, A, Murty, MV Ramana, Auciello, O, Bai, GR, and Stephenson, GB (2002). Observation of nanoscale 180° stripe domains in ferroelectric $PbTiO_3$ thin films. *Physical Review Letters*, **89**(6), 067601.

Streitz, FH and Mintmire, JW (1994). Electrostatic potentials for metal-oxide surfaces and interfaces. *Physical Review B*, **50**(16), 11996.

Strobel, Reto and Pratsinis, Sotiris E (2007). Flame aerosol synthesis of smart nanostructured materials. *Journal of Materials Chemistry*, **17**(45), 4743–4756.

Sui, Ruohong and Charpentier, Paul (2012). Synthesis of metal oxide nanostructures by direct sol–gel chemistry in supercritical fluids. *Chemical Reviews*, **112**(6), 3057–3082.

Sun, Y-N, Qin, Z-H, Lewandowski, Mikolaj, Carrasco, Esther, Sterrer, Martin, Shaikhutdinov, Shamil, and Freund, H-J (2009). Monolayer iron oxide film on platinum promotes low temperature CO oxidation. *Journal of Catalysis*, **266**(2), 359–368.

Sun, Ziqi, Liao, Ting, Dou, Yuhai, Hwang, Soo Min, Park, Min-Sik, Jiang, Lei, Kim, Jung Ho, and Dou, Shi Xue (2014). Generalized self-assembly of scalable two-dimensional transition metal oxide nanosheets. *Nature Communications*, **5**, 3813.

Sundqvist, Jonas, Lu, Jun, Ottosson, Mikael, and Hårsta, Anders (2006). Growth of SnO_2 thin films by atomic layer deposition and chemical vapour deposition: A comparative study. *Thin Solid Films*, **514**(1-2), 63–68.

Surnev, Svetlozar, Fortunelli, Alessandro, and Netzer, Falko P (2012). Structure–property relationship and chemical aspects of oxide–metal hybrid nanostructures. *Chemical Reviews*, **113**(6), 4314–4372.

Surnev, S, Kresse, G, Ramsey, MG, and Netzer, FP (2001). Novel interface-mediated metastable oxide phases: Vanadium oxides on Pd(111). *Physical Review Letters*, **87**(8), 086102.

Surnev, S, Ramsey, MG, and Netzer, FP (2003). Vanadium oxide surface studies. *Progress in Surface Science*, **73**(4-8), 117–165.

Surnev, S, Schoiswohl, J, Kresse, G, Ramsey, MG, and Netzer, FP (2002). Reversible dynamic behavior in catalyst systems: Oscillations of structure and morphology. *Physical Review Letters*, **89**(24), 246101.

Svane, A and Gunnarsson, O (1990). Transition-metal oxides in the self-interaction–corrected density-functional formalism. *Physical Review Letters*, **65**(9), 1148.

Swihart, Mark T (2003). Vapor-phase synthesis of nanoparticles. *Current Opinion in Colloid & Interface Science*, **8**(1), 127–133.

Takahashi, KS, Kawasaki, M, and Tokura, Y (2001). Interface ferromagnetism in oxide superlattices of $CaMnO_3/CaRuO_3$. *Applied Physics Letters*, **79**(9), 1324–1326.

Takeuchi, Masato, Sakamoto, Kenji, Martra, Gianmario, Coluccia, Salvatore, and Anpo, Masakazu (2005). Mechanism of photoinduced superhydrophilicity on the TiO_2 photocatalyst surface. *The Journal of Physical Chemistry B*, **109**(32), 15422–15428.

Tällberg, Rickard, Jelle, Bjørn Petter, Loonen, Roel, Gao, Tao, and Hamdy, Mohamed (2019). Comparison of the energy saving potential of adaptive and controllable smart windows: A state-of-the-art review and simulation studies of thermochromic, photochromic and electrochromic technologies. *Solar Energy Materials and Solar Cells*, **200**, 109828.

Tamura, Tomoyuki, Ishibashi, Shoji, Terakura, Kiyoyuki, and Weng, Hongming (2009). First-principles study of the rectifying properties of Pt/TiO_2 interface. *Physical Review B*, **80**(19), 195302.

Tan, Chaoliang, Cao, Xiehong, Wu, Xue-Jun, He, Qiyuan, Yang, Jian, Zhang, Xiao, Chen, Junze, Zhao, Wei, Han, Shikui, Nam, Gwang-Hyeon et al. (2017). Recent advances in ultrathin two-dimensional nanomaterials. *Chemical Reviews*, **117**(9), 6225–6331.

Tang, Qing, Li, Fengyu, Zhou, Zhen, and Chen, Zhongfang (2011). Versatile electronic and magnetic properties of corrugated V_2O_5 two-dimensional crystal and its derived one-dimensional nanoribbons: A computational exploration. *The Journal of Physical Chemistry C*, **115**(24), 11983–11990.

Tarantino, Walter, Mendoza, Bernardo S, Romaniello, Pina, Berger, JA, and Reining, Lucia (2018). Many-body perturbation theory and non-perturbative approaches: Screened interaction as the key ingredient. *Journal of Physics: Condensed Matter*, **30**(13), 135602.

Tasker, PW (1979). The stability of ionic crystal surfaces. *Journal of Physics C: Solid State Physics*, **12**(22), 4977.

Tauster, SJ, Fung, SC, Baker, RTK, and Horsley, JA (1981). Strong interactions in supported-metal catalysts. *Science*, **211**(4487), 1121–1125.

Tauster, SJ, Fung, SC, and Garten, Rl L (1978). Strong metal-support interactions: Group 8 noble metals supported on titanium dioxide. *Journal of the American Chemical Society*, **100**(1), 170–175.

Tenne, Dmitri A, Turner, P, Schmidt, JD, Biegalski, M, Li, YL, Chen, LQ, Soukiassian, A, Trolier-McKinstry, S, Schlom, DG, Xi, XX et al. (2009). Ferroelectricity in ultrathin $BaTiO_3$ films: Probing the size effect by ultraviolet Raman spectroscopy. *Physical Review Letters*, **103**(17), 177601.

Tersoff, J (1984). Schottky barrier heights and the continuum of gap states. *Physical Review Letters*, **52**(6), 465.

Tersoff, Jerry (1988). New empirical approach for the structure and energy of covalent systems. *Physical Review B*, **37**(12), 6991.

Thiel, Patricia A and Madey, Theodore E (1987). The interaction of water with solid surfaces: Fundamental aspects. *Surface Science Reports*, **7**(6-8), 211–385.

Thiel, Stefan, Hammerl, German, Schmehl, Andreas, Schneider, Christof W, and Mannhart, Jochen (2006). Tunable quasi-two-dimensional electron gases in oxide heterostructures. *Science*, **313**(5795), 1942–1945.

Thomas, IO and Fortunelli, A (2010). Analysis of the electronic structure of ultrathin NiO/Ag (100) films. *The European Physical Journal B*, **75**(1), 5–13.

Titchtmarsh, J. (2009). Transmission electron microscopy: Instrumentation, imaging modes, and analytical attachments. In *Handbook of Surface and Interface Analysis: Methods for Problem-Solving*. CRC Press.

Topsakal, Mehmet, Cahangirov, Seymur, Bekaroglu, Erman, and Ciraci, Salim (2009). First-principles study of zinc oxide honeycomb structures. *Physical Review B*, **80**(23), 235119.

Tosi, Mario P. (1964). Cohesion of ionic solids in the Born model. In *Solid State Physics* (ed. F. Seitz and D. Turnbull), Volume 16, pp. 1 – 120. Academic Press.

Tosoni, Sergio, Spinnato, Davide, and Pacchioni, Gianfranco (2015). DFT study of CO_2 activation on doped and ultrathin MgO films. *The Journal of Physical Chemistry C*, **119**(49), 27594–27602.

Tournassat, Christophe, Ferrage, Eric, Poinsignon, Christiane, and Charlet, Laurent (2004). The titration of clay minerals II. Structure-based model and implications for clay reactivity. *Journal of Colloid and Interface Science*, **273**(1), 234–246.

Towler, MD, Allan, NL, Harrison, Nicholas M, Saunders, VR, Mackrodt, WC, and Apra, E (1994). Ab initio study of MnO and NiO. *Physical Review B*, **50**(8), 5041.

Toyota, D, Ohkubo, I, Kumigashira, H, Oshima, M, Ohnishi, T, Lippmaa, M, Takizawa, M, Fujimori, A, Ono, K, Kawasaki, M et al. (2005). Thickness-dependent electronic structure of ultrathin $SrRuO_3$ films studied by in situ photoemission spectroscopy. *Applied Physics Letters*, **87**(16), 162508.

Tran, Fabien and Blaha, Peter (2009). Accurate band gaps of semiconductors and insulators with a semilocal exchange-correlation potential. *Physical Review Letters*, **102**, 226401.

Tsukada, Masaru and Hoshino, Toshiharu (1982). On the electronic structure of the polar surface of compound crystals. *Journal of the Physical Society of Japan*, **51**(8), 2562–2567.

Tsukazaki, A, Ohtomo, A, Kita, T, Ohno, Y, Ohno, H, and Kawasaki, M (2007). Quantum Hall effect in polar oxide heterostructures. *Science*, **315**(5817), 1388–1391.

Tunega, Daniel, Bučko, Tomáš, and Zaoui, Ali (2012). Assessment of ten DFT methods in predicting structures of sheet silicates: Importance of dispersion corrections. *The Journal of Chemical Physics*, **137**(11), 114105.

Turco, RP, Whitten, RC, and Toon, OB (1982). Stratospheric aerosols: Observation and theory. *Reviews of Geophysics*, **20**(2), 233–279.

Turner, John A (1999). A realizable renewable energy future. *Science*, **285**(5428), 687–689.

Tusche, C, Meyerheim, HL, and Kirschner, J (2007). Observation of depolarized ZnO (0001) monolayers: Formation of unreconstructed planar sheets. *Physical Review Letters*, **99**(2), 026102.

Tybell, Thomas, Ahn, CH, and Triscone, J-M (1999). Ferroelectricity in thin perovskite films. *Applied Physics Letters*, **75**(6), 856–858.

Ulrich, Stefan, Nilius, Niklas, Freund, Hans-Joachim, Martinez, Umberto, Giordano, Livia, and Pacchioni, Gianfranco (2009). Realization of an atomic sieve: Silica on Mo(112). *Surface Science*, **603**(8), 1145–1149.

Valeri, Sergio and Benedetti, Stefania (2012). Synthesis and preparation of oxide ultrathin films. *Oxide Ultrathin Films: Science and Technology. Wiley-VCH, Weinheim*, 1–26.

Van Delft, FCMJM and Nieuwenhuys, BE (1985). Correlation of nucleation-and growth modes with wetting, alloy segregation, catalyst preparation and strong-metal support interaction. *Solid State Ionics*, **16**, 233–240.

Van den Bossche, Maxime, Gronbeck, Henrik, and Hammer, Bjørk (2018). Tight-Binding approximation-enhanced global optimization. *Journal of Chemical Theory and Computation*, **14**(5), 2797–2807.

Van den Bossche, Maxime, Noguera, Claudine, and Goniakowski, Jacek (2020). Understanding the structural diversity of freestanding Al_2O_3 ultrathin films through DFTB-aided global optimization. *Nanoscale*, **12**, 6153–6163.

Van Driessche, Alexander ES, Kellermeier, Matthias, Benning, Liane G, and Gebauer, Denis (2016). *New Perspectives on Mineral Nucleation and Growth: From Solution Precursors to Solid Materials*. Springer, Switzerland.

Van Duin, Adri CT, Strachan, Alejandro, Stewman, Shannon, Zhang, Qingsong, Xu, Xin, and Goddard, William A (2003). ReaxFFSiO reactive force field for silicon and silicon oxide systems. *The Journal of Physical Chemistry A*, **107**(19), 3803–3811.

Van Gog, Heleen, Li, Wun-Fan, Fang, Changming, Koster, Rik S, Dijkstra, Marjolein, and van Huis, Marijn (2019). Thermal stability and electronic and magnetic properties of atomically thin 2-D transition metal oxides. *npj 2D Materials and Applications*, **3**(1), 1–12.

Vanhove, Michel A, Weinberg, William Henry, and Chan, Chi-Ming (2012). *Low-Energy Electron Diffraction: Experiment, Theory and Surface Structure Determination*. Volume 6. Springer Science & Business Media, Berlin, Heidelberg, New York, Barcelona, Hong Kong, London, Milan, Paris, Tokyo.

Variola, Fabio and Nanci, Antonio (2011). Titania thin films in biocompatible matals and medical implants. *Oxide Ultrathin Films: Science and Technology*, 309–328.

Variola, Fabio, Vetrone, Fiorenzo, Richert, Ludovic, Jedrzejowski, Pawel, Yi, Ji-Hyun, Zalzal, Sylvia, Clair, Sylvain, Sarkissian, Andranik, Perepichka, Dmitrii F, Wuest, James D et al. (2009). Improving biocompatibility of implantable metals by nanoscale modification of surfaces: An overview of strategies, fabrication methods, and challenges. *Small*, **5**(9), 996–1006.

Vayssilov, Georgi N, Lykhach, Yaroslava, Migani, Annapaola, Staudt, Thorsten, Petrova, Galina P, Tsud, Nataliya, Skála, Tomáš, Bruix, Albert, Illas, Francesc,

Prince, Kevin C et al. (2011). Support nanostructure boosts oxygen transfer to catalytically active platinum nanoparticles. *Nature Materials*, **10**(4), 310.

Védrine, Jacques C (2017). Heterogeneous catalysis on metal oxides. *Catalysts*, **7**(11), 341.

Vedrine, Jacques C (2018). *Metal Oxides in Heterogeneous Catalysis*. Elsevier, Amsterdam, Oxford, Cambridge.

Venables, JA, Spiller, GDT, and Hanbucken, M (1984). Nucleation and growth of thin films. *Reports on Progress in Physics*, **47**(4), 399.

Vidic, Jasmina, Stankic, Slavica, Haque, Francia, Ciric, Danica, Le Goffic, Ronan, Vidy, Aurore, Jupille, Jacques, and Delmas, Bernard (2013). Selective antibacterial effects of mixed ZnMgO nanoparticles. *Journal of Nanoparticle Research*, **15**(5), 1595.

Vilhelmsen, Lasse B and Hammer, Bjørk (2014). A genetic algorithm for first principles global structure optimization of supported nano structures. *The Journal of Chemical Physics*, **141**(4), 044711.

Viñes, Francesc, Lamiel-Garcia, Oriol, Illas, Francesc, and Bromley, Stefan T (2017). Size dependent structural and polymorphic transitions in ZnO: From nanocluster to bulk. *Nanoscale*, **9**(28), 10067–10074.

Vinograd, Victor L, Winkler, Björn, Putnis, Andrew, Gale, Julian D, and Sluiter, Marcel HF (2006). Static lattice energy calculations of mixing and ordering enthalpy in binary carbonate solid solutions. *Chemical Geology*, **225**(3-4), 304–313.

Viseras, C, Cerezo, P, Sanchez, R, Salcedo, I, and Aguzzi, C (2010). Current challenges in clay minerals for drug delivery. *Applied Clay Science*, **48**(3), 291–295.

Visikovskiy, Anton, Mitsuhara, K, Hazama, M, Kohyama, M, and Kido, Y (2013). The atomic and electronic structures of NiO(001)/Au(001) interfaces. *The Journal of Chemical Physics*, **139**(14), 144705.

Von Klitzing, Klaus (1993). The quantized Hall effect. *Physics 1981-1990*, **2**, 316.

Von Klitzing, Klaus (2005). 25 years of Quantum Hall Effect (QHE): A personal view on the discovery, physics and applications of this quantum effect. In *The Quantum Hall Effect*, pp. 1–21. Springer.

Voora, Vamsee K, Al-Saidi, WA, and Jordan, Kenneth D (2011). Density functional theory study of pyrophyllite and m-montmorillonites (M= Li, Na, K, Mg, and Ca): Role of dispersion interactions. *The Journal of Physical Chemistry A*, **115**(34), 9695–9703.

Wachs, Israel E (2005). Recent conceptual advances in the catalysis science of mixed metal oxide catalytic materials. *Catalysis Today*, **100**(1-2), 79–94.

Wachs, Israel E (2013). Catalysis science of supported vanadium oxide catalysts. *Dalton Transactions*, **42**(33), 11762–11769.

Wachsman, Eric D and Lee, Kang Taek (2011). Lowering the temperature of solid oxide fuel cells. *Science*, **334**(6058), 935–939.

Wagner, Carl (1933). Beitrag zur Theorie des Anlaufvorgangs. *Zeitschrift für Physikalische Chemie*, **21**(1), 25–41.

Wagner, C (1961). Theorie der Alterung von Niederschlägen durch Umlösen (Ostwald-Reifung). *Zeitschrift fur Elektrochemie*, **65**(7-8), 581–591.

Wales, David J and Doye, Jonathan PK (1997). Global optimization by Basin-

Hopping and the lowest energy structures of Lennard-Jones clusters containing up to 110 atoms. *The Journal of Physical Chemistry A*, **101**(28), 5111–5116.

Walia, Sumeet, Balendhran, Sivacarendran, Nili, Hussein, Zhuiykov, Serge, Rosengarten, Gary, Wang, Qing Hua, Bhaskaran, Madhu, Sriram, Sharath, Strano, Michael S, and Kalantar-zadeh, Kourosh (2013). Transition metal oxides– thermoelectric properties. *Progress in Materials Science*, **58**(8), 1443–1489.

Wallis, RF (1964). Surface effects on lattice vibrations. *Surface Science*, **2**, 146–155.

Wang, Baolin, Nagase, Shigeru, Zhao, Jijun, and Wang, Guanghou (2007). Structural growth sequences and electronic properties of zinc oxide clusters $(ZnO)_n$ (n= 2-18). *The Journal of Physical Chemistry C*, **111**(13), 4956–4963.

Wang, Fenggong, Di Valentin, Cristiana, and Pacchioni, Gianfranco (2012a). Doping of WO_3 for photocatalytic water splitting: Hints from density functional theory. *The Journal of Physical Chemistry C*, **116**(16), 8901–8909.

Wang, Fenggong, Di Valentin, Cristiana, and Pacchioni, Gianfranco (2012b). Rational band gap engineering of WO_3 photocatalyst for visible light water splitting. *ChemCatChem*, **4**(4), 476–478.

Wang, Hui, Zhang, Xiaodong, and Xie, Yi (2018). Recent progress in ultrathin two-dimensional semiconductors for photocatalysis. *Materials Science and Engineering: R: Reports*, **130**, 1–39.

Wang, Jianwei, Kalinichev, Andrey G, and Kirkpatrick, R James (2006). Effects of substrate structure and composition on the structure, dynamics, and energetics of water at mineral surfaces: A molecular dynamics modeling study. *Geochimica et Cosmochimica Acta*, **70**(3), 562–582.

Wang, Li, Chen, Kezheng, and Dong, Lifeng (2010a). Synthesis of exotic zigzag ZnO nanoribbons and their optical, electrical properties. *The Journal of Physical Chemistry C*, **114**(41), 17358–17361.

Wang, Lianzhou and Sasaki, Takayoshi (2014). Titanium oxide nanosheets: Graphene analogues with versatile functionalities. *Chemical Reviews*, **114**(19), 9455–9486.

Wang, Qinggao, Oganov, Artem R, Zhu, Qiang, and Zhou, Xiang-Feng (2014). New reconstructions of the (110) surface of rutile TiO_2 predicted by an evolutionary method. *Physical Review Letters*, **113**(26), 266101.

Wang, Shuqiu, Goniakowski, Jacek, Noguera, Claudine, and Castell, Martin R (2019a). Atomic and electronic structure of an epitaxial Nb_2O_3 honeycomb monolayer on Au(111). *Physical Review B*, **100**(12), 125408.

Wang, Xudong, Wang, Dan, Guo, Yali, Yang, Chengduan, Liu, Xiaoyu, Iqbal, Anam, Liu, Weisheng, Qin, Wenwu, Yan, Dan, and Guo, Huichen (2016). Fluorescent glutathione probe based on MnO_2-phenol formaldehyde resin nanocomposite. *Biosensors and Bioelectronics*, **77**, 299–305.

Wang, Yuhui, Jiang, Kai, Zhu, Jiali, Zhang, Ling, and Lin, Hengwei (2015). A FRET-based carbon dot–MnO_2 nanosheet architecture for glutathione sensing in human whole blood samples. *Chemical Communications*, **51**(64), 12748–12751.

Wang, Y, Meyer, B, Yin, X, Kunat, M, Langenberg, D, Traeger, F, Birkner, A, and Wöll, Ch (2005). Hydrogen induced metallicity on the $ZnO(10\bar{1}0)$ surface. *Physical Review Letters*, **95**(26), 266104.

Wang, Yong, Sun, Chenghua, Yan, Xiaoxia, Xiu, Faxian, Wang, Lianzhou, Smith,

Sean C, Wang, Kang L, Lu, Gao Qing, and Zou, Jin (2010*b*). Lattice distortion oriented angular self-assembly of monolayer titania sheets. *Journal of the American Chemical Society*, **133**(4), 695–697.

Wang, Yanzong, Wang, Baolin, Zhang, Qinfang, Shi, Daning, Yunoki, Seiji, Kong, Fanjie, and Xu, Ning (2012*c*). A simple capacitor model and first-principles study of carbon-doped zigzag ZnO nanoribbons. *Solid State Communications*, **152**(6), 534–539.

Wang, Zheng, Li, Can, and Domen, Kazunari (2019*b*). Recent developments in heterogeneous photocatalysts for solar-driven overall water splitting. *Chemical Society Reviews*, **48**(7), 2109–2125.

Wang, Zhong Lin (2004*a*). Nanostructures of zinc oxide. *Materials Today*, **7**(6), 26–33.

Wang, Zhong Lin (2004*b*). Zinc oxide nanostructures: Growth, properties and applications. *Journal of Physics: Condensed Matter*, **16**(25), R829.

Watanabe, Yukio, Okano, Motochika, and Masuda, Akihiro (2001). Surface conduction on insulating $BaTiO_3$ crystal suggesting an intrinsic surface electron layer. *Physical Review Letters*, **86**(2), 332.

Weaver, Charles E and Pollard, Lin D (2011). *The chemistry of clay minerals*. Volume 15. Elsevier, Amsterdam, Oxford, Cambridge.

Weirum, Gunther, Barcaro, G, Fortunelli, A, Weber, Frederik, Schennach, Robert, Surnev, S, and Netzer, FP (2010). Growth and surface structure of zinc oxide layers on a Pd(111) surface. *The Journal of Physical Chemistry C*, **114**(36), 15432–15439.

Weiss, Werner and Ranke, Wolfgang (2002). Surface chemistry and catalysis on well-defined epitaxial iron-oxide layers. *Progress in Surface Science*, **70**(1-3), 1–151.

Weissenrieder, Jonas, Kaya, Sarp, Lu, J-L, Gao, H-J, Shaikhutdinov, Shamil, Freund, H-J, Sierka, Marek, Todorova, Tanya K, and Sauer, Joachim (2005). Atomic structure of a thin silica film on a Mo(112) substrate: A two-dimensional network of SiO_4 tetrahedra. *Physical Review Letters*, **95**(7), 076103.

White, G Norman and Zelazny, LW (1988). Analysis and implications of the edge structure of dioctahedral phyllosilicates. *Clays and Clay Minerals*, **36**(2), 141–146.

Williams, David B, Carter, C Barry, and Veyssiere, P (1998). *Transmission Electron Microscopy: A Textbook for Materials Science*. Volume 10. Springer, Berlin, Heidelberg, New York, Barcelona, Hong Kong, London, Milan, Paris, Tokyo.

Williams, KJ, Boffa, AB, Lahtinen, J, Salmeron, M, Bell, AT, and Somorjai, Gabor A (1990). Hydrogenation of CO_2, acetone, and CO on a Rh foil promoted by titania overlayers. *Catalysis Letters*, **5**(4-6), 385–394.

Wimpenny, Josh (2016). Clay minerals. *Encyclopedia of Geochemistry: A Comprehensive Reference Source on the Chemistry of the Earth*, 1–11.

Wolf, SA, Awschalom, DD, Buhrman, RA, Daughton, JM, Von Molnar, S, Roukes, ML, Chtchelkanova, A Yu, and Treger, DM (2001). Spintronics: A spin-based electronics vision for the future. *Science*, **294**(5546), 1488–1495.

Wöll, Christof (2007). The chemistry and physics of zinc oxide surfaces. *Progress in Surface Science*, **82**(2-3), 55–120.

Wollschläger, J, Erdös, D, Goldbach, H, Höpken, R, and Schröder, KM (2001). Growth of NiO and MgO films on Ag(100). *Thin Solid Films*, **400**(1-2), 1–8.

Woodley, S, Battle, P, Gale, J, and Catlow, CRA (1999). The prediction of inorganic crystal structures using a genetic algorithm and energy minimisation. *Physical Chemistry Chemical Physics*, **1**(10), 2535–2542.

Woodruff, David Phillip (2016). *Modern Techniques of Surface Science*. Cambridge University Press, Cambridge.

Wu, Chen, Marshall, Matthew SJ, and Castell, Martin R (2011*a*). Surface structures of ultrathin TiO_x films on Au(111). *The Journal of Physical Chemistry C*, **115**(17), 8643–8652.

Wu, Qi-Hui, Fortunelli, Alessandro, and Granozzi, Gaetano (2009). Preparation, characterisation and structure of Ti and Al ultrathin oxide films on metals. *International Reviews in Physical Chemistry*, **28**(4), 517–576.

Wu, S Q, Ji, M, Wang, C Z, Nguyen, M C, Zhao, X, Umemoto, K, Wentzcovitch, R M, and Ho, K M (2014). An adaptive genetic algorithm for crystal structure prediction. *Journal of Physics: Condensed Matter*, **26**(3), 035402.

Wu, Wenzhi, Lu, Peng, Zhang, Zhuhua, and Guo, Wanlin (2011*b*). Electronic and magnetic properties and structural stability of BeO sheet and nanoribbons. *ACS Applied Materials & Interfaces*, **3**(12), 4787–4795.

Wu, Xianzhang, Xia, Xinnian, Chen, You, and Lu, Yanbing (2016*a*). Mesoporous Al-incorporated silica-pillared clay interlayer materials for catalytic hydroxyalkylation of phenol to bisphenol F. *RSC Advances*, **6**(78), 74028–74038.

Wu, Zhaohui, Yang, Shuanglei, and Wu, Wei (2016*b*). Shape control of inorganic nanoparticles from solution. *Nanoscale*, **8**(3), 1237–1259.

Xia, Jing, Siemons, W, Koster, Gertjan, Beasley, MR, and Kapitulnik, A (2009). Critical thickness for itinerant ferromagnetism in ultrathin films of $SrRuO_3$. *Physical Review B*, **79**(14), 140407.

Xia, Tian, Zhao, Yan, Sager, Tina, George, Saji, Pokhrel, Suman, Li, Ning, Schoenfeld, David, Meng, Huan, Lin, Sijie, Wang, Xiang et al. (2011). Decreased dissolution of ZnO by iron doping yields nanoparticles with reduced toxicity in the rodent lung and zebrafish embryos. *ACS Nano*, **5**(2), 1223–1235.

Xiao, Xu, Song, Huaibing, Lin, Shizhe, Zhou, Ying, Zhan, Xiaojun, Hu, Zhimi, Zhang, Qi, Sun, Jiyu, Yang, Bo, Li, Tianqi et al. (2016). Scalable salt-templated synthesis of two-dimensional transition metal oxides. *Nature Communications*, **7**, 11296.

Xu, Chaoqiang, Que, Yande, Zhuang, Yuan, Liu, Bin, Ma, Yaping, Wang, Kedong, and Xiao, Xudong (2019). Manipulating the edge of a two-dimensional MgO nanoisland. *The Journal of Physical Chemistry C*, **123**(32), 19619–19624.

Xu, Tao, Zhou, Xi, Jiang, Zhiyuan, Kuang, Qin, Xie, Zhaoxiong, and Zheng, Lansun (2008). Syntheses of nano/submicrostructured metal oxides with all polar surfaces exposed via a molten salt route. *Crystal Growth and Design*, **9**(1), 192–196.

Yagmurcukardes, M, Peeters, François M, Senger, Ramazan Tuğrul, and Sahin, H (2016). Nanoribbons: From fundamentals to state-of-the-art applications. *Applied Physics Reviews*, **3**(4), 041302.

Yang, Peihua, Sun, Peng, and Mai, Wenjie (2016). Electrochromic energy storage devices. *Materials Today*, **19**(7), 394–402.

Yang, Seolun, Park, H-K, Kim, J-S, Phark, S-H, Chang, Young Jun, Noh, TW, Hwang, H-N, Hwang, C-C, and Kim, H-D (2013). Reduction of charge fluctuation energies in ultrathin NiO films on Ag(001). *Surface Science*, **616**, 12–18.

Yang, Tong, Song, Ting Ting, Callsen, Martin, Zhou, Jun, Chai, Jian Wei, Feng, Yuan Ping, Wang, Shi Jie, and Yang, Ming (2019). Atomically thin 2-D transition metal oxides: Structural reconstruction, interaction with substrates, and potential applications. *Advanced Materials Interfaces*, **6**(1), 1801160.

Yang, Zongxian, Lu, Zhansheng, Luo, Gaixia, and Hermansson, Kersti (2007). Oxygen vacancy formation energy at the Pd/CeO_2(111) interface. *Physics Letters A*, **369**(1-2), 132–139.

Yao, Chang, Slamovich, Elliott B, and Webster, Thomas J (2008). Enhanced osteoblast functions on anodized titanium with nanotube-like structures. *Journal of Biomedical Materials Research Part A: An Official Journal of The Society for Biomaterials, The Japanese Society for Biomaterials, and The Australian Society for Biomaterials and the Korean Society for Biomaterials*, **85**(1), 157–166.

York, Darrin M and Yang, Weitao (1996). A chemical potential equalization method for molecular simulations. *The Journal of Chemical Physics*, **104**(1), 159–172.

Yuan, Changzhou, Wu, Hao Bin, Xie, Yi, and Lou, Xiong Wen (2014). Mixed transition-metal oxides: Design, synthesis, and energy-related applications. *Angewandte Chemie International Edition*, **53**(6), 1488–1504.

Zaanen, J, Sawatzky, GA, and Allen, JW (1985). Band gaps and electronic structure of transition-metal compounds. *Physical Review Letters*, **55**(4), 418.

Zangwill, A (1990). *Physics at surfaces*. Cambridge University Press, NewYork.

Zeuthen, Helene, Kudernatsch, Wilhelmine, Peng, Guowen, Merte, Lindsay R, Ono, Luis K, Lammich, Lutz, Bai, Yunhai, Grabow, Lars C, Mavrikakis, Manos, Wendt, Stefan et al. (2013). Structure of stoichiometric and oxygen-rich ultrathin FeO(111) films grown on Pd(111). *The Journal of Physical Chemistry C*, **117**(29), 15155–15163.

Zhang, Di, Zhou, Chun-Hui, Lin, Chun-Xiang, Tong, Dong-Shen, and Yu, Wei-Hua (2010). Synthesis of clay minerals. *Applied Clay Science*, **50**(1), 1–11.

Zhang, Shou-juan, Zhang, Chang-wen, Zhang, Shu-feng, Ji, Wei-xiao, Li, Ping, Wang, Pei-ji, Li, Sheng-shi, and Yan, Shi-shen (2017). Intrinsic Dirac half-metal and quantum anomalous Hall phase in a hexagonal metal-oxide lattice. *Physical Review B*, **96**(20), 205433.

Zhang, Yiling, Schultz, Andrew M, Li, Li, Chien, Harry, Salvador, Paul A, and Rohrer, Gregory S (2012). Combinatorial substrate epitaxy: A high-throughput method for determining phase and orientation relationships and its application to $BiFeO_3/TiO_2$ heterostructures. *Acta Materialia*, **60**(19), 6486–6493.

Zhao, Chunsong, Zhang, Haitian, Si, Wenjie, and Wu, Hui (2016). Mass production of two-dimensional oxides by rapid heating of hydrous chlorides. *Nature Communications*, **7**, 12543.

Zhao, Zilong, Fan, Huanhuan, Zhou, Gaofeng, Bai, Huarong, Liang, Hao, Wang, Ruowen, Zhang, Xiaobing, and Tan, Weihong (2014). Activatable fluorescence/MRI bimodal platform for tumor cell imaging via MnO_2 nanosheet–aptamer nanoprobe. *Journal of the American Chemical Society*, **136**(32), 11220–11223.

Zhong, Zhicheng and Hansmann, Philipp (2017). Band alignment and charge transfer in complex oxide interfaces. *Physical Review X*, **7**(1), 011023.

Zhong, Zhicheng, Zhang, Qinfang, and Held, Karsten (2013). Quantum confinement in perovskite oxide heterostructures: Tight binding instead of a nearly free electron picture. *Physical Review B*, **88**(12), 125401.

Zhou, Chun Hui (2011). An overview on strategies towards clay-based designer catalysts for green and sustainable catalysis. *Applied Clay Science*, **53**(2), 87–96.

Zhu, J, Li, Q, Li, JX, Ding, Z, Liang, JH, Xiao, X, Luo, YM, Hua, CY, Lin, H-J, Pi, TW et al. (2014). Antiferromagnetic spin reorientation transition in epitaxial NiO/CoO/MgO(001) systems. *Physical Review B*, **90**(5), 054403.

Zhu, Qiang, Li, Li, Oganov, Artem R, and Allen, Philip B (2013*a*). Evolutionary method for predicting surface reconstructions with variable stoichiometry. *Physical Review B*, **87**(19), 195317.

Zhu, Yuanyuan, Song, Chengyu, Minor, Andrew M, and Wang, Haiyan (2013*b*). Cs-corrected scanning transmission electron microscopy investigation of dislocation core configurations at a $SrTiO_3$/MgO heterogeneous interface. *Microscopy and Microanalysis*, **19**(3), 706–715.

Ziemann, Paul J and Castleman Jr, AW (1991). Stabilities and structures of gas phase MgO clusters. *The Journal of Chemical Physics*, **94**(1), 718–728.

Zwijnenburg, MA, Sokol, AA, Sousa, C, and Bromley, ST (2009). The effect of local environment on photoluminescence: A time-dependent density functional theory study of silanone groups on the surface of silica nanostructures. *The Journal of Chemical Physics*, **131**(3), 034705.

Index